Geoscience: Concepts and Applications

Geoscience: Concepts and Applications

Edited by Theodore Roa

SYRAWOOD
PUBLISHING HOUSE

New York

Published by Syrawood Publishing House,
750 Third Avenue, 9th Floor,
New York, NY 10017, USA
www.syrawoodpublishinghouse.com

Geoscience: Concepts and Applications
Edited by Theodore Roa

International Standard Book Number: 978-1-68286-602-3 (Hardback)

Cataloging-in-Publication Data

Geoscience : concepts and applications / edited by Theodore Roa.
    p. cm.
Includes bibliographical references and index.
ISBN 978-1-68286-602-3
1. Geology. 2. Earth sciences. I. Roa, Theodore.
QE26.3 .G46 2018
551--dc23

TABLE OF CONTENTS

PREFACE

Geoscience or Earth science is the holistic study of the planet Earth. It is mainly concerned with the analysis of the atmosphere, geology, hydrosphere and lithosphere of Earth. It integrates elements from chemistry, biology, geography, physics, mathematics, etc. This book traces the progress of this field and highlights some of its key concepts and applications. The topics included herein on geoscience are of utmost significance and bound to provide incredible insights to readers. The various sub-fields along with technological progress that have future implications are also glanced at. It is a complete source of knowledge on the present status of this important field.

Various studies have approached the subject by analyzing it with a single perspective, but the present book provides diverse methodologies and techniques to address this field. This book contains theories and applications needed for understanding the subject from different perspectives. The aim is to keep the readers informed about the progresses in the field; therefore, the contributions were carefully examined to compile novel researches by specialists from across the globe.

Indeed, the job of the editor is the most crucial and challenging in compiling all chapters into a single book. In the end, I would extend my sincere thanks to the chapter authors for their profound work. I am also thankful for the support provided by my family and colleagues during the compilation of this book.

Editor

Present status of soil moisture estimation by microwave remote sensing

Kousik Das[1]* and Prabir Kumar Paul[1]

*Corresponding author: Kousik Das, Department of Mining Engineering, Indian Institute of Engineering Science and Technology, Shibpur, Howrah, India

E-mail: kousik.envs@gmail.com

Reviewing editor:

Zdena Dobesova, Palacky University, Czech Republic

Abstract: The spatiotemporal distribution of soil moisture is a key variable for hydrological and meteorological applications that influences the exchange of water and energy fluxes at the land surface/atmosphere interface. Accurate estimate of the spatiotemporal variations of soil moisture is critical for numerous large-scale environmental studies. Recent technological advances in satellite remote sensing have shown that soil moisture can be measured by a variety of remote sensing techniques, each with its own strengths and weaknesses which minimizes the ill-posed conventional problems. Technical and methodological advances such as multi-configuration radar and forthcoming SAR constellations are increasingly mitigating the shortcomings of SAR with respect to soil moisture estimation at the field and catchment scale. This paper presents a comprehensive review of few selected inversion methods of soil moisture, with focus on technique in passive microwave and active microwave measurements, in addition to the factors which affect the microwave return. The theoretical and physical principles and the status of current basic retrieval methods are summarized. Limitations existing in current soil moisture estimation algorithms and the major influencing factors including radar configurations (polarization, incidence angel and frequency of bands) and soil surface characteristics on backscattering coefficient have been addressed and also discussed.

Subjects: GIS; Remote Sensing & Cartography; Remote Sensing; Soil Science

Keywords: soil moisture; radar; SAR; microwave; remote sensing

1. Introduction

Soil moisture is a quantity of water contained in soil on a volumetric or gravimetric basis (Al-Yaari et al., 2014; Zhao & Li, 2013). Soil moisture participates in the distribution of precipitation between run-off

ABOUT THE AUTHOR

The author has been working for a decade on Remote Sensing and Geographical Information System. The major thrust is focused on Remote Sensing and GIS application on natural resources and mining. To name a few we had been working on landslide zoning, location of sites for small hydel and ground water. Presently we are working on various applications of SAR data specifically on town planning, soil moisture etc.

PUBLIC INTEREST STATEMENT

This review paper is addressed towards the scientific community those are working or try to working on that field of soil moisture retrieval by microwave remote sensing. This review paper is used to describe the fundamentals of active and passive microwave remote sensing technology there models and algorithms to retrieve soil moisture. Basically it depicts limitations existing in current soil moisture estimation algorithms and the major influencing factors that affect radar configurations. New readers will be benefitted from this paper by the information of very beginning of this technology of soil moisture estimation to till date.

and infiltration (Petropoulos, Ireland, & Barrett, in press; Seneviratne et al., 2010). Soil moisture influences meteorological and climatic processes (Álvarez-Mozos, Casalí, González-Audícana, & Verhoest, 2005; Wagner et al., 2007), though surface soil moisture only constitutes 0.0012% of all water available on Earth (Chow, Maidment, & Mays, 1988). Soil moisture was recognized as an essential climate variable (ECV) in 2010 because it plays a crucial role in various processes occurring on the soil–atmosphere interface (European Space Agency [ESA], *Soil moisture network*, http://www.esa-soilmoisture-cci.org/ node) (Al-Yaari et al., 2014; Zhao & Li, 2013). The representative ground–based measurements of soil moisture are an unsolved problem because the only soil moisture data available are from point measurements (Laguardia & Niemeyer, 2008). Considering the high variability and the low degree of observed autocorrelation, it is difficult to obtain reliable estimates at the larger scale from point measurements (Engman & Gurney, 1991; Giacomelli, Bacchiega, Troch, & Mancini, 1995; Kornelsen & Coulibaly, 2013; Ulaby, Moore, & Fung, 1986). Possibility of retrieving soil moisture has been investigated using satellites, space shuttles and airborne synthetic aperture radars (Baghdadi, Holah, & Zribi, 2006a; Baghdadi, King, Chanzy, & Wigneron, 2002). Microwave remote sensing is the most effective technique for soil moisture estimation, with advantages for all-weather observations and solid physics (Engman, 1990; Kornelsen & Coulibaly, 2013; Petropoulos et al., in press). Since microwave measurements of the soil surface are affected by the water content (Engman & Gurney, 1991; Petropoulos et al., in press; Seneviratne et al., 2010; Shi et al., 2012; Ulaby et al., 1986), it is easy to see the potentiality of remote sensing in soil moisture mapping and other related applications (Batlivala & Ulaby, 1977; Giacomelli et al., 1995; Petropoulos et al., in press; Seneviratne et al., 2010).

The use of radar data to retrieve soil moisture is of considerable importance in many domains, including agriculture, hydrology and meteorology (Baghdadi, King, Chanzy, et al., 2002; Petropoulos et al., in press; Seneviratne et al., 2010). Despite many advantages that can be derived from the knowledge of soil moisture distribution, measurement of soil moisture has few limitations. However, the measurement of soil moisture is not only depended on target characteristics such as surface roughness, vegetation cover, dielectric constant and topography (Ualby, Batlivala, & Dobson, 1978) but also depends on various combinations of the radar sensor parameters including frequency, polarization and angle of incidence (θ) with respect to nadir (Anguela, Zribi, Baghdadi, & Loumagne, 2010; Bertoldi et al., 2014; Dobson & Ulaby, 1981, 1986; Kornelsen & Coulibaly, 2013; Ualby et al., 1978).

2. Microwave remote sensing

2.1. Active microwave remote sensing

Active microwave remote sensing uses the radar antenna in terms of either real or synthetic aperture, which transmits wave pulses of known energy and receives a return signal whose intensity depends on target characteristics (Kornelsen & Coulibaly, 2013). The returned signal which has been recorded by the sensor is usually expressed as backscattering coefficient (σ^0). The σ^0 is mostly dependent on soil moisture content mainly due to the dielectric constant (ε) of the soil (Prakash, Singh, & Pathak, 2012; Schmugge, Jackson, & McKim, 1980), and thus provides a method of retrieving soil water content. Therefore, microwave is the most suitable for the purpose of soil moisture because it is free from atmospheric attenuation. Number of factors like vegetation, surface roughness, measurement depth and topography affect the backscattering signal detected by the antenna (Bertoldi et al., 2014; Kornelsen & Coulibaly, 2013; Petropoulos et al., in press; Schmugge et al., 1980; Ulaby et al., 1986).

2.2. Passive microwave remote sensing

For soil moisture retrieval, passive microwave remote sensing has been considered to be superior and more reliable in terms of lower frequencies than active microwave remote sensing (Petropoulos et al., in press). A series of operational satellite-based passive microwave sensors have been available since 1978 and are represented in (Table 1) (Basist et al., 1998; Liu et al., 2011; Oza, Singh, Dadhwal, & Desai, 2006; Singh, Mishra, Sahoo, & Dey, 2005). Among all passive microwave sensors, more recently, the Advanced Microwave Scanning Radiometer—Earth observing system (AMSR-E)

Table 1. Passive microwave sensors used for the generation of the soil moisture data sets from 1978 to 2015							
Parameter	SMMR	SSM/I	TRMM	MSMR	AMSR-E	SMOS	SMAP
Launch date	1978–1987	1987/92/95	1997–2001	1999–2001	2002–2011	2009	2015
Frequency (GHz)	6.6, 10.7, 18.0, 21 and 37	19.3, 22.2 (V), 38.0 and 85.5	10.65, 19.35, 22.24, 37.0 and 85.50	6.6, 10.65, 18, 21	6.6, 10.65, 18.7, 23.8, 36.5, 89	1.4	1.41
Polarization	H and V	H and V (except 22.2 GHz)	H and V (except 22.23 GHz)	H and V	H and V	H and V	H, V and HV or VH
IFOV (km)	148 × 95, 91 × 59, 55 × 41, 46 × 30, 27 × 18	69 × 43, 60 × 40, 37 × 28, 15 × 13	11 × 8	150 × 144, 75 × 72, 50 × 36, 50 × 36	76 × 44, 49 × 28, 28 × 16, 31 × 18, 14 × 8, 6 × 4	43 × 43	40 × 40
Swath width (km)	822	1400	880	1,360	1445	–	1,000
Revisit coverage (days)	–	1		2	2	3	3
Incidence angle (deg.)	50.3 (at the surface)	53.3 (at the surface)	35.0	43.13	54 (at the surface)	0–55	35–50

on-board the Aqua satellite (since 2002), Soil moisture and ocean salinity satellite (SMOS since 2009), Multi-frequency Scanning Microwave Radiometer (MSMR since 1999) and Soil Moisture Active Passive (SMAP) (since January 2015) are presently operational, providing satellite data for the globe on a daily basis. Passive microwave remote sensing provides the temporal data of earth daily which are applicable to models like numerical weather predictions (NWP) model (Anudeep, 2013). Passive microwave instruments are typically characterized by wide swath and high temporal resolution, but also coarse spatial resolutions around 10–30 km at L-band and C-band, respectively (Anudeep, 2013; Moran, Peters-Lidard, Watts, & McElroy, 2004; Wigneron et al., 2003). The microwave ranges within 10–30 cm are not affected largely by the surface roughness, vegetation cover and soil texture rather it is highly sensitive to the soil moisture (Chai et al., 2010).

The ESA launched SMOS mission on 2 November 2009. It is also the first-ever L-band passive microwave sensor dedicated to the global measurement of the Earth's near-surface (up to 10 cm) soil moisture (Petropoulos et al., in press). The spatial resolution of SMOS is sufficient enough to retrieve soil moisture for many global applications. Combination of SMOS data with other sensors' higher resolution data can provide a potential solution for global soil moisture estimates (Petropoulos et al., in press).

SMAP mission was launched in January 2015. The SMAP sensor is designed in such a way to produce active (radar: VV, HH and HV polarizations) and passive (radiometer; V, H and high register and 4th Stokes parameter polarizations) soil moisture data simultaneously. Multiple polarizations help in accurate soil moisture estimates with corrections for vegetation, surface roughness, Faraday rotation and other perturbing factors in the 1.2–1.4 GHz range (L-band) from a sun-synchronous low Earth orbit (Petropoulos et al., in press). The SMAP project is managed for NASA by the Jet Propulsion Laboratory, with participation by the Goddard Space Flight Centre (Entekhabi et al., 2010).

3. Factors influencing microwave remote sensing

Surface characteristics, such as roughness and vegetation cover, have significant influence on the backscattering coefficient; thus, it is very difficult to retrieve the soil moisture without detailed knowledge of it (Altese, Bolognani, Mancini, & Troch, 1996; Anguela et al., 2010; Bertoldi et al., 2014; Kornelsen & Coulibaly, 2013; Petropoulos et al., in press). In addition, number of radar sensor parameters, such as frequency, ε, polarization and θ with respect to nadir, influence the microwave backscatter in several ways which are described as follows (Anguela et al., 2010; Bertoldi et al., 2014; Dobson & Ulaby, 1981, 1986; Kornelsen & Coulibaly, 2013; Petropoulos et al., in press; Ualby et al., 1978):

3.1. Dielectric constant (ε)

ε of moist soils is proportional to the number of water dipoles per unit volume (Dobson & Ulaby, 1986). Estimation of soil moisture with the use of microwave instruments is the perspective of sensitivity of ε to water content for such instances (Chen et al., 2012; Hallikainen, Ulaby, Dobson, El-rayes, & Wu, 1985). Real part of dielectric constant (ε'), which can vary from 2.5 for very dry soil to 25 for very moist soil, is a function of both the composition of the soil and the microwave frequency (Singh & Kathpalia, 2007; Ulaby, Moore, & Fung, 1982). Several investigations have been initiated in laboratory conditions to find out the effect of soil moisture, bulk density and soil texture on the net dielectric behaviour of the soil medium using either guided-wave or free-space transmission techniques (Dobson, Ulaby, Hallikainen, & El-Rayes, 1985; Gharechelou, Tateishi, & Sumantyo, 2015; Hallikainen et al., 1985; Li, Zhao, Ren, Ding, & Wu, 2014; Wang & Schmugge, 1980). Grain size of soil is another operating factor of ε due to their interstice water content (Hallikainen et al., 1985; Mironov, Dobson, Kaupp, Komarov, & Kleshchenko, 2004; Srivastava, Patel, & Navalgund, 2006). Hallikainen et al. (1985) generated empirical expressions for ε as a function of the volumetric moisture content (M_v) and soil textural (sand (s) and clay (c) percent of weight) configuration (1).

$$\varepsilon' \left(or\, \varepsilon'' \right) = \left(a_0 + a_1 s + a_2 c \right) + \left(b_0 + b_1 s + b_2 c \right) M_v + \left(c_0 + c_1 s + c_2 c \right) M_v^2 \tag{1}$$

where, s and c are the percentage of sand and clay by weight, and a_i, b_i, and c_i are the frequency dependent coefficients. ε' and ε'' are the real and imaginary parts of the dielectric constant.

Several models have been developed to correlate ε and soil moisture content. For soil moisture retrieval studies, the polynomial expressions fitted by Hallikainen et al. (1985) and the semi-empirical four-component mixing model developed by Dobson et al. (1985) are general-purpose models of ε. The latter model is valid for frequencies larger than 4 GHz and smaller than 18 GHz, which was further extended for 0.3–1.3 GHz range by Peplinski, Ulaby, and Dobson (1995a, 1995b). Generalized refractive mixing dielectric model (GRMDM) was developed by Mironov, Dobson, Kaupp, Komarov, and Kleshchenko (2002) and is used to retrieve the soil complex ε, which is a function of frequency for both free and bound soil water (Mironov et al., 2002, 2004). But, most common methods found in the literature to relate the soil moisture and the ε without direct field measurement are done using empirical curves of Hallikainen et al. (1985), which was extensively used by a number of researchers (Rao et al., 2013). A site-specific calibration procedure was developed by D'Urso and Minaecapilli (2006) for Oh, Sarabandi, and Ulaby (1992) model to derive soil ε without prior soil surface related information.

3.2. Backscattering coefficient (σ^0)

Backscattering coefficient is linearly dependent upon soil moisture at moisture levels below saturation. Near saturation, the backscattering levels off, apparently becoming less sensitive to added increments of water (Dobson et al., 1985). Soil moisture influences the backscattered quantity due to the dielectric properties of the soil (Altese et al., 1996; Callens, Verhoest, & Davidson, 2006). When the ε of a soil increases linearly, σ^0 also increases, i.e. σ^0 and soil moisture content become positively correlated (Champion, 1996). Unit of σ^0 is $m^2\,m^{-2}$, but in general, it is expressed in dB (Wagner, 1998). σ^0 increases with an increase in soil moisture until the moisture content reaches 35 volumetric % when the radar signal becomes insensitive to the soil moisture (Dobson & Ulaby, 1981; Gorrab, Zribi, Baghdadi, Lili-Chabaane, & Mougenot, 2014; Ulaby et al., 1986; Zribi, Baghdadi, Holah, & Fafin, 2005). From the numerous field experiments, the linear relationship between σ^0 and M_v content is empirically expressed in Equation (2) (e.g. Champion, 1996):

$$\sigma^0 = A + B.W \tag{2}$$

where A is the backscattering coefficient of a completely dry soil surface and B is the sensitivity of σ^0 to change with the surface soil moisture content. A and B are regression coefficients dependent on soil surface roughness, incidence angle and soil texture (Autret, Bernard, & Vidal-Madjar, 1989; Bertuzzi, Chaànzy, Vidal-Madjar, & Autret, 1992; Champion & Faivre, 1997; Dobson & Ulaby, 1986;

Ulaby & Batlivala, 1976; Wagner, 1998). A is primarily controlled by surface roughness and the incidence angle (Dobson & Ulaby, 1986; Wagner, 1998).

However, field measurements have shown that the saturation effect at high moisture contents and the supersaturated and flooded soils behave as specular surfaces, which yield lower backscattering at off-nadir angles than non-saturated (but wet) soils (Dobson & Ulaby, 1981; Dobson et al., 1985).

Theoretical research on scattering of electromagnetic waves by rough surfaces has been done extensively and studies show that backscatter is very sensitive on the r.m.s (route mean square surface height) and the autocorrelation function of the surface height variations (Fung, 1994; Tsang, Kong, & Shin, 1985). However, in case of retrieval of σ^0 from Integral Equation Method (IEM) in well-defined situations shows good agreement with experimental results (Baghdadi, King, & Bonnifait, 2002; Baghdadi, King, Chanzy, et al., 2002).

3.3. Surface roughness
Soil surface roughness is one of the main indicators for mapping potential run-off surfaces because it triggers the infiltration processes (Baghdadi, King, & Bonnifait, 2002). Surface roughness intensely affects radar return and it is much more than the presence of surface soil moisture (Srivastava, Patel, Manchanda, & Adiga, 2003; Srivastava et al., 2006; Srivastava, Yograjan, Jayaraman, Rao, & Chandrasekhar, 1997). Incident radar beam is scattered in specular direction, rather it is directly reflected back to the sensor depending on the magnitude of surface roughness, and thus the radar image is constructed by part of reflected radar beam received by the antenna (Callens et al., 2006).

Numerous studies have proved that radar signal is more sensitive to surface roughness at high incidence angles than at low incidence angles (Baghdadi, King, & Bonnifait, 2002; Baghdadi, Cerdan, et al., 2008; Fung & Chan, 1992; Ulaby et al., 1986; Zribi & Dechambre, 2002). However, in case of retrieval of σ^0, radar return is sensitive to soil surface roughness parameter, especially dependent on correlation length (L), whereas other surface parameters are not too much sensitive (Baghdadi, King, & Bonnifait, 2002; Baghdadi, King, Chanzy, et al., 2002). Furthermore, measuring the L is a problematic issue due to substantial instability of agricultural soils. Baghdadi, Paillou, Grandjean, Dubois, and Davidson (2000) have shown that roughness parameters estimated from field measurements are very sensitive to the length of the roughness profile, and also shown that the root mean square surface height (r.m.s) and the L increase with profile length (Baghdadi et al., 2004).

Bryant et al. (2007) reported that the main source of retrieval errors is due to the differences in soil roughness parameters resulting from different measurement techniques and roughness transects (Baghdadi et al., 2000, 2006a). The discrepancies found are mainly related to the uncertainty in the measured roughness parameters, especially with respect to the L (Baghdadi et al., 2000; Le Toan et al., 1999). L was removed from the practical implementation in Oh (2004) due to measurement uncertainty. To minimize the effect of L, Oh (2004) used r.m.s as a model parameter (13). But the L was retrieved empirically and applied further explicitly by Baghdadai and Zribi. Dubois, Van Zyl, and Engman (1995a) derived a backscattering model (16–17) which does not require any L as well as ERS Scatterometer data (19, 21).

$$r.m.s = \sqrt{\frac{\sum_{i=0}^{n}(Z_i - \bar{Z})^2}{n-1}} \tag{3}$$

where Z_i denotes the height of the point, $\bar{Z}$ is the mean height and n is the total number of points taken under consideration.

2.4. Bulk density
Soil bulk density (ρ_b) has a significant inverse relationship with microwave emission (Mattikalli, Engman, Jackson, & Ahuja, 1998). The increasing bulk density of soil affects the dielectric properties

of dry and moist soil (Hallikainen et al., 1985; Singh & Kathpalia, 2007; Ulaby et al., 1982). It has been evident that dielectric parameters of soil at microwave frequencies are mainly the function of various properties of soil such as texture, moisture, bulk density, temperature and soil type (Gupta & Jangid, 2011a, 2011b).

3.5. Soil texture

Determination of direct soil texture through SAR images is a very difficult task. Soil texture affects the backscattering coefficient by change in soil dielectric properties through its textural configuration in terms of its water-holding capacity (Hallikainen et al., 1985). Sandy soils contain higher amount of free water than clay soils (Kong & Dorling, 2008; Srivastava et al., 2006); thus, the Pearson correlation between backscatter and soil moisture is higher in sandy soils (Blumberg et al., 2000). Jackson and Schmugge (1989) found that water molecules are adsorbed onto the soil particles and effectively immobilize their dipoles, disallowing bound water to interact with the radar signal. Difference in texture explains the difference in surface drying rate. For this reason, difficulties can be encountered in the interpretation of radar signals in cases where the vertical moisture profile varies strongly in the first centimeters (Anguela et al., 2010). In the C-band, decreasing soil clay content increases the sensitivity of the radar signal to soil moisture (Aubert et al., 2011; Ualby et al., 1978). Because the distribution of grain sizes controls the amount of free water that interacts with the incident microwave, the amount of free water gives significant contribution to SAR backscatter (Aubert et al., 2011; Srivastava et al., 2006; Srivastava, Patel, Sharma, & Navalgund, 2009).

3.6. Vegetation cover

Presence of vegetation over the soil surface reduces the backscattering sensitivity of soil, even if in agricultural fields, soil moisture gradients near the surface can change rapidly (D'Urso & Minaecapilli, 2006). Vegetation cover also affects the soil moisture retrieval by microwave due to various factors which are vegetation biomass, canopy type and configuration and crop condition (Bertoldi et al., 2014; Kornelsen & Coulibaly, 2013; Petropoulos et al., in press). Vegetation effects become stronger in case of dense vegetation as well as with the increase in microwave frequency. Among all microwave bands, L-band measurements still yield good results under various canopy types because it has a higher penetration power to vegetation canopy or vegetation cover to reach the soil surface, whereas for C-band which is highly sensitive to vegetation cover may lead to a distorted measurement (Jagdhuber, Hajnsek, Bronstert, & Papathanassiou, 2013; Vereecken et al., 2014; Western et al., 2004). However, for X-band SAR signals of this wavelength ($\lambda \sim 3$ cm) are not able to penetrate vegetation cover due to the way that dielectric permittivity of the biomass affects radar response (Baghdadi, Aubert, & Zribi, 2012; Jagdhuber et al., 2013; Vereecken et al., 2014). The effect of vegetation cover is also dependent upon the incidence angle and polarization of the instrument (Ulaby et al., 1986).

3.7. Incident angle (θ)

Local incidence angle has an important role to inversion of soil moisture with respect to surface roughness condition of soil (Aubert et al., 2011; Baghdadi, Cerdan, et al., 2008). Numerous experimental results and simulated data showed that sensitivity of radar signal is more sensitive to surface roughness at high incidence angles than at low incidence angles. Though low incidence angles are optimal for soil moisture estimation (Autret et al., 1989; Baghdadi, King, Chanzy, et al., 2002; Holah, Baghdadi, Zribi, Bruand, & King, 2005; Ualby et al., 1978). Baghdadi, King, Chanzy, et al., 2002 have reported that high incidence angles (>45°) are suitable for the discrimination between smooth and rough areas, under this conditions the backscattered signal has an exponential dependence on surface roughness (e.g. Baghdadi, Cerdan, et al., 2008; Baghdadi, Zribi, et al., 2008; Zribi & Dechambre, 2002). The Frauenhofer criterion proposed in Ulaby et al. (1982) considers a soil surface as rough when the phase difference between two rays scattered from separate points on the surface ($\Delta\phi = 2 \cdot k.r.m.s \cdot cos\theta$) exceeds $\pi/8$ ($r.m.s * \lambda/(32 \cdot cos\theta)$), where, $\Delta\phi$ is phase difference, k is wave number, $k.r.m.s$ is electromagnetic surface roughness and λ is a wave length (Baghdadi, Zribi, et al., 2008). They also observed that the dependence of the radar signal on surface roughness in agricultural area is mainly significant for low levels of roughness and it is difficult to discriminate between roughness greater than around 0.015 m with C-band SAR sensors (Baghdadi, Zribi, et al., 2008).

3.8. Bands

The ability of radar sensor to measure soil moisture is very much hampered in the areas with high vegetation dominance such as forest, because the lower microwave bands are inefficient to escape from vegetation attenuation (Jagdhuber et al., 2013; Ulaby & El-Rayes, 1987; Vereecken et al., 2014; Western et al., 2004). The crop canopies can also influence the σ^0; because plants also affect the σ^0 by their leaves dielectric behaviour. Dielectric behaviour of leaves was determined by direct measurements of oven-dried various types of vegetation material to find out the influence on radar return on complete dry condition and presence of water. Leaves have real part of the dielectric constant (ε') between 1.5 and 2, and imaginary part of the dielectric constant (ε'') is below 0.1 (Ulaby & El-Rayes, 1987).

On the other hand soil moisture estimation by shorter wavelength than C-band is hydrologically inefficient due to the small surface penetration power (Ulaby, Dubois, & van Zyl, 1996), where as L-band measurements still yield good results under various canopy types (Jagdhuber et al., 2013; Vereecken et al., 2014; Western et al., 2004) (Figure 1). To retrieve soil moisture using C- and lower bands (Table 2) requires more accurate roughness information for retrieval studies (Mattia et al., 1997). Longer wavelength (> L-band) contains more soil profile information in the backscattered signal (Ulaby et al., 1996). Thus, to minimize the influence of vegetation cover on radar images, long wave bands and steep incident angles are preferred (Álvarez-Mozos et al., 2005).

Figure 1. Behaviour of three different wave lengths due to vegetation cover.

Note: The shorter microwave signal (X-band, ~3 cm) interacts mainly with the top of the canopy cover, while C-band (~8 cm) travel more than X-band at vegetation canopy while longer wavelengths (L-band, ~24 cm) are able to penetrate further into the canopy and reflect from the soil surface.

S. No.	Band	Frequency range
1	HF	3–30 MHz
2	VHF	30–300 MHz
3	UHF	300–1,000 MHz
4	L	1–2 GHz
5	S	2–4 GHz
6	C	4–8 GHz
7	X	8–12 GHz
8	Ku	12–18 GHz
9	K	18–27 GHz
10	Ka	27–40 GHz
11	V	40–75 GHz
12	VV	75–110 GHz
13	Mm	110–300 GHz

Table 2. Microwave bands and frequencies (Sengupta & Liepa, 2006)

3.9. Polarization

Retrieval of soil moisture by radar return is highly dependent on surface geometry. This kind of problems could be minimized using multi-polarized and/or multi-frequency sensors systems (Baghdadi, Cerdan, et al., 2008; Zribi & Dechambre, 2002). Hirosawa, Komiyama, and Matsuzaka (1978) in his work used 9-GHz and his observations of Kanto loam confirmed that the cross-polarized sensitivity to near-surface volumetric soil moisture was four times better than that of the like-polarized backscattering. The HH and HV polarizations are more sensitive to soil roughness than the VV polarization (Holah et al., 2005), but Baghdadi, Cerdan, et al. (2008) found that retrieval of soil moisture by radar signal was not influenced by polarization, using an assembled database of ERS-2, RADARSAT-1 and ENVISAT data.

4. Models for retrieval of backscattering coefficient and soil moisture

Numerous theoretical, empirical and semi-empirical methods have been developed since the beginning of SAR studies to relate the SAR backscatter coefficient to soil moisture (Baghdadi, El Hajj, et al., 2006; Bryant et al., 2007; Dubois et al., 1995a; Fung, Li, & Chen, 1992; Haider, Said, Kothyari, & Arora, 2004; Oh et al., 1992; Sanli, Kurucu, Esetlili, & Abdikan, 2008; Srivastava et al., 2009; Wang & Qu, 2009). These theoretical models were derived from the electromagnetic theory. These theories are dependent on the site and surface type on which they were developed and tested by considering the incidence angle, λ and soil parameters (Baghdadi et al., 2004; D'Urso and Minaecapilli, 2006; Wang & Qu, 2009). Some of the most used models have been described below.

4.1. Most used theoretical models for active imaging microwave data

Theoretical models are used to derive the general trend of σ^0 in respect to soil moisture content and surface roughness (Dubois & van Zyl, 1994; Wang & Qu, 2009). Number of factors, surface roughness, ε, polarization and problem of electromagnetic wave scattering from random surfaces, influence the retrieval of σ^0, which has long been studied because of its complexity (Wang & Qu, 2009).

Numerous currently used surface scattering models have been developed from the small perturbation method (SPM) (Rice, 1951) and the Kirchhoff model (Beckmann & Spizzichino, 1963), which limits the range of roughness conditions (Wang & Qu, 2009). Perturbation solutions can be used whenever the soil surface slightly deviates from smooth to rough surfaces, and in SPM (Rice, 1951; Tsang et al., 1985), the *r.m.s* height must be much smaller than the wavelength and the *r.m.s* slope should be of the same order of magnitude as the wave number times the *r.m.s* height (Rice, 1951). A perturbation method based on perturbation expansion of the phase of the surface field (PPM) was developed which extends the region of validity of SPM to higher values of the *r.m.s* height, provided the slope remains relatively small (Winebrenner & Ishimaru, 1985). The other limiting case is when surface irregularities are large compared to the wavelength, which is equivalent to having a large radius of curvature at each point on the surface. In this type of limiting conditions, the Kirchhoff approximation (KA) is applicable (Boisvert et al., 1997; Ulaby et al., 1986). Various types of modifications and improvements to this model can be found in the literature. Extended validity of the KA solution was considered by Oh et al. (1992) but in limited extent.

Oh et al. (1992) developed an empirical model for retrieving soil moisture using the multi-polarized radar signal (*HH, VV, HV* and *VH*), and tried to find out the extended validity of KA and SPM model to measure surface roughness (Oh et al., 1992). Using the multi-polarized radar signal and considering the co-polarized ($p = \sigma^0_{HH}/\sigma^0_{VV}$) and cross-polarized ($q = \sigma^0_{HH}/\sigma^0_{VV}$) ratio, this model could predict the *r.m.s* height of the surface and soil ε. Oh, Sarabandi, and Ulaby (1993) modified and developed an empirical relation between the co-polarized phase parameters and roughness and ε of rough surfaces, where α is the degree of correlation, which is a measure of width of probability density function of a co-polarized phase angle. Oh, Sarabandi, and Ulaby (1994) modified the expression of cross-polarized ratio (*q*) in respect to Oh et al. (1992) for the same purpose. The expressions for *p* and *q* were further modified in 2002, and a new expression was proposed for the cross-polarized backscatter coefficient (Oh et al., 2002). Oh (2004) further updated semi-empirical polarimetric backscattering model to retrieve both M_v and $k.r.m.s$ (electromagnetic surface roughness) height by subsequent modification of q in respect of Oh et al. (2002).

A site-specific calibration procedure was developed by D'Urso and Minaecapilli (2006) for Oh et al.'s model (1992) to derive soil moisture content without prior information of surface roughness and soil water content.

The initial version of the Oh's model was presented by Oh et al. (1992) in Equations (4–6).

$$p = \frac{\sigma_{HH}^0}{\sigma_{VV}^0} = \left[1 - \left(\frac{\theta}{90°} \right)^{1/3\Gamma_0} e^{-k.r.m.s} \right]^2 \tag{4}$$

$$q = \frac{\sigma_{HH}^0}{\sigma_{VV}^0} = 0.23 \sqrt{\Gamma_0} \left(1 - e^{-k.r.m.s} \right) \tag{5}$$

$$\Gamma_0 = \left| \frac{1 - \sqrt{\varepsilon_r}}{1 + \sqrt{\varepsilon_r}} \right|^2 \tag{6}$$

where co-polarized ratio p ($p = \sigma_{HH}^0/\sigma_{VV}^0$) and the cross-polarized ratio q ($q = \sigma_{HV}^0/\sigma_{VV}^0$) to incident angle ($\theta$), wave number ($k$) and Fresnel reflectivity of the surface at nadir (Γ_0). The parameters p and q are derived by empirical fitting to the ground-based measurements of σ_{HH}^0, σ_{VV}^0 and σ_{HV}^0.

A new expression for q was proposed by Oh et al. (1994) to incorporate the effect of the incidence angle in Equation (7).

$$q = 0.25 \sqrt{\Gamma_0(0.1 + sin^{0.9}\theta)} \left(1 - e^{-[1.4 - 1.6\Gamma_0]k.r.m.s} \right) \tag{7}$$

The expressions for p and q were again modified in 2002, and an expression was proposed for the cross-polarized backscatter coefficient, expressed in Equations (8–10) (Oh, Sarabandi, & Ulaby, 2002):

$$p = 1 - \left(\frac{\theta}{90°} \right)^{0.35 \, Mv^{-0.65}} e^{-0.4 \, (k.r.m.s)^{1.4}} \tag{8}$$

$$q = 0.1 \left(\frac{r.m.s}{L} + sin \, 1.3\theta \right)^{1.2} \left(1 - e^{-0.9 \, (k.r.m.s)^{0.8}} \right) \tag{9}$$

$$\sigma_{HV}^0 = 0.11 \, Mv^{0.7} cos^{2.2}\theta \left(1 - e^{-0.32 \, (k.r.m.s)^{1.8}} \right) \tag{10}$$

Given that the measurement of the correlation length may not be exact (Baghdadi et al., 2000; Oh & Kay, 1998) and that the ratio q is insensitive to the roughness parameter, Oh (2004) proposed a new formulation for q that ignores the correlation length (11).

$$q = 0.095(0.13 + sin \, 1.5\theta)^{1.4} \left(1 - e^{-1.3(k.r.m.s)^{0.9}} \right) \tag{11}$$

The general formula to retrieve backscattering coefficient is expressed in Equation (12).

$$\sigma_{VV}^0 = \frac{\sigma_{HV}^0}{q} \tag{12}$$

where σ_{HV}^0 was derived from Equation (10).

$$\sigma_{HH}^0 = p\, \sigma_{VV}^0 = \frac{p}{q}\sigma_{HV}^0 \tag{13}$$

The estimates of r.m.s and M_v can also be obtained from the measurements of σ_{HV}^0 and p by the simple computation in Equation (10). Solving Equation (10) for the estimate of k.r.m.s yields Equation (14).

$$k.r.m.s\left(\theta, M_v, \sigma_{VHM}^0\right) = \left[-3.125 \ln\left\{1 - \frac{\sigma_{VHM}^0}{0.11 M_v^{0.7}(cos\theta)^{2.2}}\right\}\right]^{0.556} \tag{14}$$

where σ_{VHM}^0 is the measurement of the VH-polarized scattering coefficient.

$$1 - (\frac{\theta}{90°})^{0.35 Mv^{-0.65}}.e^{-0.4}[k.r.m.s(\theta, Mv,\ \sigma_{VHM}^0]^{1.4} - p_m = 0 \tag{15}$$

From the above Equation (15), M_v can be estimated, where p_m denotes the measured co-polarized ratio of p and k.r.m.s ($\theta, M_v, \sigma_{VHM}^0$) given in Equation (14) and k.r.m.s can computed from Equation (14) and r.m.s height can also be obtained subsequently (Oh, 2004).

   Dubois et al. (1995a) developed an empirical algorithm for the retrieval of soil moisture content and r.m.s from remotely sensed scatterometer data. The algorithm was optimized for bare surfaces and developed with data for frequencies varying between 1.5 ($\lambda = 0.205$ m) and 11 GHz ($\lambda = 0.028$ m), roughness ranging from 0.003 to 0.03 m and incidence angles between 30 and 45°. Using two co-polarized signals and omitting the usually weaker HV-polarized returns made the algorithm less sensitive to system. Dubois et al. (1995a) chose to use only the co-polarized backscatter signal instead of cross-polarized signals because they are less sensitive to vegetation, easy to calibrate and less susceptible to system noise. The empirical model of Dubois et al. (1995a) was widely used in retrieval of soil moisture (Rao et al., 2013) (16, 17). Even this model could be used in sparsely vegetated surfaces but limiting towards normalized difference vegetation index (NDVI) up to 0.4 (Neusch & Sties, 1999; Sikdar & Cumming, 2004). Dubois model is not valid for P-band (Western et al., 2004). To increase the domain of applicability of this model, Dubois et al. (1995a) found that vegetation effects could be minimized by excluding areas where the L-band $\sigma_{LHV}^0 / \sigma_{LVV}^0$ ratio (an index of vegetation cover) exceeds −11 dB.

$$\sigma_{HH}^0 = 10^{-2.75}\left(\frac{cos^{1.5}\theta}{sin^5\theta}\right)10^{0.028\varepsilon\ tan\theta}(k.r.m.s\,sin\,\theta)^{1.4}\lambda^{0.7} \tag{16}$$

$$\sigma_{VV}^0 = 10^{-2.35}\left(\frac{cos^3\theta}{sin^3\theta}\right)10^{0.046\varepsilon\ tan\theta}(k.r.m.s\,sin\,\theta)^{1.1}\lambda^{0.7} \tag{17}$$

4.2. Physical model for active imaging microwave data

The physical models evolved to predict the radar σ^0 and soil characteristics using the theoretical approaches of radar return. Physical models have site-specific dependencies and are limited to the range of roughness (Baghdadi, El Hajj, et al., 2006; Paloscia, Pampaloni, Pettinato, & Santi, 2008). The integral equation model (IEM) (Fung et al., 1992) is a widely used physical model to retrieve σ^0 by considering roughness parameters. IEM's validity domain covers wide range of surface roughness values encountered on agricultural soils. IEM model (Fung, 1994) provided a good result in laboratory experiments (Mancini, Hoeben, & Troch, 1999) but was unable to shown a good retrieval capability in field base measurement (Altese et al., 1996; Baghdadi, Holah, & Zribi, 2006b; Paloscia et al., 2008). The major difficulty associated with this model was that it was highly sensitive to surface roughness in terms of L and r.m.s height (Davidson et al., 2000). To minimize the retrieval error, Baghdadi, King, Chanzy, et al. (2002), Baghdadi, King, and Bonnifait (2002) developed a calibration method to retrieve soil moisture with low retrieval error by IEM. As IEM is dependent on surface roughness, L, Baghdadi, King, Chanzy, et al. (2002), Baghdadi, King, and Bonnifait (2002) developed an optical correlation length (L_{opt}) to derive σ^0 with minimum error.

4.3. Topp model

Topp model (Topp, Davis, & Annan, 1980) has been effectively used to derive soil moisture from soil ε (Song et al., 2010) (18). Topp model is used to create a comparative study with retrieved soil moisture and σ^0 for each of the theoretical and physical models (Aqil & Schmitt, 2010; Rao et al., 2013). To derive soil moisture, this model does not require any prior knowledge of soil texture and surface roughness and M_v can be retrieved by algorithm (18).

$$M_v = -5.3 \times 10^{-2} + 2.92 \times 10^{-2}\varepsilon' - 5.5 \times 10^{-4}\varepsilon'^2 + 4.3 \times 10^{-6}\varepsilon'^3 \tag{18}$$

4.4. Relative (m_s) and profile ($W(t)$) soil moisture content

Soil moisture retrieval method for ERS Scatteromater data was presented by Magagi and Kerr (1997), Pulliainen, Manninen, and Hallikainen (1998), Wagner, Noll, Borgeaud, and Rott (1999) and Wagner, Lemoine, Borgeaud, and Rott (1999). This method of profile soil moisture content (W) retrieval was explicitly described by Wagner, Lemoine, et al. (1999). This experiment required auxiliary information on soil type, soil texture, bulk density (kg m^{-3}), wilting level of both gravimetric and volumetric units, field capacity (FC) in mm and porosity/total water capacity (TWC) in mm. FC is the saturation level of soil with water when deep percolation nearly stops, and porosity is relative pore volume of the soil and is equal to the TWC of the soil. To retrieve soil moisture, the method compared time series data of σ^0 with standard reference incident angle 40° of ERS Scatterometer data. From this time series data, highest and lowest σ^0 were determined and denoted as $\sigma^0_{wet}(40, t)$ and $\sigma^0_{dry}(40, t)$, where 40° is the reference incident angle. σ^0_{dry} is considered to be the lowest σ^0 when no liquid water is present in the soil surface layer and $\sigma^0_{wet}(40, t)$ is the highest σ^0 of the soil surface layer when it is saturated with water. $\sigma^0_{wet}(40, t)$ and $\sigma^0_{dry}(40, t)$ were calculated according to Wagner, Noll, et al. (1999).

The relative soil moisture content (m_s) was calculated according to Wagner, Lemoine, et al. (1999) (19).

$$m_s(t) = \frac{\sigma^0(40, t) - \sigma^0_{dry}(40, t)}{\sigma^0_{wet}(40, t) - \sigma^0_{dry}(40, t)} \tag{19}$$

M_v of surface layer can be estimated by multiplying the m_s with the soil porosity or TWC and this m_s is considered to be the degree of saturation (Wagner, Lemoine, & Rott, 1999).

However, several simple models are used in soil moisture estimation, but with the increase in time lag, the potentiality of measurement decreased simultaneously. Thus, to improve the retrieval potentiality, temporal variations in terms of characteristic time length (T) were taken into consideration in soil water index (SWI) as given by Equation (20).

$$SWI(t) = \frac{\sum_i m_s(t_i)e^{-(t-t_i)/T}}{\sum_i e^{-(t-t_i)/T}} \text{ for } t_i \leq t \tag{20}$$

where m_s is the surface soil moisture estimated from the ERS Scatterometer at time t_i. The SWI is calculated if there is at least one ERS Scatterometer measurement in the time interval [$t-T,t$] and at least three measurements in the interval [$t-5T,t$]. Parameter T is the characteristic time length (Wagner, Lemoine, & Rott, 1999).

The profile soil moisture content W at time t can be estimated from SWI (21).

$$W(t) = W_{min} + SWI(t).(W_{max} - W_{min}) \tag{21}$$

SWI is the trend indicator which ranges between 0 and 1. W_{min} and W_{max} are the minimum and maximum wetness values.

4.5. Passive radiometric models

4.5.1. Radiative transfer model

Radiative transfer model (Njoku, 1999) is used to retrieve M_v. The algorithm requires the values of T_e (physical temperature) and W_e (vegetation columnar water content). The baseline algorithm (Njoku, 1999) uses two lowest AMSR frequencies (6.9 and 10.7 GHz), because above ~10 GHz frequency, surface roughness and vegetation scattering effects produce complexity and uncertainty in derived products. These frequencies have better vegetation penetration and soil moisture sensitivity, although with decreased spatial resolution (Njoku, 1999). In general parameter retrieval algorithms, the land surface is modelled as absorbing vegetation layer above soil in Figure 2.

The brightness temperature (T_{Bp}) observed at the top of the atmosphere at a given incidence angle and at a given frequency can be expressed by the radiative transfer Equation (22) (Njoku, 1999).

$$T_{Bp} = T_u + \exp\left(-\tau_c\right)\left[\left\{T_d r_{sp}\exp\left(-2\tau_c\right)\right\} + \left\{e_{sp}T_s\exp\left(-\tau_c\right) + T_c\left(1 - \omega_p\right)\right.\right.$$
$$\left.\left.\left[1 - \exp\left(-\tau_c\right)\right]\left[1 + r_{sp}\exp\left(-\tau_c\right)\right]\right\}\right] \tag{22}$$

where, T_u is the upwelling atmospheric temperature (K), T_d is the downwelling atmospheric temperature (K), T_a is the atmospheric opacity, T_c is the vegetation temperature (K), τ_c is the vegetation opacity, r_{sp} is the soil reflectivity, T_s is the effective soil temperature (K) (the effective temperature is the weighted-average temperature over the microwave penetration depth in the medium) and ω_p is the vegetation single-scattering albedo.

A simplified approximation is that the vegetation and underlying soil are close to the same physical temperature T_e. This approximation does not degrade the moisture retrieval accuracy, but will result in the retrieval of a mean or "effective" radiating temperature of the composite soil/vegetation medium.

Substituting $T_s \cong T_c = T_e$ in Equation (22), we obtain Equation (23).

$$T_{Bp} = T_u + \exp\left(-\tau_a\right)\left[\left\{T_d r_{sp}\exp\left(-2\tau_c\right)\right\} + T_e\left\{(1 - r_{sp})\exp\left(-\tau_c\right) + \left(1 - \omega_p\right)\right.\right.$$
$$\left.\left.\left[1 - \exp\left(-\tau_c\right)\right]\left[1 + r_{sp}\exp\left(-\tau_c\right)\right]\right\}\right] \tag{23}$$

4.5.1.1. Atmosphere. Standard expression for T_u and T_d can be obtained from the literature (Hofer & Njoku, 1981; Njoku, 1999; Njoku & Li, 1999). At atmospheric window frequencies T_u and T_d, it can be expressed using the effective radiating temperature approximation as Equation (24) (ignoring the space contribution to T_d):

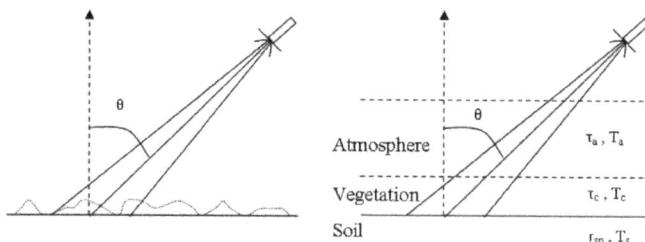

Figure 2. Model representation of a space-borne radiometer, viewing a heterogeneous earth surface (Njoku, 1999).

$$T_u \cong T_d \cong T_{ae}[1 - \exp(-\tau_a)] \tag{24}$$

where T_{ae} is the mean temperature of the microwave-emitting region of the atmosphere. This expression is valid for most atmospheric water vapour and cloud conditions. T_{ae} is frequency dependent and also depends on the distributions of temperature, humidity and liquid water. T_{ae} may be expressed simply as a function of the surface air temperature T_{as} and a frequency-dependent offset δT_a in Equation (25).

$$T_{ae} \approx T_{as} - \delta T_a \tag{25}$$

The effect of uncertainty in T_{ae} on the observed T_{Bp} is sufficiently small. The opacity τ_a along the atmospheric path is dependent on the viewing angle θ and the precipitable/vertical-column amounts of water q_v and vertical-column cloud liquid water q_l. It can be expressed (for a plane parallel atmosphere) in Equation (26).

$$\tau_a = (t_0 + a_v q_v + a_1 q_1)/cos\theta \tag{26}$$

where τ_o is the oxygen opacity at nadir and a_v and a_l are frequency-dependent coefficients and viewing angle θ.

4.5.1.2. *Surface.* The dependence of τ_c on vegetation columnar water content follows an approximately linear relationship, depicted in Equation (27) (Njoku, 1999).

$$\tau_c = bw_e/cos\theta \tag{27}$$

where $cos\theta$ accounts for the slant path through the vegetation. The coefficient b depends on canopy structure and frequency. Theory and experimental data suggest that for a given vegetation type (b), is approximately proportional to frequencies below ~10 GHz (Jackson & Schmugge, 1991; Levine & Karam, 1996).

This indicates that at higher frequencies, the frequency dependence of b decreases and its dependence on canopy structure eventually increases. This provides rationale for restricting the physically based retrieval algorithm to be in between 6.9 and 10.7 GHz.

The reflectivity of rough soil, r_{sp}, is related to that of smooth soil; r_{op} is computed by Equations (28) and (29) (Njoku & Li, 1999):

$$rsv = \left[(1 - Q)rov + Qroh\right]\exp(-h) \tag{28}$$

$$rsh = \left[(1 - Q)roh + Qrov\right]\exp(-h) \tag{29}$$

where rsv and rsh denoted the reflectivity of soil for both polarizations horizontal and vertical. Expression for h and Q is expressed in Equations (30) and (31).

$$Q = 0.35\left[1 - \exp\left(-0.6.\sigma^2.\lambda\right)\right] \tag{30}$$

$$h = (\frac{4\pi\sigma \ cos\theta}{\lambda})^2 \tag{31}$$

where λ is the wavelength of the radiometer and σ is the surface r.m.s height.

$$r_{ov} = \left| \frac{\varepsilon r \cos\theta - \sqrt{\varepsilon r - \sin^2\theta}}{\varepsilon r \cos\theta + \sqrt{\varepsilon r - \sin^2\theta}} \right|^2 \tag{32}$$

$$r_{oh} = \left| \frac{\cos\theta - \sqrt{\varepsilon r - \sin^2\theta}}{\cos\theta + \sqrt{\varepsilon r - \sin^2\theta}} \right|^2 \tag{33}$$

The Fresnel expressions relate the reflectivities r_{ov} and r_{oh} of a smooth, homogeneous soil to the complex dielectric constant of the soil ε_r. In the above equation, θ is the incidence angle relative to the surface normal. For a given frequency, the dielectric constant depends on the M_v and to a lesser extent on the soil type. This relationship can be expressed in Equation (34).

$$\varepsilon_r = f(M_v; \rho_b, s, c) \tag{34}$$

where M_v is the volumetric moisture content, ρ_b is the soil bulk density and s and c are soil and clay fraction.

4.5.2. Change detection algorithm

Global SWI is also known as the Basist wetness index (Basist, Grody, Peterson, & Williams, 1998). Surface wetness Index (SWI) is used to retrieve surface soil moisture using multi-temporal data from multi-frequency passive microwave radiometer (SSM/I). A change detection algorithm is used to compute soil moisture (SM) from SWI variations by considering field capacity and air–dry status, pixel by pixel (Oza et al., 2006; Singh, Oza, Chaudhari, & Dadhwal, 2005). Air–dry moisture status is considered to be minimum soil water content when SWI value is lowest and is considered to be maximum soil water content when SWI value is highest. Field-capacity and air–dry moisture status are obtained by considering multi temporal SWI data (Oza et al., 2006). The added advantage was that they have higher sensitivity towards soil moisture and less towards the surface geometry (Basist et al., 1998; Singh, Oza, Chaudhari, & Dhawal, 2005; Oza et al., 2006). By comparing the emissivity and wave frequency, Basist et al. (1998) developed a relation to derive SWI. SWI is proportional to the slope of emissivity as a function of frequency and is defined in Equation (35) (Basist et al., 1998).

$$SWI = \Delta \in T_s \tag{35}$$

where

$$\Delta \in = \beta_0 [\in (f_2) - \in (f_1)] + \beta_1 [\in (f_3) - \in (f_2)] \tag{36}$$

where T_s is the surface temperature, f_1, f_2 and f_3 represent operated vertical channels (frequency) and β_0 and β_1 are proportionality constants.

$$SM_i = M_{ad} + \left[\frac{(M_{fc} - M_{ad})}{SWI_{max} - SWI_{min}} \right] (SWI_i - SWI_{min}) \tag{37}$$

where SM_i is the soil moisture at pixel i, M_{fc} is the field capacity of soil at pixel i, M_{ad} is air–dry moisture level of soil at pixel i, SWI_{max} and SWI_{min} are the maximum and minimum SWI on multi-temporal data-set at pixel i and SWI_i is SWI in wetness composite image at pixel i.

Singh, Mishra, Shaoo, and Dey (2005) retrieved the soil moisture using IRS P4 (Oceansat 1) MSMR data. Algorithm of Gohill (1999) has been used to derive soil moisture at 6.6 GHz frequency because that frequency can't be affected by atmospheric attenuation or vegetation sensitivity. For radiometers working in shorter wavelength ranges, atmospheric attenuation and emission of the signal can be expressed as (Engman, 1991; Schmugge, 1985):

$$T_B = t(H)\left[rT_{sky} + (1-r)T_{soil}\right] + \{1 - t(H)\}T_{atm} \tag{38}$$

where $e = (1-r)$ is the emissivity, T_B is the microwave brightness temperature, $t(H)$ is the atmospheric transmission, r is the surface reflectivity and T_{sky}, T_{soil}, T_{atm} are temperatures of sky, soil and atmosphere respectively.

The SM has been estimated using 6.6 GHz horizontally polarized MSMR data using Equation (39) (Gohill, 1999):

$$SM = \left(-0.284T_{B6.6H}\right) + 76.2 \tag{39}$$

where SM is the soil moisture, total amount of water available (% volume) and $T_{B6.6H}$ is the brightness temperature at 6.6 GHz frequency in horizontal polarization.

4.5.3. Optimization of models

How to invert the moisture, soil texture and roughness from remote sensing data has been one of the most interesting problems to be resolved. Lots of theoretical and empirical models (Dubois et al., 1995a; Fung et al., 1992; Oh, 2004; Oh et al., 1992; Satalino, Mattia, Pasquariello, & Dente, 2005; Ulaby et al., 1982) were developed to retrieve the moisture content and roughness, but limitations of these models are that they can invert a few parameters only depending upon the number of the data-sets. Optimization techniques generally started to retrieve more of the parameters with less number of datasets. Nowadays, commonly used optimization techniques are artificial neural networks (ANN) (Lakhankar, Ghedira, & Khanbilvardi, 2006; Zhao et al., 2003) and genetic algorithms (GA) (Jin & Wang, 2000; Singh & Kathpalia, 2007).

5. Conclusion

The aim of this paper is to provide a systematic review on different basic soil moisture retrieval methods and models since 1978 until 2015 for both active and passive microwave remote sensing. Each of the methods and models has its own limitations of retrieval capacity in terms of their microwave bands (L-, C- and X-) used or in their target surface characteristics. However, the fact is that the contribution of other factors that influences the soil reflectance may not be effectively minimized.

To overcome these problems, SAR has shown its large potential for retrieving soil moisture maps at regional scales. However, since the backscattered signal is determined by several surface characteristics, the retrieval of soil moisture is an ill-posed problem when using single-configuration imagery. The advent of new, high-resolution sensors observing in X- and L-band, and C-band sensors yielding polarimetric data (HH, VV, HV and VH) allow for a better characterization of surface parameters. Along with the sensor configuration, different inversion methods have different validity regions. In hydrological perspective, L-band measurements yield good results under various canopy types due to high penetration power (Western et al., 2004). Passive microwave has more potential for large-scale soil moisture monitoring, but has a low spatial resolution. Active microwave can provide high spatial resolution, but has low revisit frequency and is more sensitive to soil roughness and vegetation. At microwave frequencies, many natural surfaces do not fall into the validity regions of the theoretical models, and even when they do, the available models fail to provide results in good agreement with experimental observations due to the following reasons (Oh et al., 1992):

(a) Dubois et al. (1995a), Dubois, VanZyl, and Engman (1995b) could yield good results of an area having normalized difference vegetation index (NDVI) up to 0.4 (Neusch & Sties, 1999; Sikdar & Cumming, 2004), but the range of this NDVI value is not valid for P-band (Western et al., 2004).

(b) The major difficulty associated with Oh model (Oh et al., 1992) is that the model is highly sensitive to surface roughness in terms of correlation length (L) and r.m.s height (Davidson et al., 2000). To minimize the effect of surface roughness, the Oh (2004) model was further modified to remove the effect of correlation length.

(c) However, ERS Scatterometer data have evolved to estimate the soil profile moisture content (W) by considering temporal variations.

(d) Soil moisture can be estimated using passive radiometer; for radiometer brightness temperature (T_{Bp}), it was shown to be sensitive to soil moisture.

(e) For IRS P4 (Oceansat 1) MSMR data, Gohill (1999) developed an algorithm to retrieve soil moisture at 6.6 GHz frequency.

Beside the active microwave remote sensing, ESA launched SMOS on 2 November 2009 which is the first L-band passive satellite sensor dedicated for global soil moisture estimation. Thereafter SMAP mission is for first-tire mission by NRC launch in January 2015, for estimation of soil moisture in L-band for both active and passive modes with each polarization (HH, VV, HV or VH). Furthermore, ESA is developing to launch Sentinel-1 mission that includes radar and multi-spectral imaging instruments for land, ocean and atmospheric monitoring (ESA, 2014). This mission can be able to provide high spatial (<1 km) and temporal (every 6 days globally) soil moisture data.

Funding
The authors received no direct funding for this research.

Author details
Kousik Das[1]
E-mail: kousik.envs@gmail.com
ORCID ID: http://orcid.org/0000-0002-7914-2276
Prabir Kumar Paul[1]
E-mail: prabirpaul59@gmail.com
ORCID ID: http://orcid.org/0000 -0002-6222-3408
[1] Department of Mining Engineering, Indian Institute of
Engineering Science and Technology, Shibpur, Howrah, India.

Cover image
Source: Authors.

References
Altese, E., Bolognani, O., Mancini, M., & Troch, P. A. (1996). Retrieving soil moisture over bare soil from ERS 1 synthetic aperture radar data: Sensitivity analysis based on a theoretical surface scattering model and field data. *Water Resources Research, 32*, 653–661. doi:10.1029/95WR03638

Álvarez-Mozos, J., Casalí, J., González-Audícana, M., & Verhoest, N. E. C. (2005). Correlation between ground measured soil moisture and RADARSAT-1 derived backscattering coefficient over an agricultural catchment of Navarre (North of Spain). *Biosystems Engineering, 92*, 119–133. doi:10.1016/j.biosystemseng.2005.06.008

Al-Yaari, A., Wigneron, J. P., Ducharne, A., Kerr, Y. H., Wagner, W., De Lannoy, G., ... Mialon, A. (2014). Global-scale comparison of passive (SMOS) and active (ASCAT) satellite based microwave soil moisture retrievals with soil moisture simulations (MERRA-Land). *Remote Sensing of Environment, 152*, 614–626. http://dx.doi.org/10.1016/j.rse.2014.07.013

Anguela, T. P., Zribi, M., Baghdadi, N., & Loumagne, C. (2010). Analysis of local variation of soil surface parameters with TerraSAR-X Radar data over bare agricultural fields. *IEEE Transactions on Geoscience and Remote Sensing, 48*, 874–881. http://dx.doi.org/10.1109/TGRS.2009.2028019

Anudeep, S. (2013). *Blending approach for soil moisture retrieval using microwave remote sensing* (MTech dissertation, Andhra University, pp. 1–95). Indian Institute of Remote Sensing, ISRO, Dehradun.

Aqil, S., & Schmitt, D. R. (2010, May 10-14). Dielectric permittivity of clay adsorbed water: Effect of salinity (AAPG Search and Discovery Article #90172 © CSPG/CSEG/ CWLS, pp. 1–4). In *GeoConvention 2010*. Calgary.

Aubert, M., Baghdadi, N., Zribi, M., Douaoui, A., Loumagne, C., Baup, F., ...Garrigues, S. (2011). Analysis of TerraSAR-X data sensitivity to bare soil moisture, roughness, composition and soil crust. *Remote Sensing of Environment, 115*, 1801– 1810. http://dx.doi.org/10.1016/j.rse.2011.02.021

Autret, M., Bernard, R., & Vidal-Madjar, D. (1989). Theoretical study of the sensitivity of the microwave backscattering coefficient to the soil surface parameters. *International Journal of Remote Sensing, 10*, 171–179. doi:10.1080/01431168908903854

Baghdadi, N., Aubert, M., & Zribi, M. (2012). Use of TerraSAR-X data to retrieve soil moisture over bare soil agricultural fields. *IEEE Geoscience and Remote Sensing Letters, 9*, 512–516. http://dx.doi.org/10.1109/LGRS.2011.2173155

Baghdadi, N., Cerdan, O., Zribi, M., Auzet, V., Darboux, F., El Hajj, M., & Kheir, R. B. (2008). Operational performance of current synthetic aperture radar sensors in mapping soil surface characteristics in agricultural environments: Application to hydrological and erosion modelling. *Hydrological Processes, 22*, 9–20. doi:10.1002/hyp.6609

Baghdadi, N., Gherboudj, I., Zribi, M. M., Sahebi, M., King, C., & Bonn, F. (2004). Semi-empirical calibration of the IEM backscattering model using radar images and moisture and roughness field measurements. *International Journal of Remote Sensing, 25*, 3593–3623. doi:10.1080/0143116 0310001654392

Baghdadi, N., Holah, N., & Zribi, M. (2006a). Soil moisture estimation using multi-incidence and multi-polarization ASAR data. *International Journal of Remote Sensing, 27*, 1907–1920. doi:10.1080/01431160500239032

Baghdadi, N., Holah, N., & Zribi, M. (2006b). Calibration of the integral equation model for SAR data in C-band and HH and VV polarizations. *International Journal of Remote Sensing, 27*, 805–816. http://dx.doi.org/10.1080/01431160500212278

Baghdadi, N., King, C., Chanzy, A., & Wigneron, J. P. (2002). An empirical calibration of the integral equation model based on SAR data, soil moisture and surface roughness measurement over bare soils. *International Journal of Remote Sensing, 23*, 4325–4340. doi:10.1080/01431160110107671

Baghdadi, N., King, C., & Bonnifait, L. (2002). An empirical calibration of the integral equation model based on SAR data and soil parameters measurements. In *Geoscience and Remote Sensing Symposium, IGARSS '02* (pp. 2646–2650). 2002 IEEE International. doi:10.1109/ IGARSS.2002.1026729

Baghdadi, N., Paillou, P., Grandjean, G., Dubois, P., & Davidson, M. (2000). Relationship between profile length and roughness variables for natural surfaces. *International Journal of Remote Sensing, 21*, 3375–3381. doi:10.1080/014311600750019994

Baghdadi, N., Zribi, M., Loumagne, C., Ansart, P., & Anguela, T. P. (2008). Analysis of TerraSAR-X data and their sensitivity to soil surface parameters over bare agricultural fields. *Remote Sensing of Environment, 112*, 4370–4379. doi:10.1016/j.rse.2008.08.004

Basist, A., Grody, N. C., Peterson, T. C., & Williams, C. N. (1998). Using the special sensor microwave/imager to monitor land surface temperatures, wetness, and snow cover. *Journal of Applied Meteorology, 37*, 888–911. doi:10.1175/1520-0450(1998)037<0888:UTSSMI>2.0.CO;2

Batlivala, P. P. & Ulaby, F. T. (1977). *Feasibility of monitoring soil moisture using active microwave remote sensing Remote Sensing Laboratory* (Technical Report 264-12). Lawrence: University of Kansas.

Beckmann, P., & Spizzichino, A. (1963). *The scattering of electromagnetic waves from rough surfaces* (Chapters 3 and 5). New York, NY: MacMillan.

Bertoldi, G., Chiesa, S. D., Notarnicola, C., Pasolli, L., Niedrist, G., & Tappeiner, U. (2014). Estimation of soil moisture patterns in mountain grasslands by means of SAR RADARSAT2 images and hydrological modeling. *Journal of Hydrology, 516*, 245–257. http://dx.doi.org/10.1016/j.jhydrol.2014.02.018

Bertuzzi, P., Chaànzy, A., Vidal-Madjar, D., & Autret, M. (1992). The use of a microwave backscatter model for retrieving soil moisture over bare soil. *International Journal of Remote Sensing, 13*, 2653–2668. doi:10.1080/01431169208904070

Blumberg, D. G., Freilikher, V., Lyalko, I. V., Vulfson, L. D., Kotlyar, A. L., Shevchenko, V. N., & Ryabokonenko, A. D. (2000). Soil moisture (water-content) assessment by an airborne scatterometer. *Remote Sensing of Environment, 71*, 309–319. doi:10.1016/S0034-4257(99)00087-5

Boisvert, J. B., Gwyn, Q. H. J., Chanzy, A., Major, D. J., Brisco, B., & Brown, R. J. (1997). Effect of surface soil moisture gradients on modelling radar backscattering from bare fields. *International Journal of Remote Sensing, 18*, 153–170. doi:10.1080/014311697219330

Bryant, R., Moran, M. S., Thoma, D. P., Collins, C. D. H., Skirvin, S., Rahman, M., ... Gonzalez-Dugo, M. P. (2007). Measuring surface roughness height to parameterize radar backscatter models for retrieval of surface soil moisture. *IEEE Geoscience and Remote Sensing Letters, 4*, 137–141. doi:/10.1109/LGRS.2006.887146

Callens, M., Verhoest, N. C. E., & Davidson, M. W. J. (2006). Parameterization of tillage-induced single scale soil roughness from 4-m profiles. *IEEE Transactions on Geoscience and Remote Sensing, 44*, 878–888. doi:10.1109/TGRS.2005.860488

Chai, S. S., Walker, J. P., Makarynskyy, O., Kuhn, M., Veenendaal, B., & West, G. (2010). Use of soil moisture variability in artificial neural network retrieval of soil moisture. *Remote Sensing, 2*, 166–190.

Champion, I. (1996). Simple modelling of radar backscattering coefficient over a bare soil: Variation with incidence angle, frequency and polarization. *International Journal of Remote Sensing, 17*, 783–800. doi:10.1080/01431169608949045

Champion, I., & Faivre, R. (1997). Sensitivity of the radar signal to soil moisture: Variation with incidence angle, frequency, and polarization. *IEEE Transactions on Geoscience and Remote Sensing, 35*, 781–783. doi:10.1109/36.582001

Chen, X. Z., Chen, S. S., Zhong, R. F., Su, Y. X., Liao, J. S., Li, D., ... Li, X. (2012). A semi-empirical inversion model for assessing surface soil moisture using AMSR-E brightness temperatures. *Journal of Hydrology, 456*, 1–11.

Chow, V. T., Maidment, D. R., & Mays, L. W. (1988). *Applied hydrology*. New York, NY: McGraw-Hill Education.

D'Urso, G., & Minacapilli, M. (2006). A semi-empirical approach for surface soil water content estimation from radar data without a priori information on surface roughness. *Journal of Hydrology, 321*, 297–310. doi:10.1016/j.jhydrol.2005.08.013

Davidson, M. W. J., Le Toan, T., Mattia, F., Satalino, G., Manninen, T., & Borgeaud, M. (2000). On the characterization of agricultural soil roughness for radar remote sensing studies. *IEEE Transactions on Geoscience and Remote Sensing, 38*, 630–640. http://dx.doi.org/10.1109/36.841993

Dobson, M. C., & Ulaby, F. T. (1981). Microwave backscatter dependence on surface roughness, soil moisture, and soil texture: Part III-soil tension. *IEEE Transactions on Geoscience and Remote Sensing, 19*, 51–61. doi:10.1109/TGRS.1981.350328

Dobson, M. C., & Ulaby, F. T. (1986). Active microwave soil moisture research. *IEEE Transactions on Geoscience and Remote Sensing, 24*, 23–36. doi:10.1109/TGRS.1986.289585

Dobson, M. C., Ulaby, F. T., Hallikainen, M. T., & El-Rayes, M. A. (1985). Microwave dielectric behavior of wet soil-Part II: Dielectric mixing models. *IEEE Transactions on Geoscience and Remote Sensing, 23*, 35–46. doi:10.1109/TGRS.1985.289498

Dubois, P., & van Zyl, J. (1994). An empirical soil moisture estimation algorithm using imaging radar. *Proceedings of IGARSS'94, IEEE 3* (pp. 1573–1575). doi: 10.1109/IGARSS.1994.399501

Dubois, P., van Zyl, J., & Engman, T. (1995a). Measuring soil moisture with imaging radars. *IEEE Transactions on Geoscience and Remote Sensing, 33*, 915–926. doi:10.1109/36.406677

Dubois, P., vanZyl, J., & Engman, T. (1995b). Corrections to "Measuring soil moisture with imaging radars". *IEEE Transactions on Geoscience and Remote Sensing, 33*, 1340. doi:10.1109/TGRS.1995.477194

Engman, E. T. (1990). Progress in microwave remote sensing of soil moisture. *Canadian Journal of Remote Sensing, 16*, 6–14. http://dx.doi.org/10.1080/07038992.1990.11487620

Engman, E. T. (1991). Applications of microwave remote sensing of soil moisture for water resources and agriculture. *Remote Sensing of Environment, 35*, 213–226. doi:10.1016/0034-4257(91)90013-V

Engman, E. T., & Gurney, R. J. (1991). *Remote sensing in hydrology*. Remote Sensing Applications Series, ISBN 0 412 24450. http://dx.doi.org/10.1007/978-94-009-0407-1

Entekhabi, D., Njoku, E. G., O'Neill, P. E., Kellogg, K. H., Crow, W. T., Edelstein, W. N., ... Van Zyl, J. (2010). The soil moisture active passive (SMAP) mission. *Proceedings of the IEEE, 98*, 704–716. http://dx.doi.org/10.1109/JPROC.2010.2043918

ESA. (2014). *Sentinel-1*. Retrieved October 27, 2014, from http://www.esa.int/Our_Activities/Observing_the_Earth/Copernicus/Sentinel-1

Fung, A. K. (1994). *Microwave scattering and emission models and their applications*. Nordwood, MA: Artech House.

Fung, A. K., & Chen, K. S. (1992). Dependence of the surface backscattering coefficients on roughness, frequency and polarization states. *International Journal of Remote Sensing, 13*, 1663–1680. doi:10.1080/01431169208904219

Fung, A. K., Li, Z., & Chen, K. S. (1992). Backscattering from a randomly rough dielectric surface. *IEEE Transactions on Geoscience and Remote Sensing, 30*, 356–369. doi:10.1109/36.134085

Gharechelou, S., Tateishi, R., & Sumantyo, J. T. S. (2015). Interrelationship analysis of L-band backscattering intensity and soil dielectric constant for soil moisture retrieval using PALSAR data. *Advances in Remote Sensing, 4*, 15–24. http://dx.doi.org/10.4236/ars.2015.41002

Giacomelli, A., Bacchiega, U., Troch, P. A., & Mancini, M. (1995). Evaluation of surface soil moisture distribution by means of SAR remote sensing techniques and conceptual hydrological modelling. *Journal of Hydrology, 166*, 445–459. doi:10.1016/0022-1694(94)05100-C

Gohill, B. S. (1999, May 19–22). *First science workshop proceedings of Megha-Tropiques* (pp. 20.1–20.15). Bangalore: ISRO.

Gorrab, A., Zribi, M., Baghdadi, N., Lili-Chabaane, Z., & Mougenot, B. (2014). Multi-frequency analysis of soil moisture vertical heterogeneity effect on radar backscatter. In *Advanced Technologies for Signal and Image Processing (ATSIP), 2014 1st International Conference* (pp. 379–384). 1653 IEEE. http://dx.doi.org/10.1109/ATSIP.2014.6834640

Gupta, V. K., & Jangid, R. A. (2011a). Microwave response of rough surfaces with auto-correlation functions, RMS heights and correlation lengths using active remote sensing. *Indian Journal of Radio and Space Physics, 40*, 137–146.

Gupta, V. K., & Jangid, R. A. (2011b). The effect of bulk density on emission behavior of soil at microwave frequencies. *International Journal of Microwave Science and Technology, 2011*, 1–6. doi:10.1155/2011/160129

Haider, S. S., Said, S., Kothyari, U. C., & Arora, M. K. (2004). Soil moisture estimation using ERS 2 SAR data: A case study in the Solani River catchment. *Hydrological Sciences Journal, 49*, 323–334. doi:10.1623/hysj.49.2.323.34832

Hallikainen, M. T., Ulaby, F. T., Dobson, M. C., El-rayes, M. A., & Wu, L. K. (1985). Microwave dielectric behavior of wet soil-Part 1: Empirical models and experimental observations. *IEEE Transactions on Geoscience and Remote Sensing, 23*, 25–34. doi:10.1109/TGRS.1985.289497

Hirosawa, H., Komiyama, S., & Matsuzaka, Y. (1978). Cross-polarized radar backscatter from moist soil. *Remote Sensing of Environment, 7*, 211–217. doi:10.1016/0034-4257(78)90032-9

Hofer, R., & Njoku, E. G. (1981). Regression techniques for oceanographic parameter retrieval using space-borne microwave radiometry. *IEEE Transactions on Geoscience and Remote Sensing, 19*, 178–189. http://dx.doi.org/10.1109/TGRS.1981.350370

Holah, N., Baghdadi, N., Zribi, M., Bruand, A., & King, C. (2005). Potential of ASAR/ENVISAT for the characterization of soil surface parameters over bare agricultural fields. *Remote Sensing of Environment, 96*, 78–86. doi:10.1016/j.rse.2005.01.008

Jackson, T. J., & Schmugge, T. J. (1989). Passive microwave remote-sensing system for soil moisture: Some supporting research. *IEEE Transactions on Geoscience and Remote Sensing, 27*, 225–235. doi:10.1109/36.20301

Jackson, T. J., & Schmugge, T. J. (1991). Vegetation effects on the microwave emission of soils. *Remote Sensing of Environment, 36*, 203–212. doi:10.1016/0034-4257(91)90057-D

Jagdhuber, T., Hajnsek, I., Bronstert, A., & Papathanassiou, K. P. (2013). Soil moisture estimation under low vegetation cover using a multi-angular polarimetric decomposition. *IEEE Transactions on Geoscience and Remote Sensing, 51*, 2201–2215. http://dx.doi.org/10.1109/TGRS.2012.2209433

Jin, Y. Q., & Wang, Y. (2000). A novel genetic algorithm to retrieve surface roughness and wetness from angular back-scattering. In *Proceedings of IEEE, IGARSS*.

Kong, X., & Dorling, S. R. (2008). Near-surface soil moisture retrieval from ASAR wide swath imagery using a principal component analysis. *International Journal of Remote Sensing, 29*, 2925–2942. doi:10.1080/01431160701442088

Kornelsen, K. C., & Coulibaly, P. (2013). Advances in soil moisture retrieval from synthetic aperture radar and hydrological applications. *Journal of Hydrology, 476*, 460–489. http://dx.doi.org/10.1016/j.jhydrol.2012.10.044

Laguardia, G., & Niemeyer, S. (2008). On the comparison between the LISFLOOD modelled and the ERS/SCAT derived soil moisture estimates. *Hydrology and Earth System Sciences, 12*, 1339–1351. http://dx.doi.org/10.5194/hess-12-1339-2008

Lakhankar, T., Ghedira, H., & Khanbilvardi, R. (2006, May 1–5). Neural network and fuzzy logic for an improved soil moisture estimation. In *ASPRS 2006 Annual Conference Reno*. Nevada.

Le Toan, T., Davidson, M., Mattia, F., Borderies, P., Chenerie, I., Manninen, T., & Borgeaud, M. (1999). Improved observation and modeling of bare surfaces for soil moisture retrieval. *Earth Observation Quarterly, 62*, 20–24.

Levine, D. M., & Karam, M. A. (1996). Dependence of attenuation in a vegetation canopy on frequency and plant water content. *IEEE Transactions on Geoscience and Remote Sensing, 34*, 1090–1096. doi:10.1109/36.536525

Li, Y. Y., Zhao, K., Ren, J. H., Ding, Y. L., & Wu, L. L. (2014). Analysis of the dielectric constant of saline-alkali soils and the effect on radar backscattering coefficient: A case study of soda alkaline saline soils in Western Jilin Province using RADARSAT-2 data. *The Scientific World Journal, 2014*, 1–14. doi:10.1155/2014/56301

Liu, Y. Y., Parinussa, R. M., Dorigo, W. A., De Jeu, R. A. M., Wagner, W., van Dijk, A. I. J. M., ... Evans, J. P. (2011). Developing an improved soil moisture dataset by blending passive and active microwave satellite-based retrievals. *Hydrology and Earth System Sciences, 15*, 425–436. doi:10.5194/hess-15-425-2011

Magagi, R. D., & Kerr, Y. H. (1997). Retrieval of soil moisture and vegetation characteristics by use of ERS-1 wind scatterometer over arid and semi-arid areas. *Journal of Hydrology, 188–189*, 361–384. doi:10.1016/S0022-1694(96)03166-6

Mancini, M., Hoeben, R., & Troch, P. A. (1999). Multifrequency radar observations of bare surface soil moisture content: A laboratory experiment. *Water Resources Research, 35*, 1827–1838. doi:10.1029/1999WR900033

Mattia, F., Le Toan, T., Souyris, J. C., De Carolis, G., Floury, N., Posa, F., & Pasquariello, G. (1997). The effect of surface roughness on multifrequency polarimetric SAR data. *IEEE Transactions on Geoscience and Remote Sensing, 35*, 915–926.

Mattikalli, N. M., Engman, E. T., Jackson, T. J., & Ahuja, L. R. (1998). Microwave remote sensing of temporal variations of brightness temperature and near-surface soil water content during a watershed-scale field experiment, and its application to the estimation of soil physical properties. *Water Resources Research, 34*, 2289–2299.

Mironov, V. L., Dobson, M. C., Kaupp, V. H., Komarov, S. A., & Kleshchenko, V. N. (2002). Generalized refractive mixing dielectric model for moist soils. *IEEE International Geoscience and Remotes Sensing Symposium, 4*, 3556–3558. http://dx.doi.org/10.1109/IGARSS.2002.1027247

Mironov, V. L., Dobson, M. C., Kaupp, V. H., Komarov, S. A., & Kleshchenko, V. N. (2004). Generalized refractive mixing dielectric model for moist soils. *IEEE Transactions on Geoscience and Remote Sensing, 42*, 773–785. http://dx.doi.org/10.1109/TGRS.2003.823288

Moran, M. S., Peters-Lidard, C. D., Watts, J. M., & McElroy, S. (2004). Estimating soil moisture at the watershed scale with satellite-based radar and land surface models. *Canadian Journal of Remote Sensing, 30*, 805–826. http://dx.doi.org/10.5589/m04-043

Neusch, T., & Sties, M. (1999). Application of the Dubois-model using experimental synthetic aperture radar data for the determination of soil moisture and surface roughness.

ISPRS Journal of Photogrammetry and Remote Sensing, 54, 273–278. doi:10.1016/S0924-2716(99)00019-2

Njoku, E. G. (1999). *AMSR land surface parameters.* Pasadena, CA: Jet Propulsion Laboratory, California Institute of Technology.

Njoku, E. G., & Li Li, L. (1999). Retrieval of land surface parameters using passive microwave measurements at 6-18 GHz. *IEEE Transactions on Geoscience and Remote Sensing, 37,* 79–93. http://dx.doi.org/10.1109/36.739125

Oh, Y. (2004). Quantitative retrieval of soil moisture content and surface roughness from multipolarized radar observations of bare soil surfaces. *IEEE Transactions on Geoscience and Remote Sensing, 42,* 596–601. doi:10.1109/TGRS.2003.821065

Oh, Y., & Kay, Y. (1998). Condition for precise measurement of soil surface roughness. *IEEE Transactions on Geoscience and Remote Sensing, 36,* 691–695. doi:10.1109/36.662751

Oh, Y., Sarabandi, K., & Ulaby, F. T. (1992). An empirical model and an inversion technique for radar scattering from bare soil surfaces. *IEEE Transactions on Geoscience and Remote Sensing, 30,* 370–381. doi:10.1109/36.134086

Oh, Y., Sarabandi, K., & Ulaby, F. T. (1993). An empirical model for phase difference statistics of rough surfaces. *Geoscience and Remote Sensing Symposium, 1993. IGARSS '93. Better Understanding of Earth Environment., International, 3,* 1003–1005. doi:10.1109/IGARSS.1993.322639

Oh, Y., Sarabandi, K., & Ulaby, F. T. (1994). An inversion algorithm for retrieving soil moisture and surface roughness from polarimetric radar observation. *Proceedings IGARSS'94, Pasadena* (IEEE catalog No. 94CH3378-7, III, pp. 1582–1584). New York, NY: IEEE.

Oh, Y., Sarabandi, K., & Ulaby, F. T. (2002). Semi-empirical model of the ensemble averaged differential Mueller matrix for microwave backscattering from bare soil surfaces. *IEEE Transactions on Geoscience and Remote Sensing, 40,* 1348–1355. doi:10.1109/TGRS.2002.800232

Oza, S. R., Singh, R. R., Dadhwal, V. K., & Desai, P. S. (2006). Large area soil moisture estimation and mapping using space-borne multi-frequency passive microwave data. *Journal of the Indian Society of Remote Sensing, 34,* 343–350. http://dx.doi.org/10.1007/BF02990919

Paloscia, S., Pampaloni, P., Pettinato, S., & Santi, E. (2008). A comparison of algorithms for retrieving soil moisture from ENVISAT/ASAR images. *IEEE Transactions on Geoscience and Remote Sensing, 46,* 3274–3284.

Peplinski, N. R., Ulaby, F. T., & Dobson, M. C. (1995a). Dielectric properties of soils in the 0.3-1.3 GHz range. *IEEE Transactions on Geoscience and Remote Sensing, 33,* 803–807. doi:10.1109/36.387598

Peplinski, N. R., Ulaby, F. T., & Dobson, M. C. (1995b). Corrections to "Dielectric properties of soils in the 0.3-1.3-GHz range". *IEEE Transactions on Geoscience and Remote Sensing, 33,* 1340. doi:10.1109/TGRS.1995.477193

Petropoulos, G. P., Ireland, G., & Barrett, B. (in press). Surface soil moisture retrievals from remote sensing: Current status, products and future trends. *Physics and Chemistry of the Earth.*

Prakash, R., Singh, D., & Pathak, N. P. (2012). A fusion approach to retrieve soil moisture with SAR and optical data. *IEEE Journal of Selected Topics in Applied Earth Observations and Remote Sensing, 5,* 196–206. doi:10.1109/JSTARS.2011.2169236

Pulliainen, J. T., Manninen, T., & Hallikainen, M. T. (1998). Application of ERS-1 wind scatterometer data to soil frost and soil moisture monitoring in boreal forest zone. *IEEE Transactions on Geoscience and Remote Sensing, 36,* 849–863. doi:10.1109/36.673678

Rao, S. S., Dinesh, S. K., Das, N., Nagaraju, M. S. S., Venugopal, M. V., Rajanar, P., ... Sharma, J. R. (2013). Modified Dubois model for estimating soil moisture with dual polarized SAR data. *Journal of the Indian Society of Remote Sensing, 41,* 865–872. doi:10.1007/s12524-013-0274-3

Rice, S. O. (1951). Reflection of electromagnetic waves from slightly rough surfaces. *Communications on Pure and Applied Mathematics, 4,* 351–378. doi:10.1002/cpa.3160040206

Sanli, F. B., Kurucu, Y., Esetlili, M. T., & Abdikan, S. (2008). Soil moisture estimation from RADARSAT -1, ASAR and PALSAR data in agricultural fields of Menemen plane of western Turkey. In *Proceedings of Commission VII ISPRS (International Society for Photogrammetry and Remote Sensing) Congress* (pp. 75–82). Beijing.

Satalino, G., Mattia, F., Pasquariello, G., & Dente, L. (2005). Soil moisture retrieval from ASAR measurements over natural surfaces with a large roughness variability (pp. 396–399). In *Procedings of IEEE, IGARSS.*

Schmugge, T. (1985). *Remote sensing of soil moisture: Hydrology forecasting.* Chichester.

Schmugge, T. J., Jackson, T. J., & McKim, H. L. (1980). Survey of methods for soil moisture determination. *Water Resources Research, 16,* 961–979. doi:10.1029/WR016i006p00961

Seneviratne, S. I., Corti, T., Davin, E. L., Hirschi, M., Jaeger, E. B., Lehner, I., ... Teuling, A. J. (2010). Investigating soil moisture–climate interactions in a changing climate: A review. *Earth-Science Reviews, 99,* 125–161. http://dx.doi.org/10.1016/j.earscirev.2010.02.004

Sengupta, D. L., & Leipa, V. V. (2006). Frequency band designation: Appendix B. *Applied Electromagnetics Compatability,* 467–472.

Shi, J., Du, Y., Du, J., Jiang, L., Chai, L., Mao, K., ... Wang, Y. (2012). Progresses on microwave remote sensing of land surface parameters. *Science China Earth Sciences, 55,* 1052–1078. http://dx.doi.org/10.1007/s11430-012-4444-x

Sikdar, M., & Cumming, I. (2004). A modified empirical model for soil moisture estimation in vegetated areas using SAR data. *Geoscience and Remote Sensing Symposium, 2004. IGARSS'04. Proceedings. 2004 IEEE International, 2,* 803–806. http://dx.doi.org/10.1109/IGARSS.2004.1368526

Singh, D., & Kathpalia, A. (2007). An efficient modeling with ga approach to retrieve soil texture, moisture and roughness from Ers-2 SAR data. *Progress In Electromagnetics Research, 77,* 121–136. doi:10.2528/PIER07071803

Singh, R. P., Mishra, D. R., Sahoo, A. K., & Dey, S. (2005). Spatial and temporal variability of soil moisture over India using IRS P4 MSMR data. *International Journal of Remote Sensing, 26,* 2241–2247. doi:10.1080/01431160500043723

Singh, R. P., Oza, S. R., Chaudhari, K. N., & Dadhwal, V. K. (2005). Spatial and temporal patterns of surface soil moisture over India estimated using surface wetness index from SSM/I microwave radiometer. *International Journal of Remote Sensing, 26,* 1269–1276. doi:10.1080/0143116041 2331330284

Song, K., Zhou, X., & Fan, Y. (2010). Retrieval of soil moisture content from microwave backscattering using a modified IEM model. *Progress In Electromagnetics Research B, 26,* 383–399. doi:10.2528/PIERB10072905

Srivastava, H. S., Patel, P., Manchanda, M. L., & Adiga, S. (2003). Use of multiincidence angle RADARSAT-1 SAR data to incorporate the effect of surface roughness in soil moisture estimation. *IEEE Transactions on Geoscience and Remote Sensing, 41,* 1638–1640. doi:10.1109/TGRS.2003.813356

Srivastava, H. S., Patel, P., & Navalgund, R. R. (2006, November 13–17). How far SAR has fulfilled its expectation for soil moisture retrieval? (SPIE Digital Library, 6410 Paper No. 64100, pp. 1–12). In *Microwave Remote Sensing of Atmosphere and Environment-II, AE107, Asia Pacific Remote Sensing Symposium.* Goa.

Srivastava, H. S., Patel, P., Sharma, Y., & Navalgund, R. R. (2009). Large-area soil moisture estimation using multi-incidence-angle RADARSAT-1 SAR data. *IEEE Transactions* on *Geoscience and Remote Sensing, 47,* 2528–2535. doi:10.1109/TGRS.2009.2018448

Srivastava, S. K., Yograjan, N., Jayaraman, V., Rao, P. P. N., & Chandrasekhar, M. G. (1997). On the relationship between ERS-1 SAR/backscatter and surface/sub-surface soil moisture variations in vertisols. *Acra Astronaurica, 40,* 693–699. doi:10.1016/S0094-5765(97)00125-2

Topp, G. C., Davis, J. L., & Annan, A. P. (1980). Electromagnetic determination of soil water content: Measurements in coaxial transmission lines. *Water Resources Research, 16,* 574–582. doi:10.1029/WR016i003p00574

Tsang, L., Kong, J. A., & Shin, R. T. (1985). *Theory of microwave remote sensing* (613 p.). New York, NY: John Wiley & Sons.

Ulaby, F. T., & Batlivala, P. P. (1976). Optimum radar parameters for mapping soil moisture. *IEEE Transactions on Geoscience Electronics, 14,* 81–93. doi:10.1109/TGE.1976.294414

Ulaby, F. T., Batlivala, P. P., & Dobson, M. C. (1978). Microwave backscatter dependence on surface roughness, soil moisture, and soil texture: Part I-bare soil. *IEEE Transactions on Geoscience Electronics, 16,* 286–295. doi:10.1109/TGE.1978.294586

Ulaby, F. T., & El-rayes, M. A. (1987). Microwave dielectric spectrum of vegetation - Part II: Dual-dispersion model. *IEEE Transactions on Geoscience and Remote Sensing, 25,* 550–557. doi:10.1109/TGRS.1987.289833

Ulaby, F. T., Dubois, P. C., & van Zyl, J. (1996). Radar mapping of surface soil moisture. *Journal of Hydrology, 184,* 57–84. doi:10.1016/0022-1694(95)02968-0

Ulaby, F. T., Moore, R. K., & Fung, A. K. (1982). *Microwave remote sensing—Active and passive* (Vols. 2–3). Addison Wesley.

Ulaby, F. T., Moore, R. K., & Fung, A. K. (1986). *Microwave remote sensing: Active and passive* (Vol. 3). Reading, MA: Addison-Wesley.

Vereecken, H., Huisman, J. A., Pachepsky, Y., Montzka, C., van der Kruk, J., Bogena, H., ... Vanderborght, J. (2014). On the spatio-temporal dynamics of soil moisture at the field scale. *Journal of Hydrology, 516,* 76–96. http://dx.doi.org/10.1016/j.jhydrol.2013.11.061

Wagner, W. (1998). *Soil moisture retrieval from ERS Scatterometer data* (Dissertation). Vienna University of Technology, Vienna. Publ. EUR 18670 EN, Office of Official Publications, Europian Community.

Wagner, W., Blöschl, G., Pampaloni, P., Calvet, J.-C., Bizzarri, B., Wigneron, J.-P., & Kerr, Y. (2007). Operational readiness of microwave remote sensing of soil moisture for hydrologic applications. *Nordic Hydrology, 38*(1), 1–20. doi:10.2166/nh.2007.029

Wagner, W., Lemoine, G., Borgeaud, M., & Rott, H. (1999). A study of vegetation cover effects on ERS Scatterometer data. *IEEE Transactions on Geoscience and Remote Sensing, 37,* 338–998. doi:10.1109/36.752212

Wagner, W., Lemoine, G., & Rott, H. (1999). A method for estimating soil moisture from ERS scatterometer and soil data. *Remote Sensing of Environment, 70,* 191–207. doi:10.1016/S0034-4257(99)00036-X

Wagner, W., Noll, J., Borgeaud, M., & Rott, H. (1999). Monitoring soil moisture over the Canadian Prairies with the ERS scatterometer. *IEEE Transactions on Geoscience and Remote Sensing, 37,* 206–216. doi:10.1109/36.739155

Wang, J. R., & Schmugge, T. J. (1980). An empirical model for the complex dielectric permittivity of soils as a function of water content. *IEEE Transactions on Geoscience and Remote Sensing, 18,* 288–295. doi:10.1109/TGRS.1980.350304

Wang, L., & Qu, J. J. (2009). Satellite remote sensing applications for surface soil moisture monitoring: A review. *Frontiers of Earth Science in China, 3,* 237–247. doi:10.1007/s11707-009-0023-7

Western, A. W., Zhou, S. L., Grayson, R. B., McMahon, T. A., Blöschl, G., & Wilson, D. J. (2004). Spatial correlation of soil moisture in small catchments and its relationship to dominant spatial hydrological processes. *Journal of Hydrology, 286,* 113–134. doi:10.1016/j.jhydrol.2003.09.014

Wigneron, J. P., Calvet, J. P., Pellarin, T., Van de Griend, A. A., Berger, M., & Ferrazzoli, P. (2003). Retrieving near-surface soil moisture from microwave radiometric observations: current status and future plans. *Remote Sensing of Environment, 85,* 489–506. http://dx.doi.org/10.1016/S0034-4257(03)00051-8

Winebrenner, D., & Ishimaru, A. (1985). Investigation of a surface field phase perturbation technique for scattering from rough surfaces. *Radio Science, 20,* 161–170. doi:10.1029/RS020i002p00161

Zhao, K. G., Shi, J. C., Zhang, L. X., Jiang, L. M., Zhang, Z. J., Qin, J., ... Hu, J. C. (2003). Retrieval of bare soil surface parameters from simulated data using neural networks combined with IEM (pp. 3881–3883). *Procedings of IEEE, IGARSS.*

Zhao, W., & Li, Z. L. (2013). Sensitivity study of soil moisture on the temporal evolution of surface temperature over bare surfaces. *International Journal of Remote Sensing, 34,* 3314–3331. http://dx.doi.org/10.1080/01431161.2012.716532

Zribi, M., Baghdadi, N., Holah, N., & Fafin, O. (2005). New methodology for soil surface moisture estimation and its application to ENVISAT-ASAR multi incidence data inversion. *Remote Sensing of Environment, 96,* 485–496. http://dx.doi.org/10.1016/j.rse.2005.04.005

Zribi, M., & Dechambre, M. (2002). A new empirical model to retrieve soil moisture and roughness from C-band radar data. *Remote Sensing of Environment, 84,* 42–52. doi:10.1016/S0034-4257(02)00069-X

2

Un-differenced precise point positioning model using triple GNSS constellations

Akram Afifi[1]* and Ahmed El-Rabbany[1]

*Corresponding author: Akram Afifi, Department of Civil Engineering, Ryerson University, Toronto, Ontario, Canada

E-mail: akram.afifi@ryerson.ca

Reviewing editor:
Shuanggen Jin, Shanghai Astronomical Observatory, Chinese Academy of Sciences, China

Abstract: This paper introduces a dual-frequency precise point positioning (PPP) model, which combines the observations of three different global navigation satellite system (GNSS) constellations, namely GPS, Galileo, and BeiDou. A drawback of a single GNSS system such as GPS, however, is the availability of sufficient number of visible satellites in urban areas. Combining GNSS observations offers more visible satellites to users, which in turn is expected to enhance the satellite geometry and the overall positioning solution. However, combining several GNSS observables introduces additional biases, which require rigorous modeling, including the GNSS time offsets and hardware delays. In this paper, un-differenced ionosphere-free linear combination PPP model is developed. The additional biases of the GPS, Galileo, and BeiDou combination are accounted for through the introduction of a new unknown parameter, which is identified as the inter-system bias, in the PPP mathematical model. Natural Resources Canada's GPSPace PPP software is modified to enable a combined GPS, Galileo, and BeiDou PPP solution and to handle the newly introduced biases. A total of four data-sets collected at four different IGS stations are processed to verify the developed PPP model. Precise satellite orbit and clock products from the International GNSS Service Multi-GNSS Experiment (IGS-MGEX) network are used to correct the GPS, Galileo, and BeiDou measurements. It is shown that the un-differenced GPS-only post-processed PPP solution indicates that the model is capable of obtaining a sub-decimeter-level accuracy. However, the solution takes about 20 min

ABOUT THE AUTHOR

Akram Afifi obtained his PhD degree in Geomatics Engineering from the Department of Civil Engineering, Ryerosn University, Canada. He is currently a postdoctoral fellow at Ryerson University, Toronto, Canada. He is acting as an adjunct professor with active membership of the graduate school at Ryerson University. Afifi's areas of interest include Satellite Navigation, Geodesy, Land survey, and Hydrographic Surveying. He published several papers and posters at various journals, conferences, and professional events. He acted as a chair of the Canadian Institute of Geomatics student affair committee from 2013 to 2014. Afifi received numerous awards in recognition of his academic achievements, including three awards from the AOLS in the annual general meeting and Dennis Mock leadership award from Ryerson University.

PUBLIC INTEREST STATEMENT

This paper introduces a dual-frequency precise point positioning (PPP) model, which combines the observations of three different global navigation satellite system (GNSS) constellations, namely GPS, Galileo, and BeiDou. A drawback of a single GNSS system such as GPS, however, is the availability of sufficient number of visible satellites in urban areas. Combining GNSS observations offers more visible satellites to users, which in turn is expected to enhance the satellite geometry and the overall positioning solution. It is shown that the un-differenced GPS-only post-processed PPP solution indicates that the model is capable of obtaining a sub-decimeter-level accuracy. However, the solution takes about 20 min to converge to decimeter-level precision. The convergence time of the combined GNSS post-processed PPP solutions takes about 15 min to reach the decimeter-level precision, which represent a 25% improvement in comparison with the GPS-only post-processed PPP solution.

to converge to decimeter-level precision. The convergence time of the combined GNSS post-processed PPP solutions takes about 15 min to reach the decimeter-level precision, which represent a 25% improvement in comparison with the GPS-only post-processed PPP solution.

Subjects: Aerospace Engineering; Civil, Environmental and Geotechnical Engineering; Earth Sciences

Keywords: PPP; GPS; Galileo; BeiDou

1. Introduction

Precise point positioning (PPP) has proven to be capable of providing positioning accuracy at the sub-decimeter and decimeter levels in static and kinematic modes, respectively. PPP accuracy and convergence time are controlled by the ability to mitigate all potential error sources in the system. Several comprehensive studies have been published on the accuracy and convergence time of undifferenced combined GPS/Galileo PPP model (see, e.g. Afifi & El-Rabbany, 2015; Collins, Bisnath, Lahaye, & Héroux, 2010; Colombo, Sutter, & Evans, 2004; Ge, Gendt, Rothacher, Shi, & Liu, 2008; Hofmann-Wellenhof, Lichtenegger, & Wasle, 2008; Kouba & Héroux, 2001; Leick, 2004; Zumberge, Heflin, Jefferson, Watkins, & Webb, 1997). PPP relies essentially on the availability and use of precise satellite products, namely orbital and clock corrections. At present, the Multi-global navigation satellite systems (GNSS) Experiment (MGEX) of the International GNSS Service (IGS) provides the precise satellite orbital and clock corrections for all the GNSS (Montenbruck et al., 2014).

Unfortunately, the use of a single constellation limits the number of visible satellites, especially in urban areas, which affects the PPP solution. Recently, a number of researchers showed that combining GPS and Galileo observations in PPP solution enhances the positioning convergence and precision in comparison with the GPS-only PPP solution (Afifi & El-Rabbany, 2016a; Melgard, Tegedor, de Jong, Lapucha, & Lachapelle, 2013). At present, the IGS-MEGX network provides the GNSS users with precise clock and orbit products to all currently available satellite systems (Montenbruck et al., 2014). This makes it possible to obtain a PPP solution by combining the observations of two or more GNSS constellations. This research focuses on combining the GPS, Galileo, and BeiDou observations in a PPP model.

Presently, there exist four operational GNSS. These include the US global positioning system (GPS), the Russian global navigation satellite system (GLONASS), the European Galileo system, and the Chinese BeiDou system. Combing the measurements of multiple systems can significantly improve the availability of a navigation solution, especially in urban areas (Afifi & EL-Rabbany, 2016b). GPS satellites transmit signals on three different frequencies, which are controlled by the GPS time frame (GPST). Currently, the GPS users can receive the modernized civil L2C and L5 signals. On the other hand, Galileo satellite constellation foresees 27 operational and three spare satellites positioned in three nearly circular medium earth orbits (MEO). Galileo system transmits six signals on different frequencies using the Galileo time system (GST). Unlike GLONASS satellite system, Galileo and GPS have partial frequency overlaps, which simplify the dual-system integration. In addition, GPS and Galileo operators have agreed to measure and broadcast a GPS to Galileo time offset (GGTO) parameter, in order to facilitate the interchangeable mode (Melgard et al., 2013). BeiDou navigation satellite system, being developed independently by China, is pacing steadily forward toward completing the constellation. China has indicated a plan to complete the second generation of Beidou satellite system by expanding the regional service into global coverage. Beidou system transmits three signals on different frequencies using the BeiDou time frame (BDT). The BeiDou-2 system is proposed to consist of 30 medium Earth orbiting satellites and five geostationary satellites (BeiDou, 2015; ESA, 2015; Hofmann-Wellenhof et al., 2008; IAC, 2015).

In this paper, a triple GNSS constellation (GPS, Galileo, and BeiDou) PPP model is developed. Four combinations are considered in the PPP modeling namely; GPS/Galileo, GPS/BeiDou, Galileo/BeiDou, and GPS/Galileo/BeiDou. All the combined PPP models results are compared with the GPS-only PPP model results. In the developed model GPS L1/L2, Galileo E1/E5a, and BeiDou B1/B2 signals are used in a dual-frequency ionosphere-free linear combination. Precise satellite corrections from the International GNSS Service multi-GNSS experiment (IGS-MEGX) network are used to account for GPS, Galileo and BeiDou satellite orbit and clock errors (Montenbruck et al., 2014). As these products are presently referenced to the GPS time and since we use mixed GNSS receivers that also use GPS time as a reference, the GGTO and the GPS to BeiDou time offset are canceled out in our model. The inter-system bias is treated as an additional unknown parameter. The hydrostatic component of the tropospheric zenith path delay is modeled through the Hopfield model, while the wet component is considered as an additional unknown parameter (Hofmann-Wellenhof et al., 2008; Hopfield, 1972). All remaining errors and biases are accounted for using existing models as shown in Kouba (2009). The inter-system bias parameter was found to be essentially constant over the one-hour observation time span and was receiver dependent. The positioning results of the developed combined GPS/Galileo, GPS/BeiDou, Galileo/BeiDou, and GPS/Galileo/BeiDou PPP models showed a sub-decimeter accuracy level and 25% convergence time improvement in comparison with the GPS-only PPP results.

2. Un-differenced PPP models

PPP has been carried out using dual-frequency ionosphere-free linear combinations of carrier-phase and pseudorange GPS measurements. Equations (1)–(6) show the ionosphere free linear combination of GPS, Galileo, and BeiDou observations (Afifi & El-Rabbany, 2016a).

$$P_{G_{IF}} = \rho_G + c[dt_{rG} - dt^s] + c[\alpha d_{P1} - \beta d_{P2}]_r + c[\alpha d_{P1} - \beta d_{P2}]^s + T_G + \varepsilon_{PG_{IF}} \tag{1}$$

$$P_{E_{IF}} = \rho_E + c[dt_{rG} - GGTO - dt^s] + c[\alpha d_{E1} - \beta d_{E5a}]_r + c[\alpha d_{E1} - \beta d_{E5a2}]^s + T_E + \varepsilon_{E_{IF}} \tag{2}$$

$$P_{B_{IF}} = \rho_B + c[dt_{rG} - GB - dt^s] + c[\alpha d_{B1} - \beta d_{B2}]_r + c[\alpha d_{B1} - \beta d_{B2}]^s + T_B + \varepsilon_{B_{IF}} \tag{3}$$

$$\Phi_{G_{IF}} = \rho_G + c[dt_{rG} - dt^s] + c[\alpha \delta_{L1} - \beta \delta_{L2}]_r + c[\alpha \delta_{L1} - \beta \delta_{L2}]^s + T_G + N_{G_{IF}} + \phi_{r0_{G_{IF}}} + \phi^s_{0_{G_{IF}}} + \varepsilon_{\Phi G_{IF}} \tag{4}$$

$$\Phi_{E_{IF}} = \rho_E + c[dt_{rG} - GGTO - dt^s] + c[\alpha \delta_{E1} - \beta \delta_{E5a}]_r + c[\alpha \delta_{E1} - \beta \delta_{E5a}]^s + T_E + N_{E_{IF}} + \phi_{r0_{E_{IF}}} + \phi^s_{0_{E_{IF}}} + \varepsilon_{\Phi E_{IF}} \tag{5}$$

$$\Phi_{B_{IF}} = \rho_B + c[dt_{rG} - GB - dt^s] + c[\alpha \delta_{B1} - \beta \delta_{B2}]_r + c[\alpha \delta_{B1} - \beta \delta_{B2}]^s + T_B + N_{B_{IF}} + \phi_{r0_{B_{IF}}} + \phi^s_{0_{B_{IF}}} + \varepsilon_{\Phi B_{IF}} \tag{6}$$

where the subscripts G, E, and B refer to the GPS, Galileo, and BeiDou satellite systems, respectively; $P_{G_{IF}}$, $P_{E_{IF}}$ and $P_{B_{IF}}$ are the ionosphere-free pseudoranges in meters for GPS, Galileo, and BeiDou systems, respectively; $\Phi_{G_{IF}}$, $\Phi_{E_{IF}}$ and $\Phi_{B_{IF}}$ are the ionosphere-free carrier phase measurements in meters for GPS, Galileo, and BeiDou systems, respectively; GGTO is the GPS to Galileo time offset; GB is the GPS to BeiDou time offset; ρ is the true geometric range from receiver at reception time to satellite at transmission time in meter; dt_r, dt^s are the clock errors in seconds for the receiver at signal reception time and the satellite at signal transmission time, respectively; d_{P1}, d_{P2}, d_{E1}, d_{E5a}, d_{B1}, d_{B2} are frequency-dependent code hardware delays for the receiver at reception time in seconds; d^s_{P1}, d^s_{P2}, d^s_{E1}, d^s_{E5a}, d^s_{B1}, d^s_{B2a} are frequency-dependent code hardware delays for the satellite at transmission time in seconds; δ_{L1r}, δ_{L2r}, δ_{E1r}, δ_{E5ar}, δ_{B1r}, δ_{B2r} are frequency-dependent carrier-phase hardware delays for the receiver at reception time in seconds; δ^s_{L1}, δ^s_{L2}, δ^s_{E1}, δ^s_{E5a}, δ^s_{B1}, δ^s_{B2} are frequency-dependent carrier-phase hardware delays for the satellite at transmission time in seconds; T is the tropospheric delay in meter; $N_{G_{IF}}$, $N_{E_{IF}}$, $N_{B_{IF}}$ are the ionosphere-free linear combinations of the ambiguity parameters for both GPS, Galileo, and BeiDou carrier-phase measurements in meters, respectively; $\phi_{r0_{G_{IF}}}$, $\phi^s_{0_{G_{IF}}}$, $\phi_{r0_{E_{IF}}}$, $\phi^s_{0_{E_{IF}}}$, $\phi_{r0_{B_{IF}}}$, $\phi^s_{0_{B_{IF}}}$ are ionosphere-free linear combinations of frequency-dependent initial fractional phase biases in the receiver and satellite channels for both GPS, Galileo, and BeiDou in meters, respectively;

c is the speed of light in vacuum in meter per second; $\varepsilon_{P_{IF}}, \varepsilon_{E_{IF}}, \varepsilon_{\Phi G_{IF}}, \varepsilon_{\Phi E_{IF}}, \varepsilon_{B_{IF}}, \varepsilon_{\Phi B_{IF}}$ are the ionosphere-free linear combinations of the relevant noise and un-modeled errors in meter; $\alpha_G, \beta_G, \alpha_E, \beta_E, \alpha_B, \beta_B$ are the ionosphere-free linear combination coefficients for GPS, Galileo, and BeiDou which are given, respectively, by: $\alpha_G = \frac{f_1^2}{f_1^2 - f_2^2}, \beta_G = \frac{f_2^2}{f_1^2 - f_2^2}, \alpha_E = \frac{f_{E1}^2}{f_{E1}^2 - f_{E5a}^2}, \beta_E = \frac{f_{E5a}^2}{f_{E1}^2 - f_{E5a}^2}, \alpha_B = \frac{f_{B1}^2}{f_{B1}^2 - f_{B2}^2}, \beta_B = \frac{f_{B2}^2}{f_{B1}^2 - f_{B2}^2}$.

where f_1 and f_2 are GPS L_1 and L_2 signals frequencies; f_{E1} and f_{E5a} are Galileo E_1 and E_{5a} signals frequencies; f_{B1} and f_{B2} are BeiDou B_1 and B_2 signals frequencies.

$$N_{G_{IF}} = \alpha_G \lambda_1 N_1 - \beta_G \lambda_2 N_2 \tag{7}$$

$$N_{E_{IF}} = \alpha_E \lambda_{E1} N_{E1} - \beta_E \lambda_{E5a} N_{E5a} \tag{8}$$

$$N_{B_{IF}} = \alpha_B \lambda_{B1} N_{B1} - \beta_B \lambda_{B2} N_{B2} \tag{9}$$

where λ_1 and λ_2 are the GPS L1 and L2 signals wavelengths in meters; λ_{E1} and λ_{E5a} are the Galileo E1 and E5a signals wavelengths in meters; λ_{B1} and λ_{B2} are the BeiDou B1 and B2 signals wavelengths in meters; N_1, N_2 are the integer ambiguity parameters of GPS signals L1 and L2, respectively; N_{E1}, N_{E5a} are the integer ambiguity parameters of Galileo signals E1 and E5a, respectively; N_{B1}, N_{B2} are the integer ambiguity parameters of BeiDou signals B1 and B2, respectively.

Precise orbit and satellite clock corrections of IGS-MGEX networks are produced for both GPS/Galileo observations and are referred to GPS time. IGS precise GPS satellite clock correction includes the effect of the ionosphere-free linear combination of the satellite hardware delays of L1/L2 P(Y) code, while the Galileo counterpart includes the effect of the ionosphere-free linear combination of the satellite hardware delays of the Galileo E1/E5a pilot code. In addition, BeiDou satellite clock correction includes the effect of the ionosphere-free linear combination of the satellite hardware delays of B1/B2 code (Montenbruck et al., 2014). By applying the precise clock products for both GPS/Galileo/BeiDou observations, Equations (1)–(6) will take the following form:

$$P_{G_{IF}} = \rho_G + c[dt_{rG} - dt_{prec}^s] + c[\alpha d_{P1} - \beta d_{P2}]_r + T_G + \varepsilon_{PG_{IF}} \tag{10}$$

$$P_{E_{IF}} = \rho_E + c[dt_{rG} - dt_{prec}^s] + c[\alpha d_{E1} - \beta d_{E5a}]_r + T_E + \varepsilon_{E_{IF}} \tag{11}$$

$$P_{B_{IF}} = \rho_B + c[dt_{rG} - dt_{prec}^s] + c[\alpha d_{B1} - \beta d_{B2}]_r + T_B + \varepsilon_{B_{IF}} \tag{12}$$

$$\Phi_{G_{IF}} = \rho_G + cdt_{rG} - c[dt_{prec}^s + [\alpha d_{P1} - \beta d_{P2}]^s] + c[\alpha \delta_{L1} - \beta \delta_{L2}]_r - c[\alpha \delta_{L1} - \beta \delta_{L2}]^s + T_G + N_{G_{IF}} \\ + \phi_{r0_{G_{IF}}} + \phi_{0_{G_{IF}}}^s + \varepsilon_{\Phi G_{IF}} \tag{13}$$

$$\Phi_{E_{IF}} = \rho_E + cdt_{rG} - c[dt_{prec}^s + [\alpha d_{E1} - \beta d_{E5a}]^s] + c[\alpha \delta_{E1} - \beta \delta_{E5a}]_r - c[\alpha \delta_{E1} - \beta \delta_{E5a}]^s \\ + T_E + N_{E_{IF}} + \phi_{r0_{E_{IF}}} + \phi_{0_{E_{IF}}}^s + \varepsilon_{\Phi E_{IF}} \tag{14}$$

$$\Phi_{B_{IF}} = \rho_B + cdt_{rG} - c[dt_{prec}^s + [\alpha d_{B1} - \beta d_{B2}]^s] + c[\alpha \delta_{B1} - \beta \delta_{B2}]_r - c[\alpha \delta_{B1} - \beta \delta_{B2}]^s \\ + T_B + N_{B_{IF}} + \phi_{r0_{B_{IF}}} + \phi_{0_{B_{IF}}}^s + \varepsilon_{\Phi B_{IF}} \tag{15}$$

For simplicity, the receiver and satellite hardware delays will take the following forms:

$$b_{r_P} = c[\alpha d_{P1} - \beta d_{P2}]_r \qquad b_P^s = c[\alpha d_{P1} - \beta d_{P2}]^s$$

$$b_{r_E} = c[\alpha d_{E1} - \beta d_{E5a}]_r \qquad b_E^s = c[\alpha d_{E1} - \beta d_{E5a}]^s$$

$$b_{r_B} = c[\alpha d_{B1} - \beta d_{B2}]_r \qquad\qquad b_B^s = c[\alpha d_{B1} - \beta d_{B2}]^s$$

$$b_{r_\Phi} = c[\alpha \delta_{L1} - \beta \delta_{L2}]_r + \phi_{r0_{G_{IF}}} \qquad\qquad b_\Phi^s = c[\alpha \delta_{L1} - \beta \delta_{L2}]^s + \phi_{0_{G_{IF}}}^s$$

$$b_{r_{E\Phi}} = c[\alpha \delta_{E1} - \beta \delta_{E5a}]_r + \phi_{r0_{E_{IF}}} \qquad\qquad b_{E\Phi}^s = c[\alpha \delta_{E1} - \beta \delta_{E5a}]^s + \phi_{0_{E_{IF}}}^s$$

$$b_{r_{B\Phi}} = c[\alpha \delta_{B1} - \beta \delta_{B2}]_r + \phi_{r0_{B_{IF}}} \qquad\qquad b_{B\Phi}^s = c[\alpha \delta_{B1} - \beta \delta_{B2}]^s + \phi_{0_{B_{IF}}}^s$$

In the combined GPS/Galileo un-differenced PPP model, the GPS receiver clock error is lumped with the GPS receiver differential code biases. In order to maintain consistency in the estimation of a common receiver clock offset, this convention is used when combining the ionosphere-free linear combination of GPS L1/L2, Galileo E1/E5a, and BeiDou B1/B2 observations in PPP solution. This, however, introduces an additional bias in the Galileo ionosphere-free PPP mathematical model, which represents the difference in the receiver differential code biases of both systems. Such an additional bias is commonly known as the inter-system bias, which is referred to as *ISB* in this paper. In our PPP mode, the Hopfield tropospheric correction model along with the Vienna mapping function are used to account for the hydrostatic component of the tropospheric delay (Boehm & Schuh, 2004; Hopfield, 1972). Other corrections are also applied, including the effect of ocean loading (Bos & Scherneck, 2011; IERS, 2010), Earth tide (Kouba, 2009), carrier-phase windup (Leick, 2004; Wu, Wu, Hajj, Bertiger, & Lichten, 1993), Sagnac (Kaplan & Heagarty, 2006), relativity (Hofmann-Wellenhof et al., 2008), and satellite and receiver antenna phase-center variations (Dow, Neilan, & Rizos, 2009). The noise terms are modeled stochastically using an exponential model, as described in Afifi and El-Rabbany (2015). With the above consideration, the GPS/Galileo ionosphere-free linear combinations of both pseudorange and carrier phase can be written as:

$$P_{G_{IF}} = \rho_G + \tilde{d}t_{rG} - dt_{prec}^s + T_G + \varepsilon_{PG_{IF}} \tag{16}$$

$$P_{E_{IF}} = \rho_E + \tilde{d}t_{rG} - dt_{prec}^s + ISB_{GE} + T_E + \varepsilon_{E_{IF}} \tag{17}$$

$$\Phi_{G_{IF}} = \rho_G + \tilde{d}t_{rG} - dt_{prec}^s + T_G + \tilde{N}_{G_{IF}} + \varepsilon_{\Phi G_{IF}} \tag{18}$$

$$\Phi_{E_{IF}} = \rho_E + \tilde{d}t_{rG} - dt_{prec}^s + T_E + \tilde{N}_{E_{IF}} + ISB_{GE} + \varepsilon_{\Phi E_{IF}} \tag{19}$$

where $\tilde{d}t_{rG}$ represents the sum of the receiver clock error and receiver hardware delay $\tilde{d}t_{rG} = cdt_{rG} + b_{r_p}$; *ISB* is the inter system bias as follows $ISB_{GE} = b_{r_E} - b_{r_p}$; $\tilde{N}_{G_{IF}}$ and $\tilde{N}_{E_{IF}}$ are given by:

$$\tilde{N}_{G_{IF}} = N_{G_{IF}} + b_{r_\Phi} + b_{r_p} - b_\Phi^s - b_P^s \tag{20}$$

$$\tilde{N}_{E_{IF}} = N_{E_{IF}} + b_{r_{E\Phi}} + b_{r_p} - b_{E\Phi}^s - b_E^s \tag{21}$$

In case of combining GPS and BeiDou observations in a PPP model ionosphere-free linear combinations of both pseudorange and carrier phase can be written as:

$$P_{G_{IF}} = \rho_G + \tilde{d}t_{rG} - dt_{prec}^s + T_G + \varepsilon_{PG_{IF}} \tag{22}$$

$$P_{B_{IF}} = \rho_B + \tilde{d}t_{rG} - dt_{prec}^s + ISB_{GB} + T_B + \varepsilon_{B_{IF}} \tag{23}$$

$$\Phi_{G_{IF}} = \rho_G + \tilde{d}t_{rG} - dt_{prec}^s + T_G + \tilde{N}_{G_{IF}} + \varepsilon_{\Phi G_{IF}} \tag{24}$$

$$\Phi_{B_{IF}} = \rho_B + \tilde{d}t_{rG} - dt^s_{prec} + T_B + \tilde{N}_{B_{IF}} + ISB_{GB} + \varepsilon_{\Phi B_{IF}}$$

(25)

where $\tilde{d}t_{rG}$ represents the sum of the receiver clock error and receiver hardware delay $\tilde{d}t_{rG} = cdt_{rG} + b_{r_p}$; ISB is the inter system bias as follows $ISB_{GB} = b_{r_B} - b_{r_p}$; $\tilde{N}_{G_{IF}}$ and $\tilde{N}_{B_{IF}}$ are given by:

$$\tilde{N}_{G_{IF}} = N_{G_{IF}} + b_{r_\Phi} + b_{r_p} - b^s_\Phi - b^s_P$$

(26)

$$\tilde{N}_{B_{IF}} = N_{B_{IF}} + b_{r_{B\Phi}} + b_{r_p} - b^s_{B\Phi} - b^s_B$$

(27)

In the combined Galileo and BeiDou un-differenced PPP model, the Galileo receiver clock error is lumped with the Galileo receiver differential code biases. In order to maintain consistency in the estimation of a common receiver clock offset, this convention is used when combining the iono-sphere-free linear combination of Galileo E1/E5a and BeiDou B1/B2 observations. This, however, in-troduces an additional bias in the BeiDou ionosphere-free PPP mathematical model, which represents the difference in the receiver differential code biases of both systems. As a result, the Galileo and BeiDou combined PPP model ionosphere-free linear combinations of both pseudorange and carrier phase can be written as:

$$P_{E_{IF}} = \rho_E + \tilde{d}t_{rE} - dt^s_{prec} + T_E + \varepsilon_{PE_{IF}}$$

(28)

$$P_{B_{IF}} = \rho_B + \tilde{d}t_{rE} - dt^s_{prec} + ISB_{EB} + T_B + \varepsilon_{B_{IF}}$$

(29)

$$\Phi_{E_{IF}} = \rho_E + \tilde{d}t_{rE} - dt^s_{prec} + T_E + \tilde{N}_{E_{IF}} + \varepsilon_{\Phi E_{IF}}$$

(30)

$$\Phi_{B_{IF}} = \rho_B + \tilde{d}t_{rG} - dt^s_{prec} + T_B + \tilde{N}_{B_{IF}} + ISB_{EB} + \varepsilon_{\Phi B_{IF}}$$

(31)

where $\tilde{d}t_{rE}$ represents the sum of the receiver clock error and receiver hardware delay $\tilde{d}t_{rE} = cdt_{rE} + b_{r_E}$; ISB is the inter system bias as follows $ISB_{EB} = b_{r_B} - b_{r_E}$; $\tilde{N}_{E_{IF}}$ and $\tilde{N}_{B_{IF}}$ are given by:

$$\tilde{N}_{E_{IF}} = N_{E_{IF}} + b_{r_{E\Phi}} + b_{r_E} - b^s_{E\Phi} - b^s_E$$

(32)

$$\tilde{N}_{B_{IF}} = N_{B_{IF}} + b_{r_{B\Phi}} + b_{r_p} - b^s_{B\Phi} - b^s_B$$

(33)

When using the combined GPS/Galileo or GPS/BeiDou or Galileo/BeiDou un-differenced PPP model, the ambiguity parameters lose their integer nature as they are contaminated by receiver and satel-lite hardware delays. It should be pointed out that the number of unknown parameters in the com-bined PPP model equals the number of visible satellites from any system plus six parameters, while the number of equations equals double the number of the visible satellites. This means that the re-dundancy equals $n_G + n_E - 6$. In other words, at least six mixed satellites are needed for the solution to exist. In comparison with the GPS-only un-differenced scenario, which requires a minimum of five satellites for the solution to exist, the addition of Galileo or BeiDou satellites increases the redun-dancy by $n_E - 1$. In other words, we need a minimum of two satellites from any GNSS system in order to contribute to the solution.

3. Least-squares estimation technique

Under the assumption that the observations are uncorrelated and the errors are normally distrib-uted with zero mean, the covariance matrix of the un-differenced observations takes the form of a diagonal matrix. The elements along the diagonal line represent the variances of the code and car-rier phase measurements. In our solution, we consider the ratio between the standard deviation of the code and carrier-phase measurements to be 100. The general linearized form for the above observation equations around the initial (approximate) vector u^0 and observables l can be written in a compact form as:

$$f(u, l) \approx A\Delta u - w - r \approx 0 \tag{34}$$

where u is the vector of unknown parameters; A is the design matrix, which includes the partial derivatives of the observation equations with respect to the unknown parameters u; Δu is the unknown vector of corrections to the approximate parameters u^0, i.e. $u = u^0 + \Delta u$; w is the misclosure vector and r is the vector of residuals. The sequential least-squares solution for the unknown parameters Δu_i at an epoch i can be obtained from Vanicek and Krakiwsky (1986):

$$\Delta u_i = \Delta u_{i-1} + M_{i-1}^{-1} A_i^T (C_{l_i} + A_i M_{i-1}^{-1} A_i^T)^{-1} [w_i - A_i \Delta u_{i-1}] \tag{35}$$

$$M_i^{-1} = M_{i-1}^{-1} - M_{i-1}^{-1} A_i^T (C_{l_i} + A_i M_{i-1}^{-1} A^T)^{-1} A_i M_{i-1}^{-1} \tag{36}$$

$$C_{\Delta u_i} = M_i^{-1} = M_{i-1}^{-1} - M_{i-1}^{-1} A_i^T (C_{l_i} + A_i M_{i-1}^{-1} A^T)^{-1} A_i M_{i-1}^{-1} \tag{37}$$

where Δu_{i-1} is the least-squares solution for the estimated parameters at epoch $i-1$; M is the matrix of the normal equations; C_l and $C_{\Delta u}$ are the covariance matrices of the observations and unknown parameters, respectively. It should be pointed out that the usual batch least-squares adjustment should be used in the first epoch, i.e. for $i = 1$. The batch solution for the estimated parameters and the inverse of the normal equation matrix are given, respectively, by Vanicek and Krakiwsky (1986):

$$\Delta u_1 = [C_{x^0}^{-1} + A_1^T C_{l_1}^{-1} A_1]^{-1} A_1^T C_{l_1}^{-1} w_1 \tag{38}$$

$$M_1^{-1} = [C_{x^0}^{-1} + A_1^T C_{l_1}^{-1} A_1]^{-1} \tag{39}$$

where C_x^0 is a priori covariance matrix for the approximate values of the unknown parameters.

In case of the combined GPS/Galileo PPP model, the design matrix A and the vector of corrections to the unknown parameters Δx take the following forms:

$$A = \begin{bmatrix} \left(\frac{x_0-X^{1_G}}{\rho_0^{1_G}}\right) & \left(\frac{y_0-Y^{1_G}}{\rho_0^{1_G}}\right) & \left(\frac{z_0-Z^{1_G}}{\rho_0^{1_G}}\right) & 1 & m_f^{1_G} & 0 & 0 & \cdots & 0 & 0 & \cdots & 0 \\ \left(\frac{x_0-X^{1_G}}{\rho_0^{1_G}}\right) & \left(\frac{y_0-Y^{1_G}}{\rho_0^{1_G}}\right) & \left(\frac{z_0-Z^{1_G}}{\rho_0^{1_G}}\right) & 1 & m_f^{1_G} & 0 & 1 & \cdots & 0 & 0 & \cdots & 0 \\ \vdots & \vdots & \vdots & \vdots & \vdots & \vdots & \vdots & \vdots & \vdots & \vdots & \ddots & \vdots \\ \left(\frac{x_0-X^{n_G}}{\rho_0^{n_G}}\right) & \left(\frac{y_0-Y^{n_G}}{\rho_0^{n_G}}\right) & \left(\frac{z_0-Z^{n_G}}{\rho_0^{n_G}}\right) & 1 & m_f^{1_{nG}} & 0 & 0 & \cdots & 0 & 0 & \cdots & 0 \\ \left(\frac{x_0-X^{n_G}}{\rho_0^{n_G}}\right) & \left(\frac{y_0-Y^{n_G}}{\rho_0^{n_G}}\right) & \left(\frac{z_0-Z^{n_G}}{\rho_0^{n_G}}\right) & 1 & m_f^{1_{nG}} & 0 & 0 & \cdots & 1 & 0 & \cdots & 0 \\ \left(\frac{x_0-X^{1_E}}{\rho_0^{1_E}}\right) & \left(\frac{y_0-Y^{1_E}}{\rho_0^{1_E}}\right) & \left(\frac{z_0-Z^{1_E}}{\rho_0^{1_E}}\right) & 1 & m_f^{1_E} & 1 & 0 & \cdots & 0 & 0 & \cdots & 0 \\ \left(\frac{x_0-X^{1_E}}{\rho_0^{1_E}}\right) & \left(\frac{y_0-Y^{1_E}}{\rho_0^{1_E}}\right) & \left(\frac{z_0-Z^{1_E}}{\rho_0^{1_E}}\right) & 1 & m_f^{1_E} & 1 & 0 & \cdots & 0 & 1 & \cdots & 0 \\ \vdots & \vdots & \vdots & \vdots & \vdots & \vdots & \vdots & \vdots & \vdots & \vdots & \ddots & \vdots \\ \left(\frac{x_0-X^{n_E}}{\rho_0^{n_E}}\right) & \left(\frac{y_0-Y^{n_E}}{\rho_0^{n_E}}\right) & \left(\frac{z_0-Z^{n_E}}{\rho_0^{n_E}}\right) & 1 & m_f^{1_{nE}} & 1 & 0 & \cdots & 0 & 0 & \cdots & 0 \\ \left(\frac{x_0-X^{n_E}}{\rho_0^{n_E}}\right) & \left(\frac{y_0-Y^{n_E}}{\rho_0^{n_E}}\right) & \left(\frac{z_0-Z^{n_E}}{\rho_0^{n_E}}\right) & 1 & m_f^{1_{nE}} & 1 & 0 & \cdots & 0 & 0 & \cdots & 1 \end{bmatrix}_{2n \times (n+6)} \quad \Delta x = \begin{bmatrix} \Delta x \\ \Delta y \\ \Delta z \\ c\, dt_{r_G} \\ zpd_w \\ ISB_{GE} \\ \tilde{N}_G^1 \\ \vdots \\ \tilde{N}_G^{n_G} \\ \tilde{N}_E^1 \\ \vdots \\ \tilde{N}_E^{n_E} \end{bmatrix}_{n+6} \tag{40}$$

where n_G refers to the number of visible GPS satellites; n_E refers to the number of visible Galileo satellites; $n = n_G + n_E$ is the total number of the observed satellites for both GPS/Galileo systems; x_0, y_0, and z_0 are the approximate receiver coordinates; $X^{j_G}, Y^{j_G}, Z^{j_G}, j = 1, 2, \ldots, n_G$ are the known GPS satellite coordinates; $X^{k_E}, Y^{k_E}, Z^{k_E}, k = 1, 2, \ldots, n_E$ are the known Galileo satellite coordinates; ρ_0 is the approximate receiver–satellite range. The unknown parameters in the above system are the corrections to the receiver coordinates, Δx, Δy, and Δz, the wet component of the tropospheric zenith path delay zpd_w, the inter-system bias ISB, and the non-integer ambiguity parameters $\widetilde{N}$.

In case of the combined GPS/BeiDou PPP model, the design matrix A and the vector of corrections to the unknown parameters Δx take the following forms:

$$
A = \begin{bmatrix}
\left(\frac{x_0-X^{1_G}}{\rho_0^{1_G}}\right) & \left(\frac{y_0-Y^{1_G}}{\rho_0^{1_G}}\right) & \left(\frac{z_0-Z^{1_G}}{\rho_0^{1_G}}\right) & 1 & m_f^{1_G} & 0 & 0 & \cdots & 0 & 0 & \cdots & 0 \\
\left(\frac{x_0-X^{1_G}}{\rho_0^{1_G}}\right) & \left(\frac{y_0-Y^{1_G}}{\rho_0^{1_G}}\right) & \left(\frac{z_0-Z^{1_G}}{\rho_0^{1_G}}\right) & 1 & m_f^{1_G} & 0 & 1 & \cdots & 0 & 0 & \cdots & 0 \\
\vdots & \vdots & \vdots & \vdots & \vdots & \vdots & \vdots & \vdots & \vdots & \vdots & \ddots & \vdots \\
\left(\frac{x_0-X^{n_G}}{\rho_0^{n_G}}\right) & \left(\frac{y_0-Y^{n_G}}{\rho_0^{n_G}}\right) & \left(\frac{z_0-Z^{n_G}}{\rho_0^{n_G}}\right) & 1 & m_f^{1_{nG}} & 0 & 0 & \cdots & 0 & 0 & \cdots & 0 \\
\left(\frac{x_0-X^{n_G}}{\rho_0^{n_G}}\right) & \left(\frac{y_0-Y^{n_G}}{\rho_0^{n_G}}\right) & \left(\frac{z_0-Z^{n_G}}{\rho_0^{n_G}}\right) & 1 & m_f^{1_{nG}} & 0 & 0 & \cdots & 1 & 0 & \cdots & 0 \\
\left(\frac{x_0-X^{1_B}}{\rho_0^{1_B}}\right) & \left(\frac{y_0-Y^{1_B}}{\rho_0^{1_B}}\right) & \left(\frac{z_0-Z^{1_B}}{\rho_0^{1_B}}\right) & 1 & m_f^{1_B} & 1 & 0 & \cdots & 0 & 0 & \cdots & 0 \\
\left(\frac{x_0-X^{1_B}}{\rho_0^{1_B}}\right) & \left(\frac{y_0-Y^{1_B}}{\rho_0^{1_B}}\right) & \left(\frac{z_0-Z^{1_B}}{\rho_0^{1_B}}\right) & 1 & m_f^{1_B} & 1 & 0 & \cdots & 0 & 1 & \cdots & 0 \\
\vdots & \vdots & \vdots & \vdots & \vdots & \vdots & \vdots & \vdots & \vdots & \vdots & \ddots & \vdots \\
\left(\frac{x_0-X^{n_B}}{\rho_0^{n}}\right) & \left(\frac{y_0-Y^{n_B}}{\rho_0^{n}}\right) & \left(\frac{z_0-Z^{n_B}}{\rho_0^{n}}\right) & 1 & m_f^{1_{nB}} & 1 & 0 & \cdots & 0 & 0 & \cdots & 0 \\
\left(\frac{x_0-X^{n_B}}{\rho_0^{n_B}}\right) & \left(\frac{y_0-Y^{n_B}}{\rho_0^{n_B}}\right) & \left(\frac{z_0-Z^{n_B}}{\rho_0^{n_B}}\right) & 1 & m_f^{1_{nB}} & 1 & 0 & \cdots & 0 & 0 & \cdots & 1
\end{bmatrix}_{2n \times (n+6)}
\qquad
\Delta x = \begin{bmatrix}
\Delta x \\
\Delta y \\
\Delta z \\
c\, dt_{r_G} \\
zpd_w \\
ISB_{GB} \\
\widetilde{N}_G^1 \\
\vdots \\
\widetilde{N}_G^{n_G} \\
\widetilde{N}_B^1 \\
\vdots \\
\widetilde{N}_B^{n_B}
\end{bmatrix}_{n+6}
\qquad (41)
$$

where n_G refers to the number of visible GPS satellites; n_B refers to the number of visible BeiDou satellites; $n = n_G + n_B$ is the total number of the observed satellites for both GPS/BeiDou systems; x_0, y_0, and z_0 are the approximate receiver coordinates; $X^{j_G}, Y^{j_G}, Z^{j_G}, j = 1, 2, \ldots, n_G$ are the known GPS satellite coordinates; $X^{k_B}, Y^{k_B}, Z^{k_B}, k = 1, 2, \ldots, n_B$ are the known BeiDou satellite coordinates; ρ_0 is the approximate receiver–satellite range. The unknown parameters in the above system are the corrections to the receiver coordinates, Δx, Δy, and Δz, the wet component of the tropospheric zenith path delay zpd_w, the inter-system bias ISB_{GB}, and the non-integer ambiguity parameters $\widetilde{N}$.

In case of the combined Galileo/BeiDou PPP model, the design matrix A and the vector of corrections to the unknown parameters Δx take the following forms:

$$
A = \begin{bmatrix}
\left(\frac{x_0-X^{1_E}}{\rho_0^{1_E}}\right) & \left(\frac{y_0-Y^{1_E}}{\rho_0^{1_E}}\right) & \left(\frac{z_0-Z^{1_E}}{\rho_0^{1_E}}\right) & 1 & m_f^{1_E} & 0 & 0 & \cdots & 0 & 0 & \cdots & 0 \\
\left(\frac{x_0-X^{1_E}}{\rho_0^{1_E}}\right) & \left(\frac{y_0-Y^{1_E}}{\rho_0^{1_E}}\right) & \left(\frac{z_0-Z^{1_E}}{\rho_0^{1_E}}\right) & 1 & m_f^{1_E} & 0 & 1 & \cdots & 0 & 0 & \cdots & 0 \\
\vdots & \vdots & \vdots & \vdots & \vdots & \vdots & \vdots & \vdots & \vdots & \vdots & \ddots & \vdots \\
\left(\frac{x_0-X^{n_E}}{\rho_0^{n_E}}\right) & \left(\frac{y_0-Y^{n_E}}{\rho_0^{n_E}}\right) & \left(\frac{z_0-Z^{n_E}}{\rho_0^{n_E}}\right) & 1 & m_f^{1_{nE}} & 0 & 0 & \cdots & 0 & 0 & \cdots & 0 \\
\left(\frac{x_0-X^{n_E}}{\rho_0^{n_E}}\right) & \left(\frac{y_0-Y^{n_E}}{\rho_0^{n_E}}\right) & \left(\frac{z_0-Z^{n_E}}{\rho_0^{n_E}}\right) & 1 & m_f^{1_{nE}} & 0 & 0 & \cdots & 1 & 0 & \cdots & 0 \\
\left(\frac{x_0-X^{1_B}}{\rho_0^{1_B}}\right) & \left(\frac{y_0-Y^{1_B}}{\rho_0^{1_B}}\right) & \left(\frac{z_0-Z^{1_B}}{\rho_0^{1_B}}\right) & 1 & m_f^{1_B} & 1 & 0 & \cdots & 0 & 0 & \cdots & 0 \\
\left(\frac{x_0-X^{1_B}}{\rho_0^{1_B}}\right) & \left(\frac{y_0-Y^{1_B}}{\rho_0^{1_B}}\right) & \left(\frac{z_0-Z^{1_B}}{\rho_0^{1_B}}\right) & 1 & m_f^{1_B} & 1 & 0 & \cdots & 0 & 1 & \cdots & 0 \\
\vdots & \vdots & \vdots & \vdots & \vdots & \vdots & \vdots & \vdots & \vdots & \vdots & \ddots & \vdots \\
\left(\frac{x_0-X^{n_B}}{\rho_0^{n}}\right) & \left(\frac{y_0-Y^{n_B}}{\rho_0^{n}}\right) & \left(\frac{z_0-Z^{n_B}}{\rho_0^{n}}\right) & 1 & m_f^{1_{nB}} & 1 & 0 & \cdots & 0 & 0 & \cdots & 0 \\
\left(\frac{x_0-X^{n_B}}{\rho_0^{n_B}}\right) & \left(\frac{y_0-Y^{n_B}}{\rho_0^{n_B}}\right) & \left(\frac{z_0-Z^{n_B}}{\rho_0^{n_B}}\right) & 1 & m_f^{1_{nB}} & 1 & 0 & \cdots & 0 & 0 & \cdots & 1
\end{bmatrix}_{2n \times (n+6)}
\qquad
\Delta x = \begin{bmatrix}
\Delta x \\
\Delta y \\
\Delta z \\
c\, dt_{r_G} \\
zpd_w \\
ISB_{GB} \\
\widetilde{N}_G^1 \\
\vdots \\
\widetilde{N}_G^{n_G} \\
\widetilde{N}_B^1 \\
\vdots \\
\widetilde{N}_B^{n_B}
\end{bmatrix}_{n+6}
\qquad (42)
$$

where n_E refers to the number of visible Galileo satellites; n_B refers to the number of visible BeiDou satellites; $n = n_G + n_B$ is the total number of the observed satellites for both Galileo/BeiDou systems; x_0, y_0, and z_0 are the approximate receiver coordinates; $X^{j_E}, Y^{j_E}, Z^{j_E}, j = 1, 2, \ldots, n_E$ are the known

Galileo satellite coordinates; $X^{k_B}, Y^{k_B}, Z^{k_B}, k = 1, 2, \ldots, n_B$ are the known BeiDou satellite coordinates; ρ_0 is the approximate receiver–satellite range. The unknown parameters in the above system are the corrections to the receiver coordinates, Δx, Δy, and Δz, the wet component of the tropospheric zenith path delay zpd_w, the inter-system bias ISB_{EB}, and the non-integer ambiguity parameters $\widetilde{N}$.

4. Results and discussion

To verify the developed combined PPP models, GPS, Galileo, and BeiDou observations at four globally distributed stations (Figure 1) were selected from the IGS tracking network (Dow et al., 2009). Those stations are occupied by GNSS receivers, which are capable of simultaneously tracking the GNSS constellations. Only one hour of observations with maximum possible number of Galileo and BeiDou satellites of each data-set is considered in our analysis. All data-sets have an interval of 30 s.

The positioning results for stations DLF1 are presented below. Similar results are obtained for the other stations. However, a summary of the convergence times and the three-dimensional PPP solution standard deviations are presented below for all stations. Natural Resources Canada's GPSPace PPP software was modified to handle data from GPS, Galileo, and BeiDou systems, which enables a combined PPP solution as detailed above (Afifi & El-Rabbany, 2016b). In addition to the combined PPP model, we also obtained the solutions of the un-differenced ionosphere-free GPS-only which is

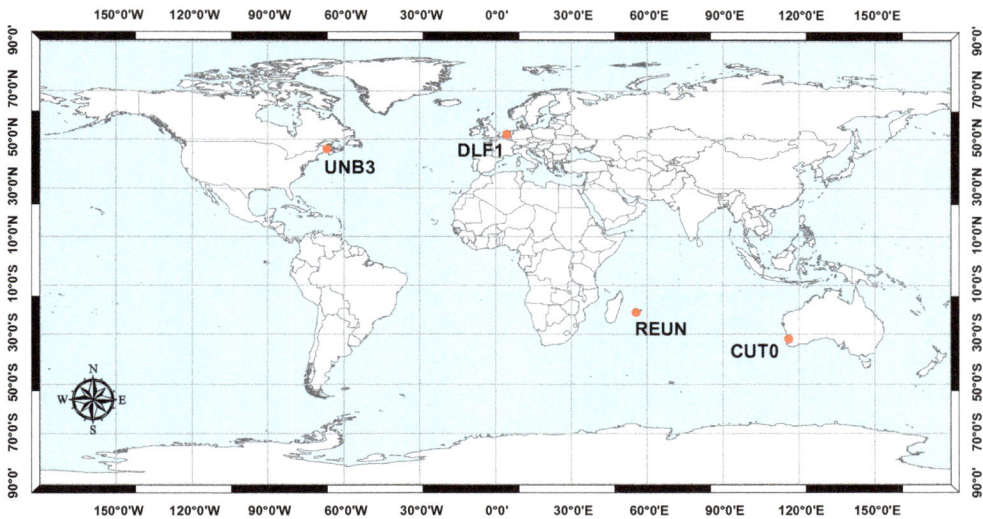

Figure 1. Analysis stations.

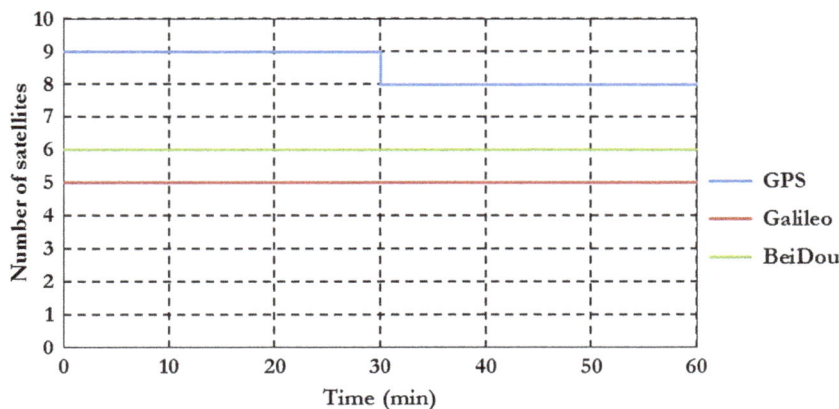

Figure 2. DLF1 station GNSS satellite availability.

used to assess the performance of the newly developed PPP model. Figure 2 summarizes the satellite availability during the analysis time (one hour) for each system at DLF1 station.

As shown in Figure 2, the GPS system offers eight visible satellites for one hour, however by adding the Galileo system the number of visible satellites will be 13 satellites. In case of combining GPS and BeiDou the number of visible satellites will be 14 satellites, however by combining the three satellite systems the number of visible satellites will be 19 satellites. Figure 3 shows the positioning results in

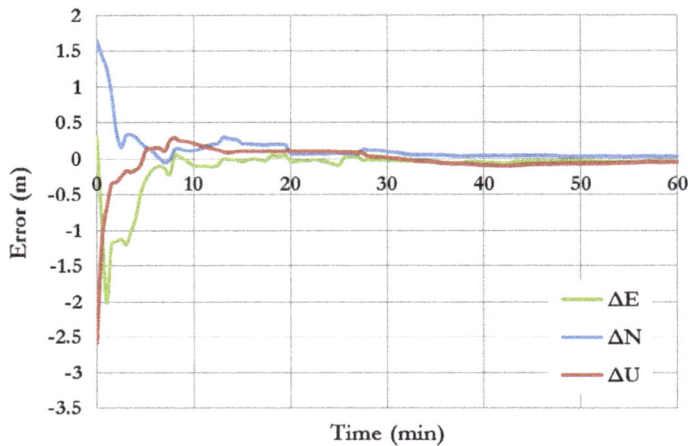

Figure 3. The positioning results of the GPS-only PPP model.

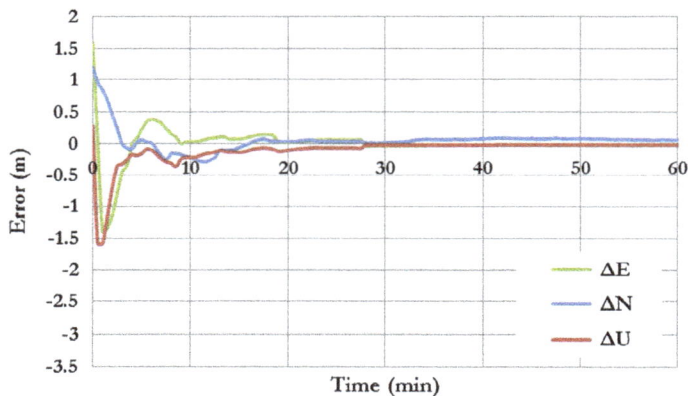

Figure 4. The positioning results of the GPS/Galileo PPP model.

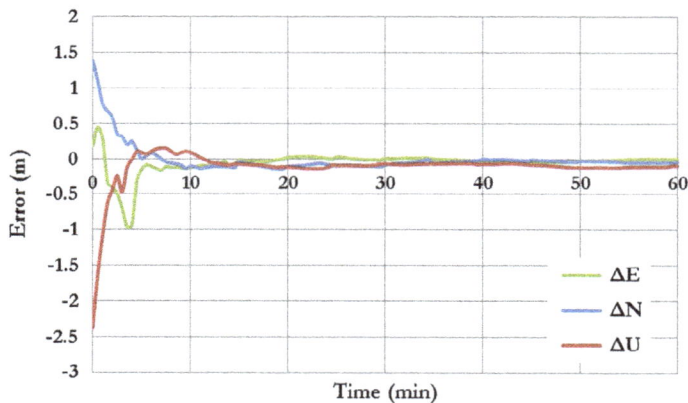

Figure 5. The positioning results of the GPS/BeiDou PPP model.

the East, North, and Up directions, respectively, for the GPS-only PPP model. As can be seen, the un-differenced GPS-only PPP solution indicates that the model is capable of obtaining a sub-decimeter-level accuracy. However, the solution takes about 20 min to converge to decimeter-level accuracy.

Figure 4 shows the positioning results combined GPS/Galileo PPP model. As shown in Figure 4, the positioning results of the combined GPS/Galileo traditional PPP model have a convergence time of 15 min to reach the decimeter-level of accuracy.

Figure 5 shows the combined GPS/BeiDou PPP model positioning results. As shown in Figure 5 the convergence time of the combined GPS/BeiDou PPP model is similar to the combined GPS/Galileo PPP model which is 15 min to reach the decimeter level of accuracy.

Figure 6 shows the combined Galileo/BeiDou PPP model positioning results. Similar to the previous combined PPP models the convergence time of the combined Galileo/BeiDou PPP model is 15 min to reach the decimeter level of accuracy.

Figure 7 shows the combined GPS/Galileo/BeiDou PPP model positioning results. As shown in Figure 7 the convergence time of the combined GPS/Galileo/BeiDou PPP model has a convergence time of 15 min to reach the decimeter level of accuracy.

Figure 8 summarizes the convergence times for all combined PPP models, which confirm the PPP solution consistency at all stations.

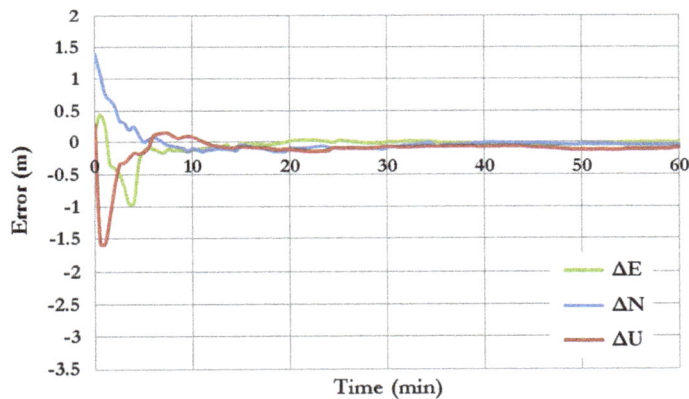

Figure 6. The positioning results of the Galileo/BeiDou PPP model.

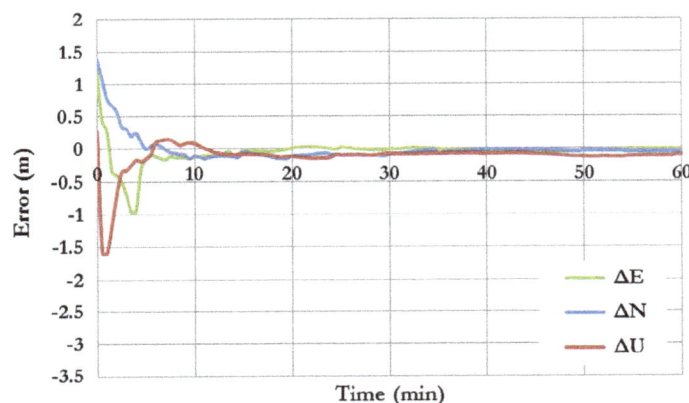

Figure 7. The positioning results of the GPS/Galileo/BeiDou PPP model.

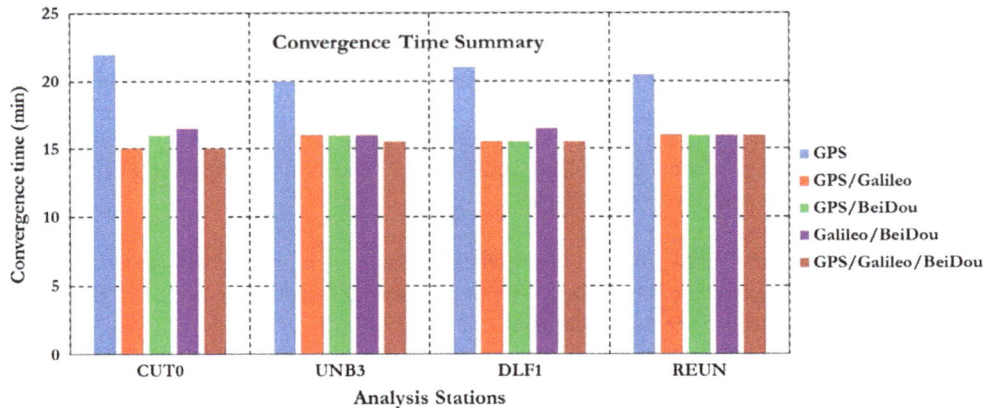

Figure 8. Summary of convergence times of all stations and PPP models.

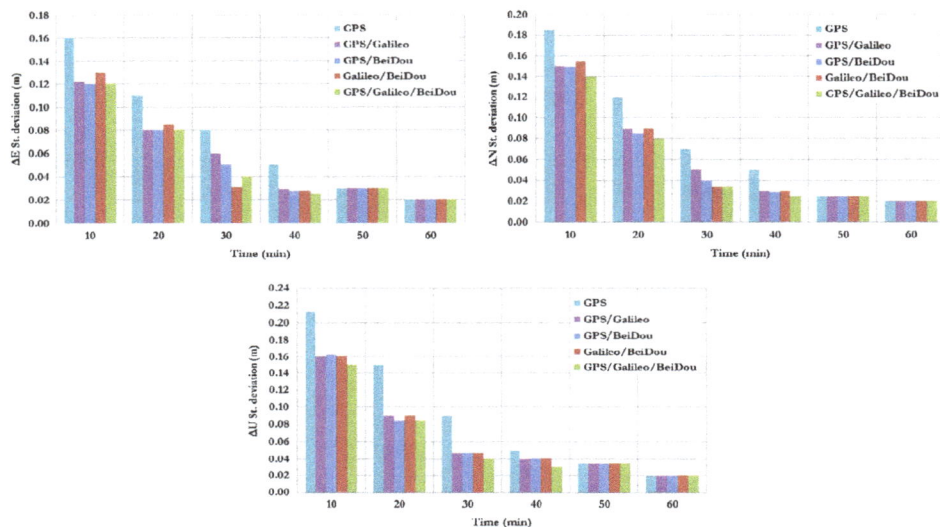

Figure 9. Summary of positioning standard deviations in East, North, and Up directions of all PPP models.

To further assess the performance of the various PPP models, the solution output is sampled every 10 min and the standard deviation of the computed station coordinates is calculated for each sample. Figure 9 shows the position standard deviations in the East, North, and Up directions, respectively. Examining the standard deviations of the combined PPP models is almost comparable to the GPS-only PPP model. As the number of epochs, and consequently the number of measurements, increases the performance of the various models tends to be comparable.

5. Conclusions

This paper presented a PPP model, which combines GPS/Galileo, GPS/BeiDou, Galileo/BeiDou, and GPS/Galileo/BeiDou observations in the un-differenced mode. The developed PPP model accounts for the combined effects of the different GNSS time offsets and hardware delays through the introduction of a new unknown parameter, the inter-system bias, in the PPP mathematical model. Four combinations are considered in the PPP modeling namely; GPS/Galileo, GPS/BeiDou, Galileo/BeiDou, and GPS/Galileo/BeiDou. All the combined PPP models results are compared with the GPS-only PPP model results. In the developed model GPS L1/L2, Galileo E1/E5a, and BeiDou B1/B2 are used in a dual-frequency ionosphere-free linear combination. It has been shown that the positioning results of the GPS-only and GPS/Galileo/BeiDou PPP are comparable and are at the sub-decimeter-level accuracy. However, the convergence time of the combined PPP models improved by about 25% in comparison with the GPS-only PPP.

Acknowledgments

This research was partially supported by the Natural Sciences and Engineering Research Council (NSERC) of Canada. The authors would like to thank the International GNSS service-Multi-GNSS Experiment (IGS-MEGX) network.

Funding

The authors received no direct funding for this research.

Author details

Akram Afifi[1]
E-mail: akram.afifi@ryerson.ca
Ahmed El-Rabbany[1]
E-mail: rabbaany@ryerson.ca
[1] Department of Civil Engineering, Ryerson University, Toronto, Ontario, Canada.

References

Afifi, A., & El-Rabbany, A. (2015). Performance analysis of several GPS/Galileo precise point positioning models. *Sensors, 15,* 14701–14726. doi:10.3390/s150614701

Afifi, A., & El-Rabbany, A. (2016a). Precise point positioning using triple GNSS constellations in various modes. *Sensors, 16,* 779. doi:10.3390/s16060779

Afifi, A., & El-Rabbany, A. (2016b). Improved between-satellite single-difference precise point positioning model using triple GNSS constellations: GPS, Galileo, and BeiDou. *Positioning, 7,* 63–74. doi:10.4236/pos.2016.72006

BeiDou. (2015). *BeiDou navigation satellite system.* Retrieved June 28, 2015, from http://en.beidou.gov.cn/

Boehm, J., & Schuh, H. (2004). Vienna mapping functions in VLBI analyses. *Geophysical Research Letters, 31,* 1601–1604. doi:10.1029/2003gl018984

Bos, M. S., & Scherneck, H.-G. (2011). *Ocean tide loading provider.* Retrieved December 1, 2014, from http://holt.oso.chalmers.se/loading/

Collins, P., Bisnath, S., Lahaye, F., & Héroux, P. (2010). Undifferenced GPS ambiguity resolution using the decoupled clock model and ambiguity datum fixing. *Navigation, 57,* 123–135. http://dx.doi.org/10.1002/navi.2010.57.issue-2

Colombo, O. L., Sutter, A. W., & Evans, A. G. (2004, September 21–24). *Evaluation of precise, kinematic GPS point positioning.* ION GNSS 17th International Technical Meeting of the Satellite Division, Long Beach, CA.

Dow, J. M., Neilan, R. E., & Rizos, C. (2009). The International GNSS service in a changing landscape of global navigation satellite systems. *Journal of Geodesy, 83,* 191–198. doi:10.1007/s00190-008-0300-3

ESA. (2015). *European space agency.* Retrieved June 28, 2015, from http://www.esa.int/ESA

Ge, M., Gendt, G., Rothacher, M., Shi, C., & Liu, J. (2008). Resolution of GPS carrier-phase ambiguities in precise point positioning (PPP) with daily observations. *Journal of Geodesy, 82,* 401. doi:10.1007/s00190-007-0208-3

Hofmann-Wellenhof, B., Lichtenegger, H., & Wasle, E. (2008). *GNSS global navigation satellite systems: GPS, Glonass, Galileo & more* (p. 501). Wien: Springer.

Hopfield, H. S. (1972). Tropospheric refraction effects on satellite range measurements. *APL Technical Digest, 11,* 11–19.

IAC. (2015). *Fedral space agency, the Information-analytical centre.* Retrieved June 28, 2015, from https://glonass-iac.ru/en/

IERS. (2010). *International earth rotation and reference system services conventions (2010)* (IERS Technical Note 36). Retrieved from http://www.iers.org/IERS/EN/Publications/TechnicalNotes/tn36.html/

Kaplan, E., & Heagarty, C. (2006). *Understanding GPS principles and applications* (650 p.). Boston, MA: Artech House.

Kouba, J. (2009). *A guide to using international GNSS service (IGS) products.* Retrieved from http://igscb.jpl.nasa.gov/igscb/resource/pubs/UsingIGSProductsVer21.pdf

Kouba, J., & Héroux, P. (2001). Precise point positioning using IGS orbit and clock products. *GPS Solutions, 5,* 12–28. http://dx.doi.org/10.1007/PL00012883

Leick, A. (2004). *GPS satellite surveying* (3rd ed., p. 435). New York, NY: Wiley.

Melgard, T., Tegedor, J., de Jong, K., Lapucha, D., & Lachapelle, G. (2013, September 16–20). Interchangeable integration of GPS and Galileo by using a common system clock in PPP. In *ION GNSS+,* Nashville TN: Institute of Navigation.

Montenbruck, O., Steigenberger, P., Khachikyan, R., Weber, G., Langley, R. B., Mervart, L., & Hugentobler, U. (2014). IGS-MGEX: Preparing the ground for multi-constellation GNSS science. *Inside GNSS, 9,* 42–49.

Vanicek, P., & Krakiwsky, E. J. (1986). *Geodesy: The concepts* (2nd ed.). Amsterdam: North-Holland.

Wu, J. T., Wu, S. C., Hajj, G. A., Bertiger, W. I., & Lichten, S. M. (1993). Effects of antenna orientation on GPS carrier phase. *Manuscripta Geodetica, 18,* 91–98.

Zumberge, J. F., Heflin, M. B., Jefferson, D. C., Watkins, M. M., & Webb, F. H. (1997). Precise point positioning for the efficient and robust analysis of GPS data from large networks. *Journal of Geophysical Research: Solid Earth, 102,* 5005–5017. http://dx.doi.org/10.1029/96JB03860

3

Assessing the potential of remote sensing to discriminate invasive *Asparagus laricinus* from adjacent land cover types

Bambo Dubula[1]*, Solomon Gebremariam Tesfamichael[1] and Isaac Tebogo Rampedi[1]
*Corresponding author: Bambo Dubula, Department of Geography, Environmental Management and Energy Studies, University of Johannesburg, Johannesburg, South Africa
E-mail: bambodubula@gmail.com
Reviewing editor:
Louis-Noel Moresi, University of Melbourne, Australia

Abstract: The utility of remote sensing technique to discriminate *Asparagus laricinus* from adjacent land cover types using a field spectrometer data was explored in this study. Analysis made use of original spectra and spectra simulated based on Landsat and SPOT 5 bands. Comparisons were made at individual and plot levels using original spectra, and individual and group level using simulated spectra. The near-infrared region showed consistent significant differences between *A. laricinus* and adjacent land cover types at the individual level analysis. In particular, Landsat- and SPOT 5-simulated spectra showed significant differences in only the NIR band. The findings suggest the potential of upscaling field-based data into airborne or spaceborne remote sensing techniques with more emphasis on the NIR band. However, more studies need to be undertaken that will make up for the shortcomings encountered in this study. In this regard, improvements can be made using large number of samples, stratifying target plants according to phenologies, and taking spectral measurements at ideal times as much as possible. Furthermore, laboratory measurements would help in drawing up conclusive statements on the discriminability of the species.

Subjects: Biodiversity; Botany; Earth Sciences

Keywords: *Asparagus laricinus*; remote sensing; spectral reflectance; spectral bands; field spectrometer

1. Introduction

Invasive alien plants are a growing global concern (Richardson & Van Wilgen, 2004; Rouget, Hui, Renteria, Richardson, & Wilson, 2015; Schor, Farwig, & Berens, 2015; Vicente et al., 2013). These

ABOUT THE AUTHORS

Bambo Dubula is a postgraduate student in environmental management, Solomon Gebremariam Tesfamichael is a senior lecturer in GIS and remote sensing, and Isaac Tebogo Rampedi is a senior lecturer in environmental management in the Department of Geography, Environmental Management and Energy Studies, University of Johannesburg. We are conducting research work on plant invasions in the Kliprivierberg Nature Reserve and we aim toward developing distribution maps of invasive plant infestation that will be helpful in the design of better management strategies by the reserve managers.

PUBLIC INTEREST STATEMENT

This study gives insights into remote sensing capabilities in discriminating *A. laricinus* from adjacent land cover types. The information will be helpful in developing spatial distribution maps of the species. Such maps are valuable to land managers for the timely information they provide on the distribution of the species, thus aid in development of better management strategies of the species.

plants hold special characters that make them outcompete and replace indigenous vegetation, and have the potential of spreading to other areas (Bradley & Marvin, 2011; Mgidi et al., 2007; Van Wilgen, 2006). As a result, they compromise ecosystem stability, delivery of ecosystem goods and services, and threaten biodiversity and economic productivity (van Wilgen, Reyers, Le Maitre, Richardson, & Schonegevel, 2008; van Wilgen et al., 2012; Van Wilgen, 2006). Mitigating these effects is costly; South Africa, for example, spends considerable amounts of money in programs such as the Working for Water (WfW) which is mandated to control invasive alien plants.

Most invasive plant control measures focus primarily on established invasions and less attention is given to new infestations (Mgidi et al., 2007). The success of this practice is unsatisfactory, since an effective management of invasive alien plants should depend on early detection and eradication (Mgidi et al., 2007). One method of achieving early detection of plant invasions is through the use of spatial and temporal maps that show the distribution of invasive plants (Dorigo, Lucieer, Podobnikar, & Čarni, 2012). Traditional methods can be used to provide spatial and temporal distribution of invasive plants, but the methods often rely on field inventories that are limited in spatial coverage, time-consuming, and relatively expensive (Dewey, Price, & Ramsey, 1991; Dorigo et al., 2012; Rodgers, Pernas, & Hill, 2014).

Remote sensing methods make up for most inefficiencies of the traditional mapping methods and are used to characterize the spatial and temporal distribution of plants (Alparone, Aiazzi, Baronti, & Andrea, 2015; Campbell & Wynne, 2011; Galvão, Epiphanio, Breunig, & Formaggio, 2011; Jensen, 2014; Lillesand, Kiefer, & Chipman, 2015). Remote sensing is the science of deriving information from electromagnetic energy reflected from objects on the ground (Alparone et al., 2015; Campbell & Wynne, 2011; Jensen, 2014). The method differentiates earth features using varying sensitivity of ground objects to electromagnetic radiation often acquired within the visible, infrared, and microwave regions of the spectrum (Campbell & Wynne, 2011; Lillesand et al., 2015). Several studies have used a variety of remote sensing techniques to study invasive alien plants (e.g. Abdel-Rahman, Mutanga, Adam, & Ismail, 2014; Adam & Mutanga, 2009; Adelabu, Mutanga, Adam, & Sebego, 2014; Bentivegna et al., 2012; Berg, Kotze, & Beukes, 2013; Manevski, Manakos, Petropoulos, & Kalaitzidis, 2011; Martín, Barreto, & Fernández-Quintanilla, 2011; Mirik et al., 2014; Narumalani, Mishra, Wilson, Reece, & Kohler, 2009; Prasad & Gnanappazham, 2014).

Plants have been mapped using multispectral remote sensing techniques in a number of studies (e.g. Dronova, Gong, Wang, & Zhong, 2015; Johansen, Phinn, & Witte, 2010; Laba et al., 2008; Lemke, Hulme, Brown, & Tadesse, 2011; Vancutsem, Pekel, Evrard, Malaisse, & Defourny, 2009). This method is good particularly for large spatial area mapping purposes (Azong, Malahlela, & Ramoelo, 2015; Cuneo, Jacobson, & Leishman, 2009; Dronova et al., 2015; Vancutsem et al., 2009). In comparison, hyperspectral remote sensing offers better accuracy levels of vegetation characterization due to the high spectral resolution and continuous hyperspectral bands they possess (Alparone et al., 2015; Carroll, Glaser, Hunt, & Sappington, 2008; Gavier-pizarro, Kuemmerle, & Stewart, 2012; Huang & Asner, 2009; Jensen, 2014). For example, Bentivegna et al. (2012) detected invasive cutleaf teasel (*Dipsacus laciniatus* L.) in Missouri, USA using high spatial resolution (1 m) hyperspectral images (63 bands in visible to near-infrared spectral region). Mirik et al. (2013) explored the ability of hyperspectral imagery for mapping infestation of musk thistle (*Carduus nutans*) on a native grassland during the pre-and peak-flowering stages using support vector machine classifier in Friona, Parmer County, USA. Ouyang et al. (2013) used a field spectrometer data to find the most appropriate period for mapping invasive *Spartina alterniflora* by measuring its community and major victims at different phonological stages in Chongming Island, China. Similarly, Rudolf, Lehmann, Große-stoltenberg, Römer, and Oldeland (2015) developed a classification model to spectrally discriminate between invasive shrub *Acacia longifolia* from other non-native and native species using field-based spectra and condensed leaf tannin content in Portuguese dune ecosystems, Portugal.

However, discrimination of plant species using hyperspectral data often places emphasis on identification of the optimal specific bands for discrimination. These bands are narrow and cannot be separated from within the broader bandwidth of multispectral data. Hyperspectral remote sensing has grown significantly in the past few decades. However, its application in operational characterization is rather limited. Although there is a promise to translate research efforts of hyperspectral remote sensing into operational tools, current advances in data availability show that multispectral remote sensing remains the most important source of information in vegetation monitoring. For example, DeVries, Verbesselt, Kooistra, and Herold (2015) monitored small-scale forest disturbances in a tropical montane forest of southern Ethiopia using Landsat time series. Gu and Wylie (2015) developed a 30-m grassland productivity estimation map for central Nebraska in USA using 250 m MODIS and 30 m Landsat 8 observations, United States. Johansen, Phinn, and Taylor (2015) mapped woody vegetation clearing in Queensland, Australia from Landsat imagery using the Google Earth Engine. Kennedy et al. (2015) described factors attributing to disturbance change from Landsat time-series in support of habitat monitoring in the Puget Sound region, USA. Therefore, research efforts involving hyperspectral remote sensing analysis need to consider extending the technique into multispectral remote sensing techniques.

This study uses a continuum of hyperspectral bands to identify best wavelength regions for discriminating *Asparagus laricinus* from adjacent land cover types. As such, it focuses on spectral regions rather than identifying individual bands in an attempt to simulate multispectral remote sensing systems. Specific objectives of the study are (1) determining whether or not *A. laricinus* can be differentiated from adjacent land cover types using a field spectrometer data and (2) to investigate the performance of spectra simulated according to Landsat and SPOT 5 images in discriminating *A. laricinus* from adjacent land cover types. There have been little or no studies that focused on discriminating *A. laricinus* from other vegetation or land cover types. *A. laricinus* is a plant belonging to the Asparagaceae family and occurs in different parts of South Africa. However, the plant is not indigenous in South Africa and has a status of "list concern" in the South African National Biodiversity Institute (SANBI) national Red List of South African plants (Foden & Potter, 2005). Knowledge on the spectral and spatial characteristics of the species assists the development of better management strategies in areas where it invades. Such maps can also help traditional health practitioners and pharmaceutical industries to locate stands of the plant for medicinal purposes, as it also has medicinal uses (Fuku, Al-Azzawi, Madamombe-Manduna, & Mashele, 2013; Mashele & Kolesnikova, 2010; Ntsoelinyane & Mashele, 2014).

2. Methods

2.1. Study area
The study was conducted in the Klipriviersberg Nature Reserve, in Johannesburg, South Africa (Figure 1). It covers an area of approximately 680 hectares in extent and is managed by the City of Johannesburg. The reserve lies in the Klipriviersberg area, a transition zone between the grass land and the savannah biome in the northern edge of the Highveld (Faiola & Vermaak, 2014). Climatic conditions experienced in the reserve vary from warm to hot summer (17–26°C) and cool to cold winter (5–7°C) (Kotze, 2002). Three geology types occur in the reserve, namely basalt and andesite volcanic rocks that underlay the reserve; quartzites and conglomerates of the upper Witwatersrand system underneath the lavas in north of the reserve; and dolomites of the Transvaal system south of the reserve (Kotze, 2002). The flora of the reserve is categorized into two broad vegetation types, the Andesite Mountain Bushveld and a section of Tsakane Clay Grassland at its flatter southern end (Faiola & Vermaak, 2014). There is relatively rich biodiversity with approximately 650 indigenous plant species, 215 bird species, 16 reptile species, and 32 butterfly species. Mammals that occur in the reserve include lesser spotted genet, African civet, zebra, red hartebeest, blesbok, springbok, duiker, black wildebeest, porcupines, meerkats, and otters (Faiola & Vermaak, 2014).

2.2. Field data
Field surveys were conducted between the 2 and 14 December 2014 during summer season of the area with the aim of characterizing the vegetation under relatively high vigor condition. *A. laricinus* is found extensively in one part of the reserve, while other occurrences are scattered in small spatial

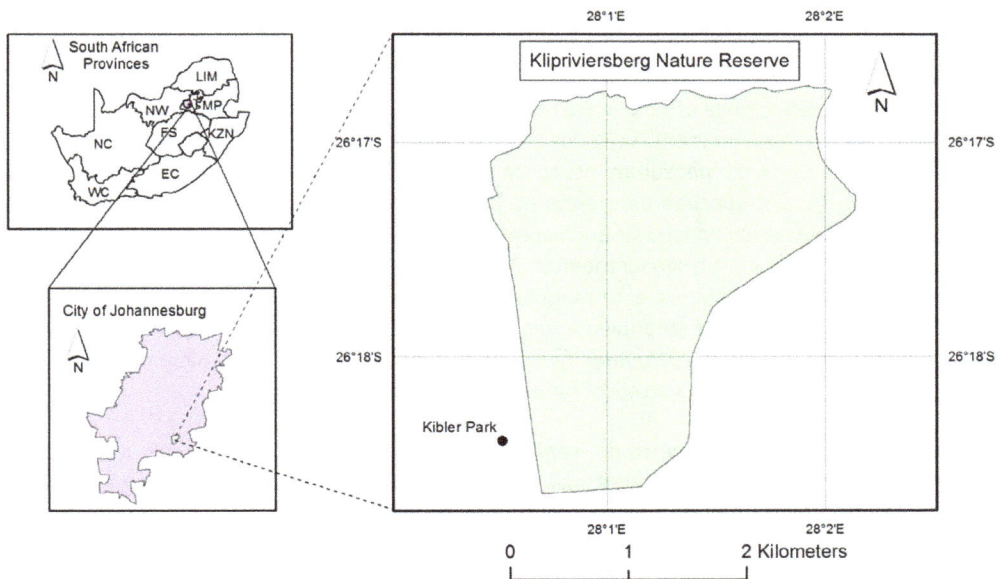

Figure 1. Map showing the Klipriviersberg Nature Reserve.

extents. Such a rather limited distribution resulted in delineation of 10 plots of 15 m radius each (Figure 2). The plot size was chosen with the anticipation of extending the investigation to space-borne remote sensing techniques. Each plot therefore accommodates at least one pixel of Landsat imagery (30 m resolution) and a number of SPOT 5 imagery pixels (1.5–10 m resolutions). The center of each plot was recorded using GPS (Garmin GPSmap® 76) within 3 m accuracy. A total of 13 sample plants were taken randomly within the 15 m radius plot area. Although random, sampling was attempted to follow a systematic design as shown in Figure 2. Therefore, samples were taken at 5-m intervals along perpendicular transects that intersect at the center of plot (Appendix A). However, this was rarely achieved as it was difficult walking through the thorny and dense stands of *A. laricinus*, prompting use of random sampling. *A. laricinus* individuals varied between six and eight plants within each plot.

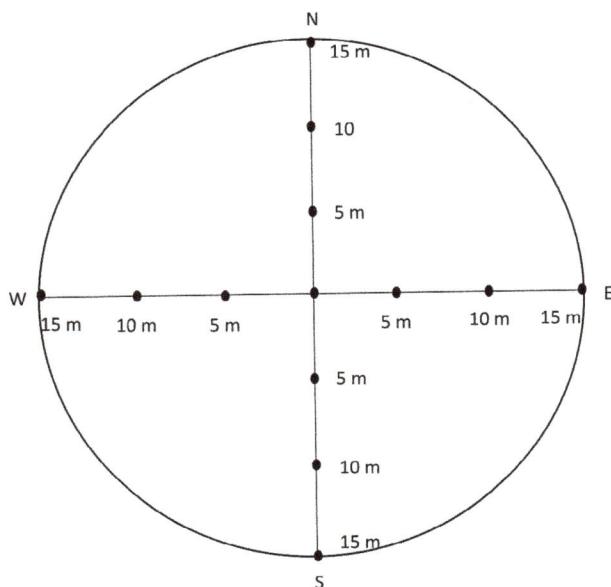

Figure 2. A layout of sampling design for spectral measurements of individual target plant.

Spectral data were collected using Spectral Evolution® SR-3500 Remote Sensing Portable Spectroradiometer (Spectral Evolution Inc., Lawrence, MA, USA). The spectrometer has a 1.6-nm spectral resolution ranging between 340 and 2,503 nm. Target radiance in energy unit was converted into percent reflectance using a white reference measurement (Prospere, McLaren, & Wilson, 2014). Three spectral measurements were taken for each *A. laricinus* plant from different leaf canopy parts of the plant with all measurements taken at 5 cm above leaf canopy to mimic a remotely sensed data (airborne and spaceborne) viewpoint. The three spectral measurements were averaged to represent the reflectance spectra of each sample plant. Spectral measurements from adjacent land cover types were taken in a similar manner. These measurements should ideally be taken when the sun is overheard to acquire electromagnetic radiation reflectance optimally (Cho, Sobhan, Skidmore, & de Leeuw, 2008; Fernandes, Aguiar, Silva, Ferreira, & Pereira, 2013; Mansour, 2013; Olsson, van Leeuwen, & Marsh, 2011; Rudolf et al., 2015). However, time constraints did not necessarily allow the application of this protocol for all measurements.

2.3. Analysis of spectral reflectance per region

Analysis was limited to the regions that showed consistent spectral differences between *A. laricinus* and adjacent land cover types. In order to identify these regions an average spectrum was computed from the three spectral measurements taken from each target (*A. laricinus* and adjacent land cover type, respectively). The resultant average values were pooled per land cover type and averaged to generate "global" spectral curves representing *A. laricinus* and each adjacent land cover type in the study area as illustrated (Figure 3). The global spectra of *A. laricinus* was compared against each adjacent land cover types, as illustrated in an example comparing *A. laricinus* and grass in Figure 4. Please note, not all global comparisons are presented in here for the sake of brevity. The global spectra of adjacent land cover types were computed to determine the potential discrimination of *A. laricinus* from them, since the species can co-exist with a mixture of land cover types in a natural environment. Comparison using global pairs is deemed a better representation of the study area than comparison of individual pairs that most likely yields results that are unable to converge to a compromise generic conclusion.

A visual assessment of the global spectra was used to determine regions that were considered unnecessary for differentiating *A. laricinus* and adjacent land cover types. Two rules were used to determine these regions. The first rule included regions that returned random reflectance properties commonly known as atmospheric noise (*A. laricinus* vs. Grass: 1,873–1,954 and 2,351–2,503 nm; *A. laricinus* vs. Acacia: 1,821–1,956 nm and 2,282–2,503 nm; *A. laricinus* vs. Herbaceous: 1,838–1,942 nm and 2,272–2,503 nm; *A. laricinus* vs. mixture of herbaceous and bare ground: 1,831–1,970 nm and 2,351–2,503 nm). The second rule included regions that did not show spectral reflectance difference between *A. laricinus* and adjacent land cover types (*A. laricinus* vs. Grass: 340–343, 684–750 nm and 1,350–1,824 nm; *A. laricinus* vs. Acacia: 650–749 and 1,331–1,448 nm: *A. laricinus* vs. Herbaceous: 340–387, 641–748 nm and 1,316–1,448 nm; *A. laricinus* vs. Mixture of herbaceous and bare ground: 340–467, 685–745 nm and 1357–1,455 nm). These exclusions resulted in four discontinuous regions (Table 1, Figure 5) based on which spectra of individual targets (individuals of *A. laricinus* and adjacent land cover types) were used in further analyses.

Analysis involved comparison of reflectance between *A. laricinus* and adjacent land cover types at two levels, namely individual and plot levels. Individual level comparison was made between *A. laricinus* and adjacent land cover type at a sampling point within each plot. On the other hand, plot level comparison was made between plot level mean reflectance of *A. laricinus* against plot level mean reflectance of dominant adjacent land cover type. Differences at both levels were assessed graphically and using statistical tests such as the analysis of variance (ANOVA) and *t*-test (Weiss, 2012). All the tests were calculated using 95% confidence level ($\alpha = 0.05$).

2.4. Simulation of Landsat and SPOT 5 imagery bands

Wavelength regions corresponding to Landsat and SPOT 5 bands were extracted from the original reflectance spectra for all *A. laricinus* and adjacent land cover types. This was an initial step to

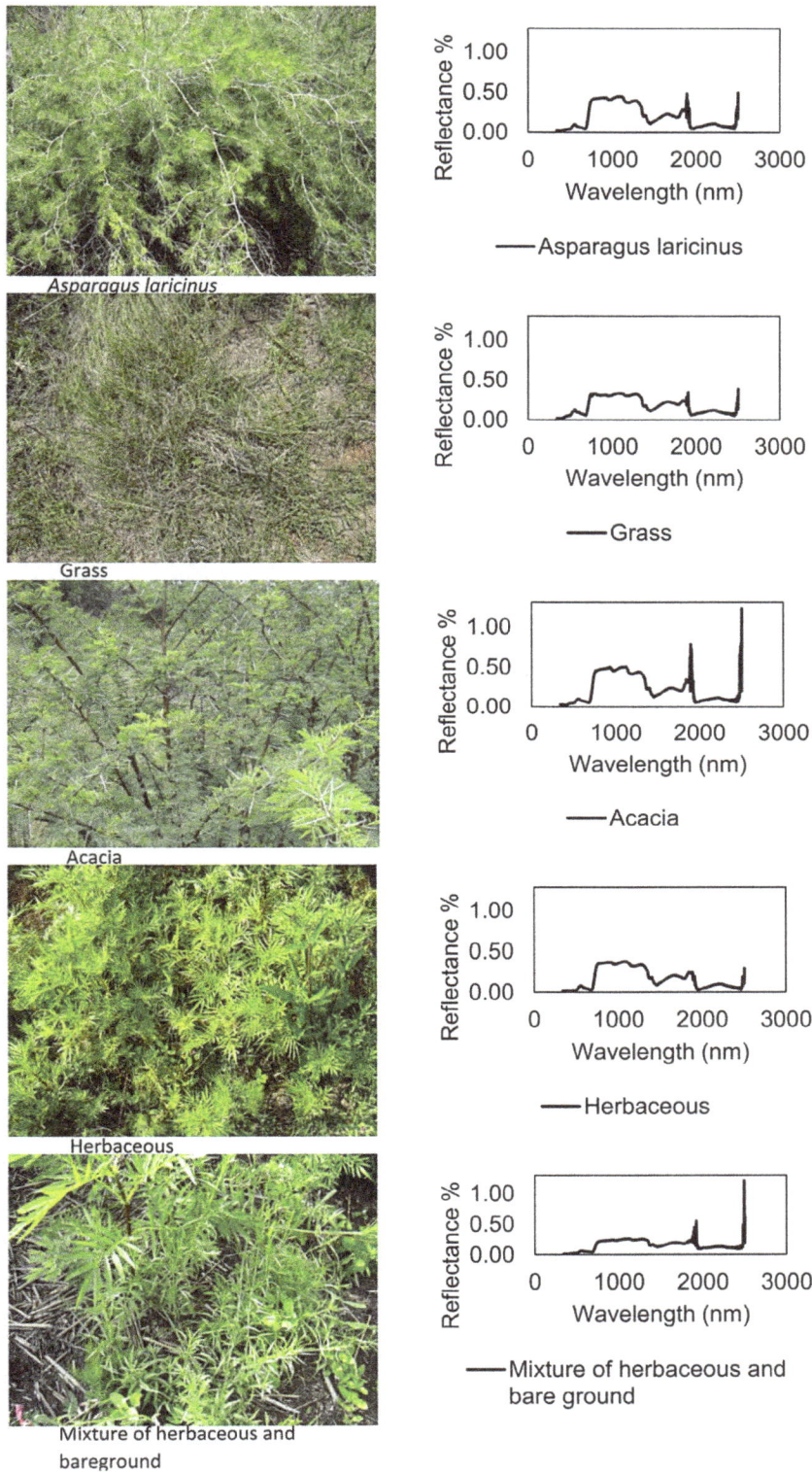

Figure 3. Global spectra of A. laricinus and adjacent land cover types.

testing the potential of upscaling field-based remote sensing information to airborne or satellite-based remote sensing. Only blue, green, red, and NIR bands were simulated or Landsat, while green, red, and NIR spectral bands were simulated for SPOT 5 imagery. These elected bands are widely used

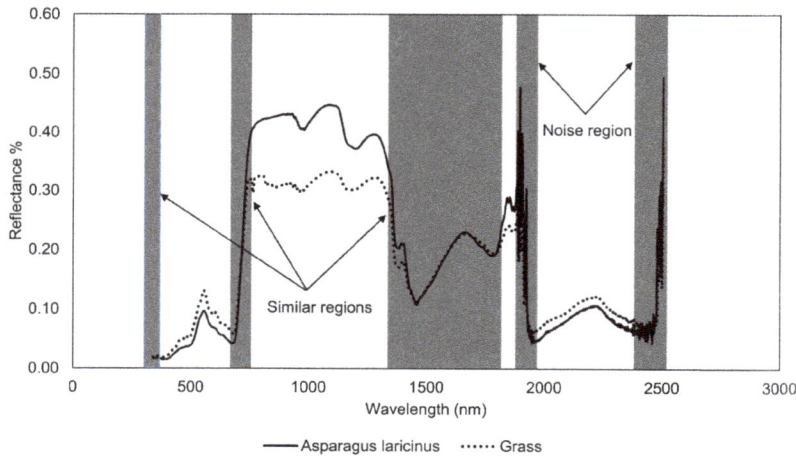

Figure 4. Global reflectance of A. *laricinus* and grass across the full spectrum.

Note: Highlighted regions show spectral parts excluded from further analysis.

Table 1. Wavelength regions used for analysis				
Comparison pairs	**Wavelength regions**			
	Region 1 (Ultraviolet & Visible), nm	**Region 2 (NIR), nm**	**Region 3 (NIR & SWIR), nm**	**Region 4 (SWIR), nm**
A. *laricinus* vs. Grass	345–683	752–1,346	1,828–1,872	1,956–2,349
A. *laricinus* vs. Acacia	340–648	750–1,327	1,452–1,817	1,959–2,279
A. *laricinus* vs. Herbaceous	389–640	749–1,312	1,452–1,835	1,945–2,269
A. *laricinus* vs. Mixture of herbaceous and bare ground	468–684	747–1,354	1,459–1,828	1,973–2,349

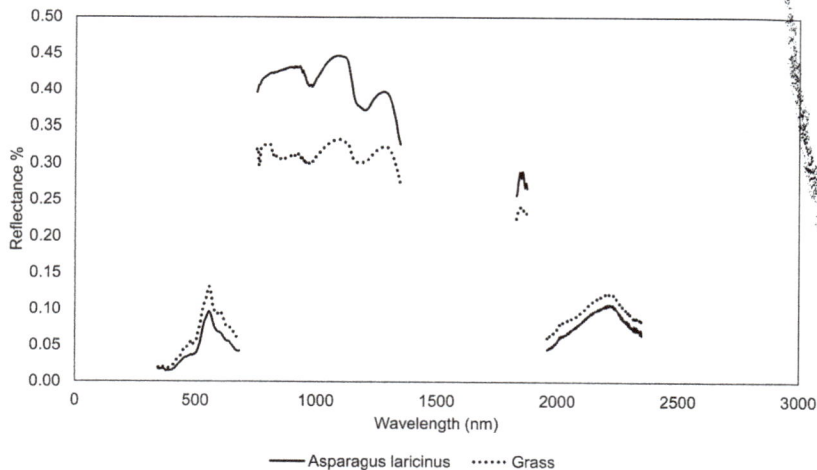

Figure 5. An example of global reflectance of A. *laricinus* and grass, representing wavelength regions used in further analysis.

in the assessment of vegetation characteristics (e.g. Manevski et al., 2011; Mirik, Ansley, et al., 2013; Mirik, Emendack, et al., 2014). Five separate pools representing A. *laricinus*, grass, acacia, herbaceous, and mixture of herbaceous and bare ground were created. Reflectance comparisons were done at individual and group level. Individual level compared the pool of A. *laricinus* against separate pools of grass, acacia, herbaceous, and mixture of herbaceous and bare ground. The group level compared *A. laricinus* pool against combined pool of adjacent land cover types. Spectral differences were assessed using ANOVA and *t*-test.

3. Results

Individual-level comparisons between *A. laricinus* and adjacent land cover types resulted in an over-all significant difference in all plots for each spectral region, based on ANOVA results. However, sepa-rate reflectance comparisons of each of the individuals per plot showed inconsistent significant differences. Distinct spectral separability between *A. laricinus* and adjacent land cover types was observed mostly in the NIR region (region 2), with seven of 10 plots. In contrast, only two in the ul-traviolet–visible (region 1), three in the NIR–SWIR (region 3), and five in the SWIR (region 4) regions showed clear separation. These differences are illustrated in Figure 6 which show spectral reflec-tance differences between *A. laricinus* and grass of one plot. Significant differences are presented using different letters, whereas same letters represent insignificant differences. Distinct separation between *A. laricinus* and adjacent land cover types in the NIR region (region 2) is shown by higher reflectance of *A. laricinus* than other land cover types (Figure 6).

Grasses represented majority of land cover types at plot level analysis (7 of 10 plots), while Herbaceous, Acacia, and Mixture of ground and herbaceous were dominant in each of the remaining plots. Comparisons at this level resulted in significant differences in all plots based on *t*-test results as illustrated in Figure 7. In most cases, *A. laricinus* had higher reflectance than adjacent land cover types in the NIR region (region 2), with 8 of 10 plots. The species had higher reflectance in five plots in the ultraviolet–visible (region 1), six plots in the NIR–SWIR (region 3) and five plots in the SWIR region (region 4).

3.1. Landsat simulation

Comparisons between *A. laricinus* and adjacent land cover types at the individual level resulted in an overall significant difference in all Landsat simulated bands (blue, green, red, and NIR), based on the ANOVA results. Individual pair comparisons using least significance difference (LSD) resulted in sig-nificant difference between *A. laricinus* and all land cover types, in most cases (Figure 8). Similarities were, however, observed between *A. laricinus* and grass in the blue and red bands, and between *A. laricinus* and herbaceous vegetation in the green band (Figure 8). *A. laricinus* had higher reflectance than other adjacent land cover types with exceptions of Acacia in the blue band, Acacia and herba-ceous in the green and NIR bands, and Acacia and grass in the red band.

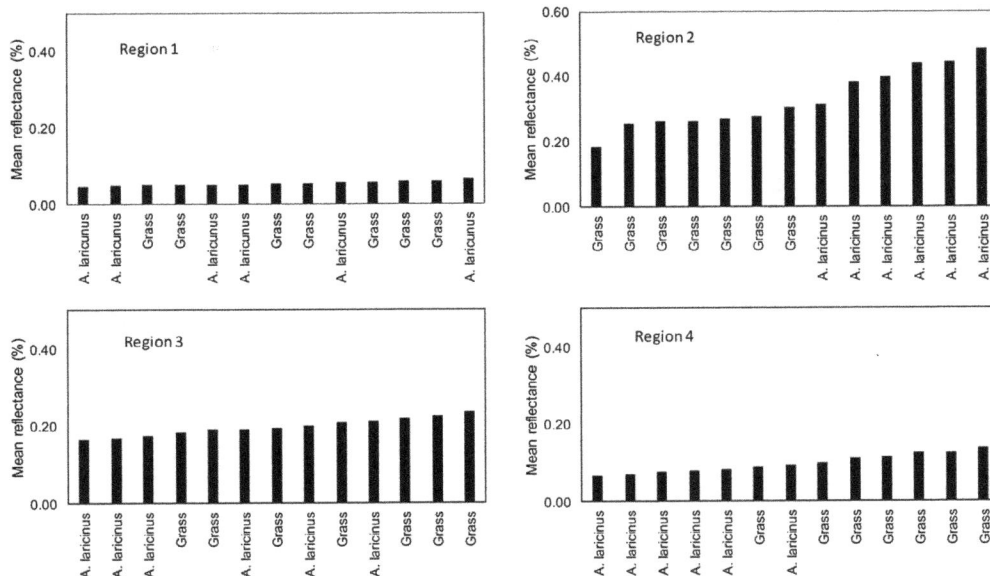

Figure 6. Reflectance of the regions used for analysis at individual plant level for a typical plot.

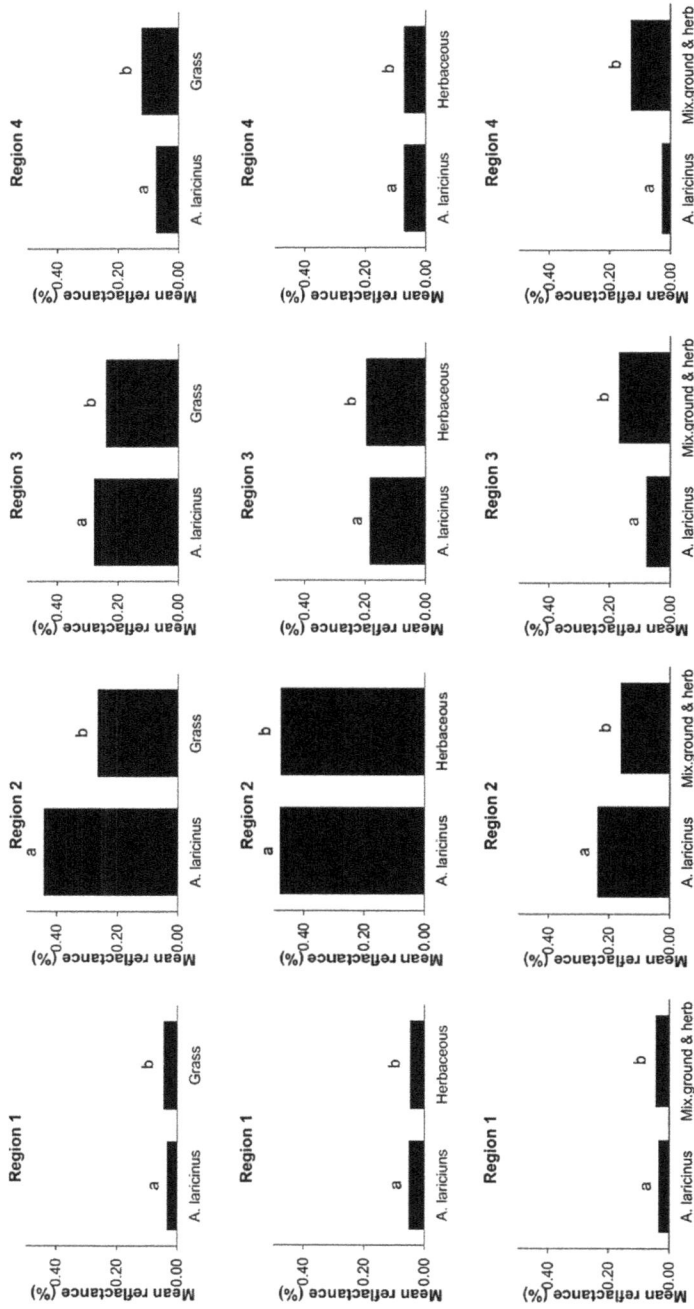

Figure 7. Plot-level mean reflectance of A. laricinus and adjacent land cover types.

Note: The comparisons are per region and per plot. (Mix. Ground & herb = Mixture of herbaceous and bare ground).

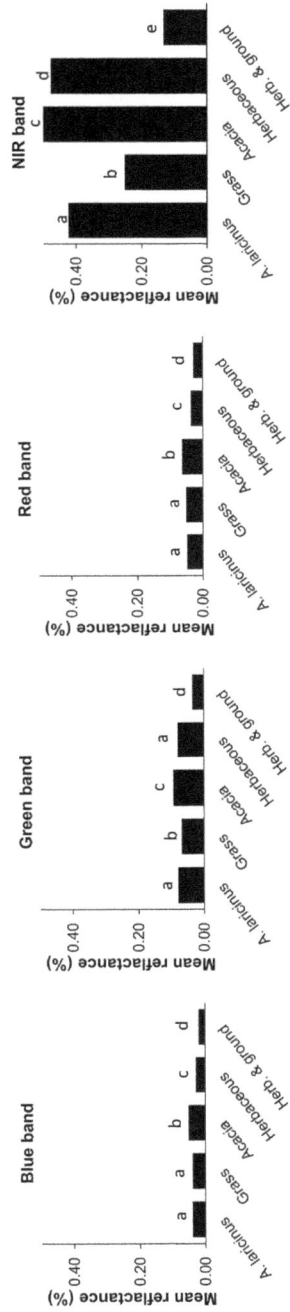

Figure 8. Mean reflectance of simulated Landsat bands per land cover type (individual level).

Note: The comparison is per spectral band. (Herb. & ground = Mixture of herbaceous and bare ground).

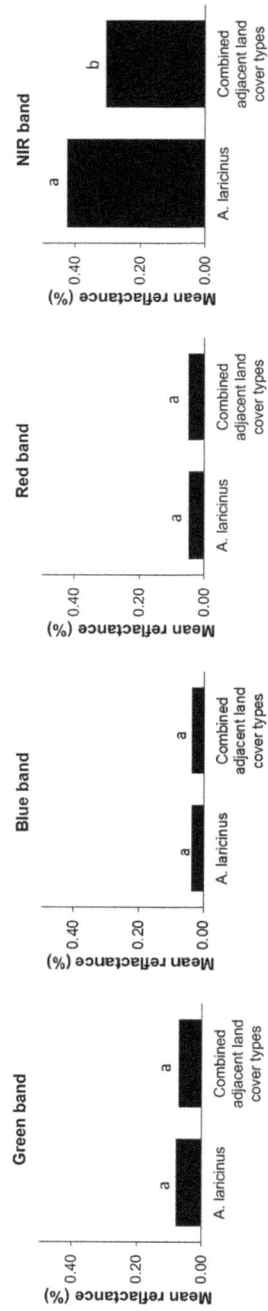

Figure 9. Mean reflectance of simulated Landsat bands per land cover type (Group level).

Note: The comparison is per spectral band.

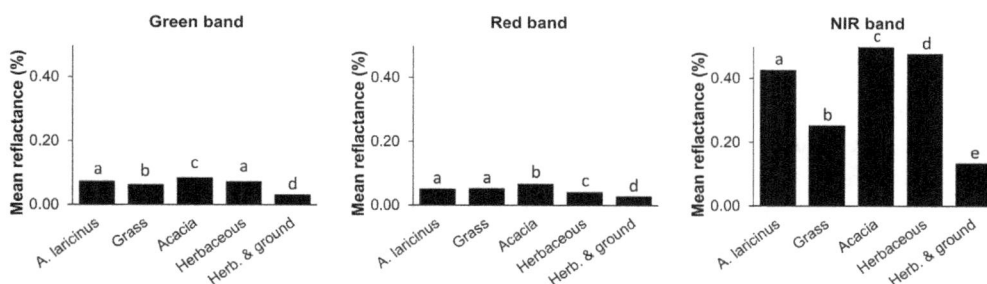

Figure 10. Mean reflectance of simulated SPOT 5 bands per land cover type (individual level).
Note: The comparison is per spectral band. (Herb. & ground = Mixture of herbaceous and bare ground).

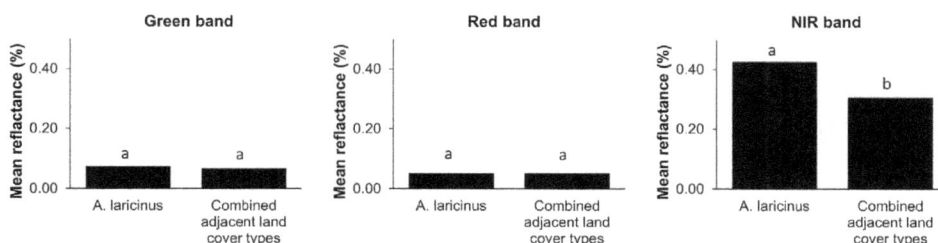

Figure 11. Mean reflectance of simulated SPOT 5 bands per land cover type (Group level).

Note: The comparison is per spectral band.

Comparison of reflectance at the group level between *A. laricinus* and combined adjacent land cover types resulted in insignificant difference in the blue, green, and red bands, while the difference was significant in the NIR (Figure 9). *A. laricuns* had higher reflectance than combined adjacent land cover types in the green and NIR band, while it had lower reflectance in the blue and red bands (Figure 9).

3.2. SPOT 5 simulation
Reflectance comparisons of SPOT 5 simulated bands resulted in overall significant differences in all bands, based on ANOVA. Individual pair comparisons using LSD showed significant differences between *A. laricinus* and adjacent land cover types in all bands, except for comparison between *A. laricinus* and herbaceous vegetation in the green band as well as between *A. laricinus* and grass in the red band (Figure 10). *A. laricinus* had a relatively high reflectance in all bands. However, it had lower reflectance than Acacia plants in all bands and herbaceous vegetation in the green and NIR bands, and grass in the red band (Figure 10).

Group-level comparisons between *A. laricinus* and combined adjacent land cover types showed significant difference in only the NIR band (Figure 11). *A. laricinus* had higher reflectance than combined adjacent land cover types in the green and NIR bands, while it had negligible reflectance in the red band (Figure 11).

4. Discussion
The utility of a field-based spectral data to discriminate *A. laricinus* from adjacent land cover types was investigated in this study. Investigations were made using original spectra and spectra simulated based on bands of Landsat and SPOT 5 images. These simulations were intended to assess the potential of upscaling the technique to spaceborne remote sensing techniques. Analyses were done at individual and plot levels using original spectra, and individual and group level for the simulated spectra. Visual comparisons using global pair reflectance of *A. laricinus* and each adjacent land cover type showed differentiation in the ultraviolet–visible (region 1), NIR (region 2), NIR–SWIR (region 3), and SWIR (region 4) spectral regions, but the difference was considerable in the NIR region (e.g. Figure 5). *A. laricinus* had high reflectance in NIR (region 2) and NIR–SWIR (region 3) and low reflectance in ultraviolet–visible region (region 1) and SWIR region (region 4) when compared with grass. *A. laricinus* reflectance was high in all regions when compared with herbaceous, while it was high in

ultraviolet–visible (region 1), NIR (region 2), and NIR–SWIR (region 3) when compared with mixture of bare ground and herbaceous plants, but it was low in all regions when compared with Acacia. All these wavelength regions are considered best at characterizing vegetation types (e.g. Manevski et al., 2011; Mirik, Ansley, et al., 2013; Mirik, Emendack, et al., 2014). The far-SWIR region on the other hand is considered best at discriminating between photosynthetic, non-photosynthetic vegetation components, and ground due to spectral absorption attributable to presence of cellulose in healthy vegetation (Daughtry et al., 2006; Guerschman et al., 2009; Nagler, Daughtry, & Goward, 2000; Serbin, Daughtry, Hunt, Reeves, & Brown, 2009).

The overall significant differences observed for individual-level comparisons per plot are not at-tributable to reflectance difference between A. laricinus and adjacent land cover types. This is be-cause significant differences were observed even within individuals of same land cover types, based on pairwise comparisons using LSD. There were further inconsistent significant differences when comparing individuals per plot separately. As such, distinct separation between A. laricinus and ad-jacent land cover types was mostly achieved in the NIR region, for 7 of 10 plots, while only a few plots showed clear separation in the ultraviolet–visible region, NIR–SWIR, and SWIR regions (Figure 6). Consistent significant difference observed in the NIR region was somewhat expected, given the distinct reflectance differences between A. laricinus and adjacent land cover types from the global spectra comparisons (e.g. Figure 5).

The plot-level differences between A. laricinus and dominant adjacent land cover types were con-siderable particularly between A. laricinus and grass as well as A. laricinus and mixture of herba-ceous vegetation and bare ground (e.g. Figure 7). The differences were somewhat expected given different global reflectance patterns of A. laricinus, grass, and mixture of herbaceous vegetation and bare ground (Figure 3). In contrast, the differences between A. laricinus and herbaceous were lower, although they were significant in the visible, NIR, and lower end of SWIR regions. This can as well be explained by the global reflectance resemblance of A. laricinus and herbaceous (Figure 3). Another noteworthy observation at the plot level was the fact that the magnitude of reflectance of A. larici-nus was greater than for herbaceous vegetation in the ultraviolet–visible (regions 1), NIR region (region 2), and SWIR (region 4), and smaller in NIR–SWIR (region 3). This is the opposite of what were observed in comparisons between A. laricinus and grass as well as A. laricinus and a mixture of her-baceous vegetation and bare ground. This dissimilarity can be attributed to the relatively heteroge-neous species composition of herbaceous plants within a plot. In contrast, grass and bare ground can be comparatively considered homogenous land cover types, respectively, having marked spec-tral difference with A. laricinus.

The significant difference between A. laricinus and adjacent land cover types using the Landsat- and SPOT 5-simulated bands achieved at the individual-level analysis (Figures 8–11) was anticipat-ed, given the distinct homogeneous setup of A. laricinus and adjacent land cover types. This setting does, however, occur rarely in an ideal natural environment where plant of different species co-exist. Unlike individual level analysis which showed significant differences in all bands (Figure 8 and 9), only the NIR band showed significant difference at group level (Figure 10 and 11). These results showed the potential of discriminating A. laricinus from adjacent land cover types using this band which is available in most remotely sensed data. This agrees with a study that classified Asparagus officinalis (a species that belongs to the same family as A. laricinus) successfully using Landsat im-agery (Tatsumi, Yamashiki, Canales Torres, & Taipe, 2015).

The NIR band was most useful in discriminating between A. laricinus and adjacent land cover types. This is not surprising as the band has been widely used in discriminating between plant spe-cies in a number of studies. For example, A. officinalis was successfully identified using NIR reflec-tance spectroscopy by Perez and Sanchez (2001). This region was used in studies on plants not related to A. laricinus, too, such as by Xu, Yu, Fu, and Ying (2009) who successfully discriminated between two tomato varieties in China using visible–near-infrared reflectance spectroscopy. Thenkabail et al. (2013) on the other hand identified individual bands that included the visible–NIR bands as well as vegetation indices that best characterize, classify, model, and map the world's main agricultural crops. Bentivegna et al. (2012) detected cutleaf teasel (*D. laciniatus*) with hyperspectral imagery using visible–NIR spectral region along Missouri Highway, USA. Calvini, Ulrici, and Amigo (2015) tested sparse methods for classifying Arabica and Robusta coffee species using near-infrared hyperspectral images.

5. Conclusion

This study aimed at determining the potential of discriminating between *A. laricinus* and adjacent land cover types in the Klipriviersberg Nature Reserve using a field spectrometer data. Analysis of spectral reflectance was done at individual and plot levels using the original spectra. Although different spectral wavelength regions showed the ability to differentiate the species from other land cover types, the NIR region was found to be the most consistent of all. This finding is in line with other vegetation studies, although such studies on asparagus are rare.

A comparative similarity between *A. laricinus* and herbaceous plants was noteworthy. This similarity can make identification of the plant challenging in such co-existence. In contrast, the species can be discriminated from grass and mixed land cover (ground and herbaceous vegetation) at relative ease. The separability from grass is particularly important if the species favors to co-exist more with grass than with other species (7 of 10 plots were dominated by *A. laricinus* and grass in this study). The ability to discriminate these pecies from mixed land cover types that include bare ground, among others, is useful since it enables early detection in sparsely vegetated areas. Further studies are however needed to determine the relative contribution of different land cover types in the mixture to spectral reflectance.

Analysis of spectra simulated based on Landsat and SPOT 5 imagery bands showed the NIR to be consistent in discriminating *A. laricinus* from other land cover types. This finding is encouraging in that it shows the potential of upscaling the application to airborne and spaceborne remote sensing that mostly include the NIR region of electromagnetic energy. This study, however, used limited number of samples and thus should rather be considered a preliminary indicator that needs further studies. Future studies should attempt to utilize large number of samples. Such sample size can be achieved with the use of small sampling units and high spatial resolution imagery (e.g. SPOT 5, 6/7), particularly in areas where the spatial extent of invasion is small relative to imagery with lower spatial resolution (e.g. Landsat). In addition, limiting spectral measurements within ideal time frames when there is enough illumination would need to be considered. Furthermore, it is vital to profile the biochemical contents of the species so that relationships can be built between the inherent contents of the plant and their effects on spectral signatures. In connection to this, it is important to take into consideration spectral properties at different phenological stages of the species.

Acknowledgements
The authors are thankful to the Klipriviersberg Nature Reserve for permitting use of the reserve as a study site. The authors also acknowledge Bongeka Wendy Mbatha and Minenhle Gumede for helping in data collection.

Funding
This research was funded by the University of Johannesburg.

Author details
Bambo Dubula[1]
E-mail: bambodubula@gmail.com
ORCID ID: http://orcid.org/0000-0002-4718-8026
Solomon Gebremariam Tesfamichael[1]
E-mail: gtesfamichael@uj.ac.za
Isaac Tebogo Rampedi[1]
E-mail: isaacr@uj.ac.za
[1] Department of Geography, Environmental Management and Energy Studies, University of Johannesburg, Johannesburg, South Africa.

References
Abdel-Rahman, E. M., Mutanga, O., Adam, E., & Ismail, R. (2014). Detecting Sirex noctilio grey-attacked and lightning-struck pine trees using airborne hyperspectral data, random forest and support vector machines classifiers. *ISPRS Journal of Photogrammetry and Remote Sensing, 88*, 48–59. doi:10.1016/j.isprsjprs.2013.11.013
Adam, E., & Mutanga, O. (2009). Spectral discrimination of papyrus vegetation (*Cyperus papyrus* L.) in swamp wetlands using field spectrometry. *ISPRS Journal of Photogrammetry and Remote Sensing, 64*, 612–620. doi:10.1016/j.isprsjprs.2009.04.004
Adelabu, S., Mutanga, O., Adam, E., & Sebego, R. (2014). Spectral discrimination of insect defoliation levels in

mopane woodland using hyperspectral data. *IEEE Journal of Selected Topics in Applied Earth Observations and Remote Sensing, 7,* 177–186. doi:10.1109/JSTARS.2013.2258329

Alparone, L., Aiazzi, B., Baronti, S., & Andrea G. (2015). *Remote sensing image fusion.* CRC Press. doi:10.1201/b18189-9

Azong, M., Malahlela, O., & Ramoelo, A. (2015). Assessing the utility WorldView-2 imagery for tree species mapping in South African subtropical humid forest and the conservation implications: Dukuduku forest patch as case study. *International Journal of Applied Earth Observations and Geoinformation, 38,* 349–357. doi:10.1016/j.jag.2015.01.015

Bentivegna, D. J., Smeda, R. J., Wang, C., Bentivegna, D. J., Smeda, R. J., & Wang, C. (2012). Detecting cutleaf teasel (*Dipsacus laciniatus*) along a Missouri highway with hyperspectral imagery. *Invasive Plant Science and Management, 5,* 155–163. doi:10.1614/IPSM-D-10-00053.1

Berg, V. Den, Kotze, I., & Beukes, H. (2013). Detection, quantification and monitoring of prosopis in the Northern Cape Province of South Africa using remote sensing and GIS. *South African Journal of Geomatics, 2,* 68–81.

Bradley, B. A., & Marvin, D. C. (2011). Using expert knowledge to satisfy data needs: Mapping invasive plant distributions in the western United States. *Western North American Naturalist, 71,* 302–315. doi:10.3398/064.071.0314

Calvini, R., Ulrici, A., & Amigo, J. (2015). Practical comparison of sparse methods for classification of Arabica and Robusta coffee species using near infrared hyperspectral imaging. *Chemometrics and Intelligent Laboratory Systems, 146,* 503–511. doi:10.1016/j.chemolab.2015.07.010

Campbell, J. B., & Wynne, R. H. (2011). *Introduction to remote sensing* (5th ed.). New York, NY: The Guilford Press.

Carroll, M. W., Glaser, J. A., Hunt, T. E., & Sappington, T. W. (2008). Use of spectral vegetation indices derived from airborne hyperspectral imagery for detection of European Corn Borer infestation in Iowa Corn plots. *Journal of Economic Entomology, 101,* 1614–1623.

Cho, M., Sobhan, I., Skidmore, A., & de Leeuw, J. (2008). Discriminating species using hyperspectral indices at leaf and canopy scales. *The International Archives of the Photogrammetry, Remote Sensing and Spatial Information Sciences, 37,* 369–376.

Cuneo, P., Jacobson, C., & Leishman, M. (2009). Landscape-scale detection and mapping of invasive African olive (*Olea europaea* L. ssp. cuspidata Wall ex G. Don Ciferri) in SW Sydney, Australia using satellite remote sensing. *Applied Vegetation Science, 12,* 145–154. http://dx.doi.org/10.1111/avsc.2009.12.issue-2

Daughtry, C. S. T., Doraiswamy, P. C., Hunt, E. R., Stern, A. J., McMurtrey, J. E., & Prueger, J. H. (2006). Remote sensing of crop residue cover and soil tillage intensity. *Soil and Tillage Research, 91,* 101–108. doi:10.1016/j.still.2005.11.013

DeVries, B., Verbesselt, J., Kooistra, L., & Herold, M. (2015). Robust monitoring of small-scale forest disturbances in a tropical montane forest using Landsat time series. *Remote Sensing of Environment, 161,* 107–121. doi:10.1016/j.rse.2015.02.012

Dewey, S. A., Price, K. P., & Ramsey, D. (1991). Satellite remote sensing to predict potential distribution of Dyers woad (*Isatis tinctoria*). *Weed Technology, 5,* 479–484.

Dorigo, W., Lucieer, A., Podobnikar, T., & Čarni, A. (2012). Mapping invasive Fallopia japonica by combined spectral, spatial, and temporal analysis of digital orthophotos. *International Journal of Applied Earth Observation and Geoinformation, 19,* 185–195. doi:10.1016/j.jag.2012.05.004

Dronova, I., Gong, P., Wang, L., & Zhong, L. (2015). Mapping dynamic cover types in a large seasonally flooded wetland using extended principal component analysis and object-based classification. *Remote Sensing of Environment, 158,* 193–206. doi:10.1016/j.rse.2014.10.027

Faiola, J., & Vermaak, V. (2014). Klipriviersberg. *Veld and Flora, 100,* 68–71.

Fernandes, M. R., Aguiar, F. C., Silva, J. M. N., Ferreira, M. T., & Pereira, J. M. C. (2013). Spectral discrimination of giant reed (*Arundo donax* L.): A seasonal study in riparian areas. *ISPRS Journal of Photogrammetry and Remote Sensing, 80,* 80–90. doi:10.1016/j.isprsjprs.2013.03.007

Foden, W., & Potter, L. (2005). Asparagus laricinus Burch. National assessment: Red list of South African plants version 2015.1. Retrieved September 18, 2015, from http://redlist.sanbi.org/species.php?species=728-59

Fuku, S., Al-Azzawi, A., Madamombe-Manduna, I., & Mashele, S. (2013). Phytochemistry and free radical scavenging activity of Asparagus laricinus. *International Journal of Pharmacology, 9,* 312–317.

Galvão, L., Epiphanio, J., Breunig, F., & Formaggio, A. (2011). Crop type discrimination using hyperspectral data. In *Hyperspectral remote sensing of vegetation* (pp. 397–422). CRC Press. doi:10.1201/b11222-25

Gavier-pizarro, G. I., Kuemmerle, T., & Stewart, S. I. (2012). Monitoring the invasion of an exotic tree (*Ligustrum lucidum*) from 1983 to 2006 with Landsat TM/ETM + satellite data and support vector machines in Córdoba, Argentina. *Remote Sensing of Environment,* 1–12.

Gu, Y., & Wylie, B. K. (2015). Developing a 30-m grassland productivity estimation map for central Nebraska using 250-m MODIS and 30-m Landsat-8 observations. *Remote Sensing of Environment, 171,* 291–298. doi:10.1016/j.rse.2015.10.018

Guerschman, J. P., Hill, M. J., Renzullo, L. J., Barrett, D. J., Marks, A. S., & Botha, E. J. (2009). Estimating fractional cover of photosynthetic vegetation, non-photosynthetic vegetation and bare soil in the Australian tropical savanna region upscaling the EO-1 Hyperion and MODIS sensors. *Remote Sensing of Environment, 113,* 928–945. doi:10.1016/j.rse.2009.01.006

Huang, C., & Asner, G. P. (2009). Applications of remote sensing to alien invasive plant studies. *Sensors, 9,* 4869–4889. doi:10.3390/s90604869

Jensen, J. R. (2014). *Remote sensing of the environment: An earth resource perspective* (2nd ed.). Harlow: Pearson.

Johansen, K., Phinn, S., & Witte, C. (2010). Mapping of riparian zone attributes using discrete return LiDAR, QuickBird and SPOT-5 imagery: Assessing accuracy and costs. *Remote Sensing of Environment, 114,* 2679–2691. doi:10.1016/j.rse.2010.06.004

Johansen, K., Phinn, S., & Taylor, M. (2015). Mapping woody vegetation clearing in Queensland, Australia from landsat imagery using the Google Earth Engine. *Remote Sensing Applications: Society and Environment, 1,* 36–49. doi:10.1016/j.rsase.2015.06.002

Kennedy, R. E., Yang, Z., Braaten, J., Copass, C., Antonova, N., Jordan, C., & Nelson, P. (2015). Attribution of disturbance change agent from landsat time-series in support of habitat monitoring in the Puget Sound region, USA. *Remote Sensing of Environment, 166,* 271–285. doi:10.1016/j.rse.2015.05.005

Kotze, P. (2002). *The ecological integrity of the Klip River and the development of a sensitivity weighted fish index of biotic integrity (SIBI). ujdigispace.* Johannesburg: University of Johannesburg.

Laba, M., Downs, R., Smith, S., Welsh, S., Neider, C., White, S., … Baveye, P. (2008). Mapping invasive wetland plants in the Hudson River National Estuarine Research Reserve using quickbird satellite imagery. *Remote Sensing of Environment, 112,* 286–300. doi:10.1016/j.rse.2007.05.003

Lemke, D., Hulme, P. E., Brown, J. A., & Tadesse, W. (2011). Distribution modelling of Japanese honeysuckle (Lonicera japonica) invasion in the Cumberland Plateau and

Mountain Region, USA. *Forest Ecology and Management, 262*, 139–149. doi:10.1016/j.foreco.2011.03.014

Lillesand, T., Kiefer, R., & Chipman, J. (Eds.). (2015). *Remote sensing and image interpretation* (7th ed.). Hoboken, NJ: John Wiley & Sons.

Manevski, K., Manakos, I., Petropoulos, G. P., & Kalaitzidis, C. (2011). Discrimination of common Mediterranean plant species using field spectroradiometry. *International Journal of Applied Earth Observation and Geoinformation, 13*, 922–933. doi:10.1016/j.jag.2011.07.001

Mansour, K. (2013). Comparing the new generation world view-2 to hyperspectral image data for species discrimination. *International Journal of Development Research, 3*, 8–13.

Martín, M., Barreto, L., & Fernández-Quintanilla, C. (2011). Discrimination of sterile oat (*Avena sterilis*) in winter barley (*Hordeum vulgare*) using QuickBird satellite images. *Crop Protection, 30*, 1363–1369. doi:10.1016/j.cropro.2011.06.008

Mashele, S., & Kolesnikova, N. (2010). In vitro anticancer screening of Asparagus laricinus extracts. *Pharmacologyonline, 2*, 246–252.

Mgidi, T. N., Le Maitre, D. C., Schonegevel, L., Nel, J. L., Rouget, M., & Richardson, D. M. (2007). Alien plant invasions—Incorporating emerging invaders in regional prioritization: A pragmatic approach for Southern Africa. *Journal of Environmental Management, 84*, 173–187. doi:10.1016/j.jenvman.2006.05.018

Mirik, M., Ansley, R. J., Steddom, K., Jones, D. C., Rush, C. M., Michels, G. J., & Elliott, N. C. (2013). Remote distinction of a noxious weed (Musk Thistle: Carduus Nutans) using airborne hyperspectral imagery and the support vector machine classifier. *Remote Sensing, 5*, 612–630. doi:10.3390/rs5020612

Mirik, M., Emendack, Y., Attia, A., Chaudhuri, S., Roy, M., Backoulou, G. F., & Cui, S. (2014). Detecting musk thistle (*Carduus nutans*) infestation using a target recognition algorithm. *Advances in Remote Sensing, 3*, 95–105. http://dx.doi.org/10.4236/ars.2014.33008

Nagler, P. L., Daughtry, C. S. T., & Goward, S. N. (2000). Plant litter and soil reflectance. *Remote Sensing of Environment, 71*, 207–215. doi:10.1016/S0034-4257(99)00082-6

Narumalani, S., Mishra, D. R., Wilson, R., Reece, P., & Kohler, A. (2009). Detecting and mapping four invasive species along the floodplain of North Platte River, Nebraska. *Weed Technology, 23*, 99–107. doi:10.1614/WT-08-007.1

Ntsoelinyane, P. H., & Mashele, S. (2014). Phytochemical screening, antibacterial and antioxidant activities of asparagus laricinus leaf and stem extracts. *Bangladesh Journal of Pharmacology, 9*, 10–14. doi:10.3329/bjp.v9i1.16967

Olsson, A. D., van Leeuwen, W. J. D., & Marsh, S. E. (2011). Feasibility of invasive grass detection in a desertscrub community using hyperspectral field measurements and landsat TM imagery. *Remote Sensing, 3*, 2283–2304. doi:10.3390/rs3102283

Ouyang, Z.-T., Gao, Y., Xie, X., Guo, H.-Q., Zhang, T.-T., & Zhao, B. (2013). Spectral discrimination of the invasive plant spartina alterniflora at multiple phenological stages in a saltmarsh wetland. *PLoS ONE, 8*, e67315. doi:10.1371/journal.pone.0067315

Perez, D., & Sanchez, M. (2001). Authentication of green asparagus varieties by near-infrared reflectance spectroscopy. *Journal of Food Science, 66*, 323–327. doi:10.1111/j.1365-2621.2001.tb11340.x

Prasad, K., & Gnanappazham, L. (2014). Discrimination of mangrove species of Rhizophoraceae using laboratory spectral signatures. In *IEEE Geoscience and Remote Sensing Symposium*, Quebec City.

Prospere, K., McLaren, K., & Wilson, B. (2014). Plant species discrimination in a tropical wetland using in situ hyperspectral data. *Remote Sensing, 6*, 8494–8523. doi:10.3390/rs6098494

Richardson, D. M., & Van Wilgen, B. W. (2004). Invasive alien plants in South Africa: How well do we understand the ecological impacts? *South African Journal of Science, 100*, 45–52.

Rodgers, L., Pernas, T., & Hill, S. D. (2014). Mapping invasive plant distributions in the Florida everglades using the digital aerial sketch mapping technique. *Invasive Plant Science and Management, 7*, 360–374. doi:10.1614/IPSM-D-12-00092.1

Rouget, M., Hui, C., Renteria, J., Richardson, D. M., & Wilson, J. R. U. (2015). Plant invasions as a biogeographical assay: Vegetation biomes constrain the distribution of invasive alien species assemblages. *South African Journal of Botany*. doi:10.1016/j.sajb.2015.04.009

Rudolf, J., Lehmann, K., Große-stoltenberg, A., Römer, M., & Oldeland, J. (2015). Field spectroscopy in the VNIR-SWIR region to discriminate between Mediterranean native plants and exotic-invasive shrubs based on leaf tannin content. *Remote Sensing, 7*, 1225–1241. doi:10.3390/rs70201225

Schor, J., Farwig, N., & Berens, D. G. (2015). Intensive land-use and high native fruit availability reduce fruit removal of the invasive Solanum mauritianum in South Africa. *South African Journal of Botany, 96*, 6–12. doi:10.1016/j.sajb.2014.11.004

Serbin, G., Daughtry, C. S. T., Hunt, E. R., Reeves, J. B., & Brown, D. J. (2009). Effects of soil composition and mineralogy on remote sensing of crop residue cover. *Remote Sensing of Environment, 113*, 224–238. doi:10.1016/j.rse.2008.09.004

Tatsumi, K., Yamashiki, Y., Canales Torres, M. A., & Taipe, C. L. R. (2015). Crop classification of upland fields using Random forest of time-series Landsat 7 ETM+ data. *Computers and Electronics in Agriculture, 115*, 171–179. doi:10.1016/j.compag.2015.05.001

Thenkabail, P. S., Mariotto, I., Gumma, M. K., Middleton, E. M., Landis, D. R., & Huemmrich, K. F. (2013). Selection of hyperspectral narrowbands (HNBs) and composition of hyperspectral twoband vegetation indices (HVIs) for biophysical characterization and discrimination of crop types using field reflectance and hyperion/EO-1 data. *IEEE Journal of Selected Topics in Applied Earth Observations and Remote Sensing, 6*, 427–439. http://dx.doi.org/10.1109/JSTARS.2013.2252601

Van Wilgen, B. (2006). Invasive alien species—An important aspect of global change. *CSIR Science Scope, 1*, 8–11.

van Wilgen, B. W., Forsyth, G. G., Le Maitre, D. C., Wannenburgh, A., Kotzé, J. D. F., van den Berg, E., & Henderson, L. (2012). An assessment of the effectiveness of a large, national-scale invasive alien plant control strategy in South Africa. *Biological Conservation, 148*, 28–38. doi:10.1016/j.biocon.2011.12.035

van Wilgen, B., Reyers, B., Le Maitre, D., Richardson, D., & Schonegevel, L. (2008). A biome-scale assessment of the impact of invasive alien plants on ecosystem services in South Africa. *Journal of Environmental Management, 89*, 336–349. doi:10.1016/j.jenvman.2007.06.015

Vancutsem, C., Pekel, J. F., Evrard, C., Malaisse, F., & Defourny, P. (2009). Mapping and characterizing the vegetation types of the Democratic Republic of Congo using SPOT VEGETATION time series. *International Journal of Applied Earth Observation and Geoinformation, 11*, 62–76. doi:10.1016/j.jag.2008.08.001

Vicente, J. R., Fernandes, R. F., Randin, C. F., Broennimann, O., Gonçalves, J., Marcos, B., ... Honrado, J. P. (2013). Will climate change drive alien invasive plants into areas of high protection value? An improved model-based regional assessment to prioritise the management of invasions. *Journal of Environmental Management, 131*, 185–195. doi:10.1016/j.jenvman.2013.09.032

Weiss, N. A. (2012). *Introductory statistics* (9th ed.). Boston, MA: Pearson Education.

Xu, H., Yu, P., Fu, X., & Ying, Y. (2009). On-site variety discrimination of tomato plant using visible-near infrared reflectance spectroscopy. *Journal of Zhejiang University SCIENCE B, 10*, 126–132. doi:10.1631/jzus.B0820200

Appendix A

Center coordinates of sample plots used in the analysis

Plots	Latitude	Longitude
1	−26.30169	28.01205
2	−26.30117	28.01164
3	−26.30085	28.01121
4	−26.30076	28.01141
5	−26.3002	28.01127
6	−26.30018	28.01058
9	−26.30063	28.01058
8	−26.30148	28.01138
9	−26.30257	28.01096
10	−26.30291	28.01106

Characteristics of soil exchangeable potassium according to soil color and landscape in Ferralsols environment

Brahima Koné[1]*, Traoré Lassane[2], Sehi Zokagon Sylvain[1] and Kouassi Kouassi Jacques[1]

*Corresponding author: Brahima Koné, Earth Science Unit, Soil Science Department, Felix Houphouet-Boigny University, 22 BP 582 Abidjan, Côte d'Ivoire

E-mail: kbrahima@hotmail.com

Reviewing editor:

Craig O'Neill, Macquarie University, Australia

Abstract: The use of soil color as Munsell data was explored for *in situ* indication of soil potassium (K) availability toward a friendly method of agricultural land survey. Soil contents of K, calcium, and magnesium were determined for 998 upland soil samples from Côte d'Ivoire (7–10°N). Soil depths (0–20, 20–60, 60–80, and 80–150 cm), redness ratio (RR), and redness factor (RF) were considered. Significant association was observed between K-levels (high, medium, and low) in topsoil and its color hue, and the highest cumulative frequency of 2.5YR in high and medium levels was characterizing the hill slope position (summit and upper slope). Deep horizon, foot slope, and yellowish color (7.5YR and 10YR) were more relevant to low K-level. Significant linear regressions of soil content of K were observed according to both redness indices indifferently to the topographic positions and soil depths in some extend. Of these finding in the line of folk knowledge, RR and RF are recommended for *in situ* measurement of soil K, and 2.5YR as color hue may be use as indicator of K-enriched soil at hill slope position.

Subjects: Earth Sciences; Environment & Agriculture; Environmental Studies & Management

Keywords: indicator of soil fertility; potassium; catena; soil color; folk knowledge

ABOUT THE AUTHORS

The data reported in the current paper were drawn from the thesis of Brahima Koné as first author. The work was carried out in Côte d'Ivoire in order to develop a tool for soil survey in Ferralsol environment. The co-authors of current paper have contributed to data analysis and interpretation as well as for the writing of the manuscript.

PUBLIC INTEREST STATEMENT

Soil use and management are important factors affecting agricultural production, while conventional standard methods of soil assessment are not well understood, especially by non-educated farmers of tropical Africa. In contrast, there were social and economical perceptions of farming including the land use as long as populations are practicing agriculture. So call folk knowledge, local populations around the word have friendly methods of soil classification in relation with its fertility and productivity. The soil color accounts for criteria in folk knowledge. Regarding to the importance of soil content of potassium in Ferralsol, the current study emphasized the relation between soil content of this nutrient and soil color in interaction with the topographic position of landscapes. The hill slope position characterized by reddish soils was found to be the most enriched land in potassium.

1. Introduction

Sustaining agriculture requires sound soil evaluation methods in concordance with morphopedology standards (Bertrand et al., 1985; Loukili et al., 2000). However, minimum data-set requirements are differing according to authors: soil attributes or both soil and plant parameters may be concerned (Larson and Pierce, 1991; Pearson et al., 1995; Pieri, 1992). This variance in methodology as a weakness was tackled by Riquier et al. (1970) when initiating the use of soil productivity index which was improved with data relevant to crop potential yielding (root development) as function of soil environment (Burger, 1996; Gale et al., 1991; Kiniry et al., 1983; Milner et al., 1996; Pierce et al., 1983). In the meantime, ecological specificity was considered by including soil contents of phosphorus (P) and organic matter (Neill, 1979) and typical model was suggested for tropical environment (Sys and Frankart, 1971). Of existing models, soil content of potassium (K) was missing as parameter though; this nutrient is among the most limiting of crop growth in tropical agro-ecologies (Koné, Fatogoma, Chérif, 2013).

The most recent approach of soil management is the fertility capability classification (FCC) system (Sanchez et al., 2003) based on five data-sets of soil as modifiers including soil content of K. However, this system is not popularly adopted yet. The wide number of required parameters, specially the chemical analytical data and the skill required for FCC may be of concerned. Therefore, a friendly method of soil chemical parameters estimation in field can contribute to wide adoption of such fertility classification of soil. For this purpose, the color of soil as a component of folk classification (Krasilnikov & Tabor, 2003), may have consistent contribution as friendly prediction method of Ferralsols K availability.

In fact, there is evidence of Ferralsols inherent fertility classification as poor for yellow and richer for reddish colors at a given topographic position, especially for soil contents of P, K, and magnesium (Mg) in addition to soil particle sizes (Koné, Yao-Kouamé, et al., 2009). Moreover, the opportunity of soil K supplying capacity was successfully explored using soil color by Koné, Bongoua-Devisme, Kouadio, Kouadio, and Traoré (2014).

Indeed, the colors of Ferralsols are relevant to difference in their mineralogy, organic matter content, and texture (Koné, Diatta, et al., 2009; Stoner et al., 1980) as major descriptors of their history.

The use of soil redness (Torrent et al., 1980, 1983; Santana, 1984) for estimating soil K availability on the basis of linear regression may be a fast and cheapest method, hence limiting constraint in agricultural soil capacity evaluation where soil content of K may be critical.

Soil survey was conducted in Côte d'Ivoire above the latitude 7°N applying randomized and unequal stratified soil sampling method. Soil color was determined by Munsell chart and soil exchangeable K was analyzed for exploring the accuracy of soil redness rate (RR) and redness factor (RF) in a linear regression of soil K. The aim was (i) to identify a soil color hue for a given level (high, medium, and low) of soil K in spatial distribution according to soil depth and topographic section, (ii) to defined the relation between soil exchangeable K and soil color, and (iii) to identify fit model of soil K among that using RR and RF. Definitively, a model of soil exchangeable K content should be recommended for landscape section and soil depth as a tool for most friendly method of soil survey.

2. Material and methods

2.1. Studied zone description

The study was carried out across 19 sites characterized by Ferralsols encountered between the latitudes 7–10°N in Côte d'Ivoire. Four major agro-ecologies were described in there by Koné (2007) as Sudan savannah with grassland, Guinea savannah with woodland, derived savannah (a transition between savannah and forest agro-ecologies), and mountainous zone located in the west of the country. Annual average rainfall amount ranged from 1,200 to 2,000 mm.

Dismantled or unaffected summit (SUM) ferruginous cuirass of plateau landscapes further characterized by slightly concave or convex sides were frequently encountered in the studied area beside of hills with bedrock outcroppings. A few inselbergs were also observed and, the landside and length of landscapes were variable accordingly. Upland soils were essentially Ferralsols plinthic belonging to hyperdystric or dystric groups.

2.2. Soil sampling

Two hundred and eighty-nine soil profiles were surveyed along the toposequences of various landscapes at 19 sites. The soil profiles were unequally distributed (Webster and Oliver, 1990) on three sub-groups of Ferralsols encountered in the studied area (Koné, Diatta, et al., 2009). The identified horizons in the soil profiles were coded according to depth classes—H1 (0–20 cm), H2 (20–60 cm), H3 (60–80 cm), and H4 (80–150 cm) dividing the soil profile into organic horizon (Diatta, 1996), minimum, medium, and maximum crop rooting depths (Böhn, 1976; Chopart, 1985), respectively. A total of 995 samples (2 kg for each) were taken from soil horizons up to a maximum depth of 1.5 m when possible (Table 1).

2.3. Soil sample characteristics

Soil sample size of a given color hue was variable according to soil depth across the studied zone.

The corresponding numbers of soil samples were 274, 325, 279, and 117 for the SUM, upper slope (US), middle slope (MS), and foot slope (FS), respectively.

2.4. Laboratory analysis and classification of soil K contents

Soil samples were dried under forced air at room temperature. Then, they were crushed before sieving through a 2.0-mm stainless steel sieve. Exchangeable K of soil was extracted by shaking 1 g in 10 ml of 1 M NH4OAc during 5 min before the use of atomic emission in a Perkin Elmer Analyst 100 spectrometer (Page, 1982). Three classes of soil contents of exchangeable K were defined as done by Berryman et al. (1984) for tropical soils:

L = Low soil content of K ranging below 0.15 cmol kg^{-1}.

M = Moderate soil content of K ranging between 0.15 and 0.30 cmol kg^{-1}.

H = High soil content of K ranging over 0.30 cmol kg^{-1}.

Soil contents of calcium (Ca) and Mg were also analyzed using the same extraction method described above.

2.5. Soil color identification

The year 2000 revised washable edition of Munsell soil color charts (Gretagmacьeth, 2000) composed of 322 different standard color ships was used in field during the survey for soil color identification. Wet soil samples were compared with the standard color ships, respectively, and the three components of the color were recorded as "Hue (He); Chroma (C)/Value (V)." The RF defined as RF by Santana (1984) was calculated for each of the soil samples likewise the redness ratio (RR) according to Torrent et al. (1980, 1983):

Table 1. Soil sample size of a given soil color hue as identified in soil depths					
	Number of soil sample				
	H1	**H2**	**H3**	**H4**	**Total**
2.5YR	77	94	68	54	293
5YR	121	134	86	46	387
7.5YR	81	74	35	16	206
10YR	40	36	18	15	109
Total	319	338	207	131	995

$$RF = (10 - He) + C/V \qquad\qquad (1)$$

$$RR = (10 - He) \times C/V \qquad\qquad (2)$$

2.6. Statistical analysis

By descriptive analysis, average frequency of soil content of exchangeable K was determined in topsoil (H1 and H2) for each topographic section (S, US, MS, and FS) and for each soil color hue as encountered (2.5, 5, 7.5, and 10YR) using SPSS 10 package. Cross-table analysis was done to determine the frequency of soil K levels (H = high, M = moderate, and L = low) according to soil depths (H1, H2, H3, and H4) and the topographic sections for the identified soil color hue. Pearson correlation analysis was also performed between RR and soil contents of K, Ca, and Mg and likely for RF, respectively. Furthermore, soil content of exchangeable K was predicted by RR and RF separately using linear regression analysis step by step with constant term or not, and the most significant ($p < 0.05$) model was reported. The thickness of elementary soil horizon was considered as weighted variable. SAS (version 8) was used for these statistical analyses and the critical level of probability was fixed as 0.05 (α).

3. Results

3.1. Characterization of soil K levels

The frequencies of soil color hues (2.5, 5, 7.5, and 10YR) as determined for different soil depths (H1, H2, H3, and H4) according to K-levels are presented in Figure 1. The highest frequencies are related to the low K-level (L) throughout the soil profile with outstanding values for 5 and 7.5YR in the topsoil layers (H1 and H2), while similarly observed for 2.5 and 7.5YR in the subsoil (H3 and H4). Nevertheless, the highest cumulative frequencies of 2.5YR in medium (M) and high (H) levels of soil K is observed likewise for 5YR as soil color hue.

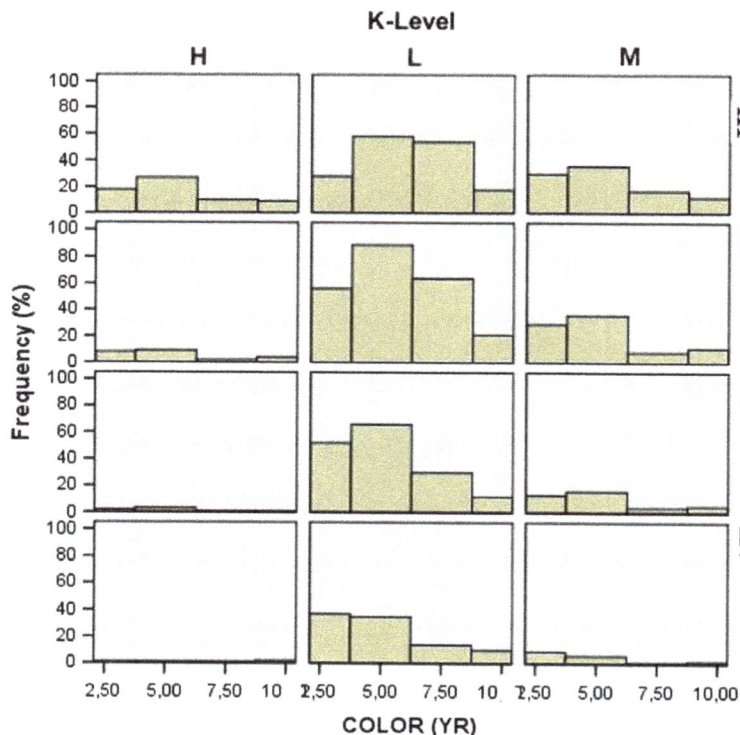

Figure 1. Frequency of soil potassium levels (H, M, and L) according encountered Munsell color hues (2.5, 5, 7.5, and 10YR) in different soil depths (H1, H2, H3, and H4).

In fact, there are 40, 37.3, and 22.6% of chance to observed 2.5YR as soil color hue in topsoil 0–20 cm for medium, low, and high soil K-levels, respectively (Figure 2a). Hence, the cumulative frequency of 2.5YR referring to medium and high K-levels is about 62.67% over the occurrence as low K-level. Similar results also account for the soil samples of 5YR (52.06% vs. 47.29%) and 10YR (53.15% vs. 46.15%) in color hues contrasting with the results observed for 7.5YR. Further contrast is observed in 20–60 cm soil depth showing the highest frequencies of low K-level (L) compared to the cumulative frequency (high (H) and medium (M) K-levels) indifferently to soil color (Figure 2b).

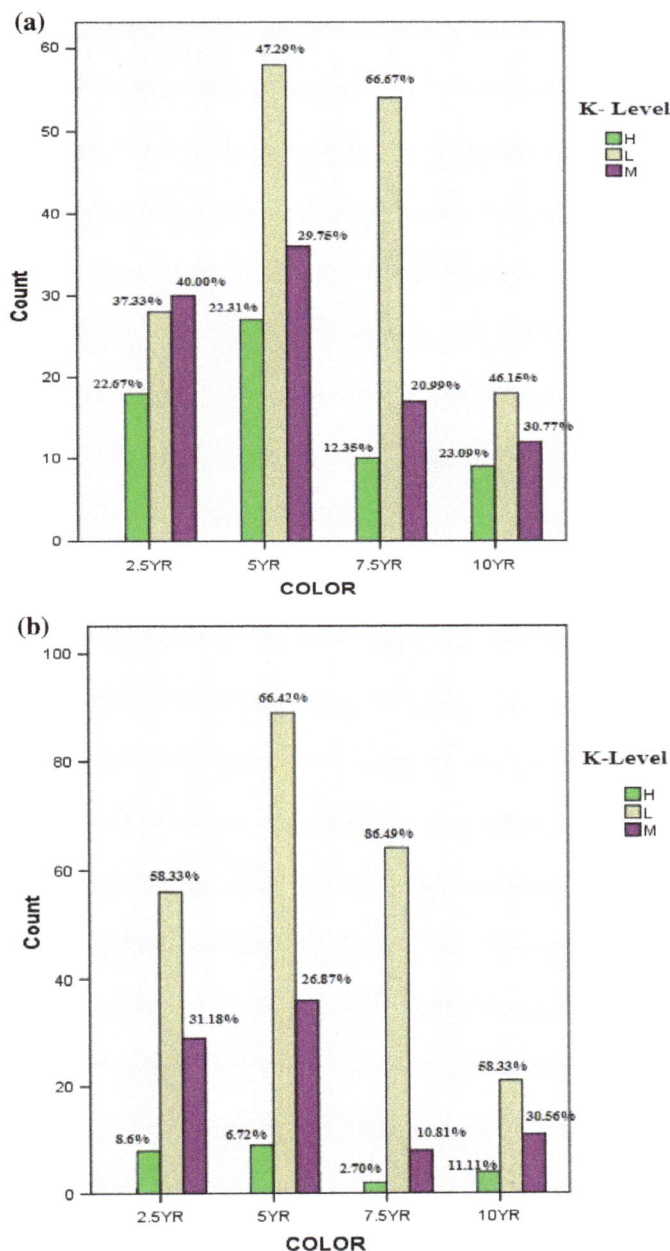

Figure 2. Frequency of potassium level in H1 (a) and H2 (b) according to encountered Munsell color hue (2.5, 5, 7.5, and 10YR).

More details of these results are presented in Figure 3 considering the topographic positions. The soils colored in 2.5YR are outstanding with the highest frequencies of K-levels in both soil depths (0–20 and 20–60 cm) at the SUM position, while similar observations account for 5, 7.5, and 10YR at the US, MS, and FS positions, respectively (Figure 3a): highest cumulative frequency of 2.5 YR in H and M levels of soil K is characterized by the soil samples of SUM and US, while almost equal chances are observed between this cumulative frequency and that related to L when referring to a soil sampled at US with 5YR in color. In turn, the highest frequencies characterizing the soils of 7.5 and 10YR in L are observed when sampled in 0–20 cm at MS and FS positions, respectively. Highest frequencies of L are also observed in 20–60 cm depth indifferently to soil color and topographic section even when compared with the cumulative frequency relative to H and M (Figure 3b). Hence, reddish (2.5YR) topsoil (0–20 cm) appeared to be most enriched (H and M) in exchangeable K, especially for the SUM and UP positions.

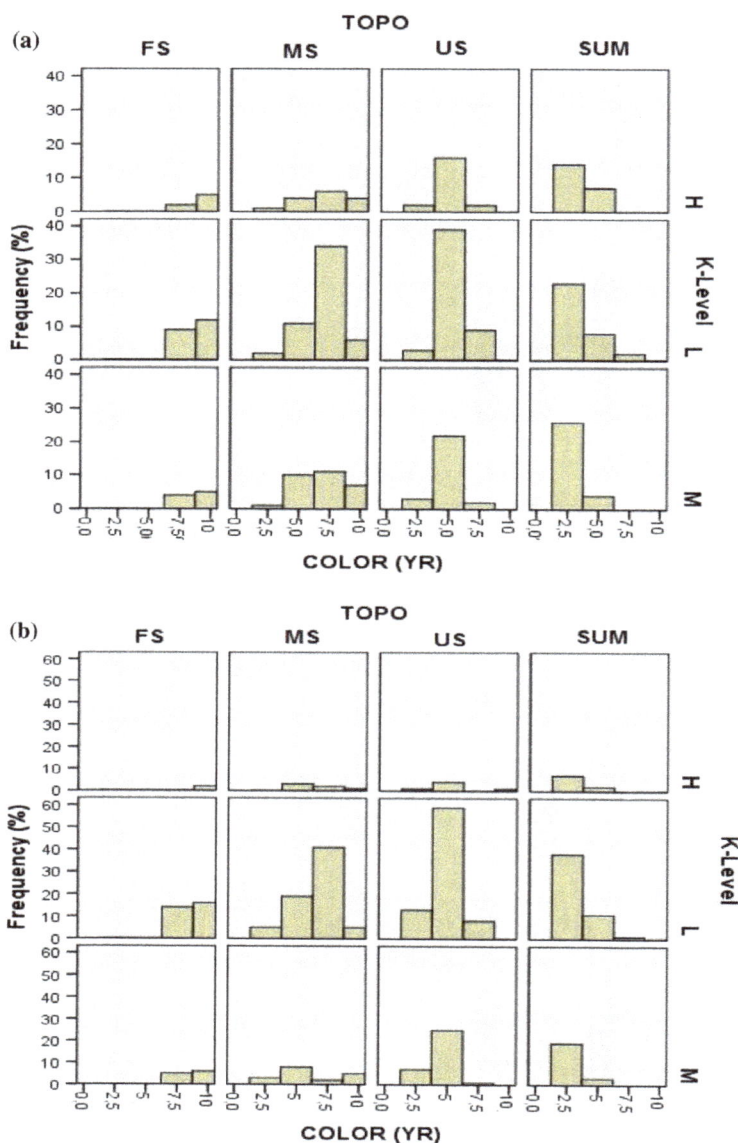

Figure 3. Frequency of potassium level in H1 (a) and H2 (b) depth according to topographic section (TOPO) and encountered Munsell color hue (2.5, 5, 7.5, and 10YR).

Table 2. Frequency of soil color hue as associated to soil K-level (H, M, and L) in 0–20 cm soil depth at different topographic positions				
		Frequency (%)		
		H	M	L
Summit	2.5YR	22.22	41.27	36.51
	5YR	36.84	21.05	42.11
	7.5YR	0.00	0.00	100
	10YR	0.00	0.00	0.00
χ^2 probability				0.521
Upper slope	2.5YR	25.00	37.50	37.50
	5YR	20.78	28.57	50.65
	7.5YR	15.38	15.38	69.24
	10YR	0.00	0.00	0.00
χ^2 probability				0.054
Middle slope	2.5YR	25.00	25.00	50.00
	5YR	16.00	40.00	44.00
	7.5YR	11.00	21.57	66.43
	10YR	23.53	41.17	35.30
χ^2 probability				0.298
Foot slope	2.5YR	0.00	0.00	0.00
	5YR	0.00	0.00	0.00
	7.5YR	13.33	26.67	60.00
	10YR	22.73	22.73	54.56
χ^2 probability				0.412

These observations are strengthened by crossing case occurrences according to the topographic sections and soil color hues as determined in the topsoil 0–20 cm soil depth (Table 2): although not always significant, the highest frequencies of reddish soils are accounting for H and M levels of K at the SUM, US, and MS. Greater cumulative frequency of reddish soil relative to H and M levels of soil K is characterizing the S and US when compared to that of L. Arguably, soil K-level of L is more associated to 7.5 and 10YR at FS position of landscape.

3.2. Relationship between soil cations and the redness of soil

Tables 3 and 4 are showing the relations (correlation coefficient) between soil contents of cations and its redness referring to the RR and the RF, respectively. Except the negatives correlation coefficients observed in 80–150 cm at the SUM (−0.48) and 20–60 cm at the FS (−0.27), no significant relation is noticed between soil content of K and RR (Table 3). This statement is contrasting with the positive correlations observed for soil contents of Mg in 20–60 cm (R = 0.22; p = 0.020) and 60–80 cm (R = 0.22; p = 0.09) depths at MS position. Significant (p = 0.040) correlation between soil content of Ca and RR only accounts for 20–60 cm soil depth at foot slope. No significant correlation accounts for the topsoil (0–20 cm) indifferently to soil contents of cations and topographic sections. Overall, almost the significant correlations are observed in 20–60 cm depth at down slope (MS and FS) of landscape involving all the studied cations. In some extend, similar results are observed in Table 4 relative to the correlation between RF and soil contents of cations (K, Ca, and Mg).

Table 3. Correlation between soil contents of cations (K, Ca, and Mg) and its (RR) in soil depths according to topographic sections

		Pearson correlation for RR							
		0–20 cm		20–60 cm		60–80 cm		80–150 cm	
		R	p-value	R	p-value	R	p-value	R	p-value
SUM	K	−0.14	0.199	0.02	0.828	−0.10	0.460	−0.48	0.001
	Ca	−0.12	0.272	0.04	0.689	−0.10	0.440	−0.003	0.818
	Mg	0.006	0.981	−0.14	0.218	−0.08	0.535	−0.22	0.127
US	K	−0.003	0.975	−0.08	0.344	−0.09	0.426	0.22	0.195
	Ca	0.020	0.837	0.03	0.747	0.009	0.961	0.07	0.666
	Mg	0.010	0.310	−0.003	0.967	−0.000	0.996	0.000	0.995
MS	K	0.01	0.877	−0.03	0.742	−0.025	0.851	−0.24	0.189
	Ca	0.008	0.931	0.14	0.828	−0.04	0.722	0.02	0.909
	Mg	−0.02	0.845	0.22	0.020	0.22	0.09	−0.04	0.789
FS	K	−0.14	0.398	−0.27	0.060	−0.17	0.404	−0.30	0.399
	Ca	−0.03	0.644	0.30	0.040	0.09	0.648	−0.07	0.822
	Mg	0.03	0.831	0.24	0.126	−0.02	0.922	−0.09	0.791

Table 4. Correlation between soil contents of cations (K, Ca, and Mg) and its redness factor (RF) in soil depths according to topographic sections

		Pearson correlation for RF							
		0–20 cm		20–60 cm		60–80 cm		80–150 cm	
		R	p-value	R	p-value	R	p-value	R	p-value
SUM	K	0.06	0.560	0.07	0.497	−0.11	0.112	−0.54	0.0002
	Ca	−0.04	0.653	0.13	0.227	0.02	0.861	0.03	0.811
	Mg	0.06	0.535	0.05	0.627	0.035	0.790	−0.09	0.503
US	K	0.06	0.50	−0.06	0.459	−0.09	0.455	0.30	0.070
	Ca	−0.01	0.862	0.04	0.612	0.005	0.961	0.15	0.342
	Mg	0.09	0.329	0.05	0.585	−0.000	0.998	0.08	0.617
MS	K	0.01	0.870	−0.02	0.828	0.08	0.548	−0.29	0.105
	Ca	−0.02	0.831	0.22	0.030	0.05	0.692	0.007	0.967
	Mg	−0.09	0.341	0.29	0.004	0.27	0.044	−0.06	0.729
FS	K	−0.19	0.260	−0.17	0.255	−0.12	0.542	−0.13	0.699
	Ca	−0.10	0.844	0.31	0.044	0.13	0.529	0.04	0.898
	Mg	0.036	0.830	0.32	0.038	−0.04	0.835	−0.08	0.808

However, significant ($p < 0.0001$) linear regressions of soil content of K are observed according to RR and RF, respectively, except the soil sampled at 80–150 cm depth at middle (RF and RR) and foot (RR) slope positions (Table 5). The coefficients of these regressions are low (1/100 times) and characterized by an increasing trend with soil depth indifferently to the descriptive variables.

Table 5. Linear regression of soil exchangeable K according to RF and RR, respectively, according to topographic sections and soil depth

		\ Linear regression of soil K												
		0–20 cm			20–60 cm			60–80 cm			80–150 cm		
		Coef.	p-value	Err.	Coef.	p-value	Err.	Coef.	p-value	Err.	Coef.	p-value	Err.
SUM	RF	0.03	<0.0001	0.003	0.02	<0.0001	0.02	0.01	<0.0001	0.001	0.02	<0.0001	0.003
	RR	0.02	<0.0001	0.002	0.01	<0.0001	0.001	0.01	<0.0001	0.001	0.01	<0.0001	0.002
US	RF	0.05	<0.0001	0.004	0.02	<0.0001	0.024	0.01	<0.0001	0.001	0.01	<0.0001	0.001
	RR	0.04	<0.0001	0.004	0.02	<0.0001	0.001	0.01	<0.0001	0.001	0.01	<0.0001	0.001
MS	RF	0.05	<0.001	0.008	0.03	<0.0001	0.004	0.02	<0.0001	0.003	0.03	0.500	0.05
	RR	0.05	<0.0001	0.009	0.02	<0.0001	0.003	0.02	<0.0001	0.003	0.02	0.711	0.046
FS	RF	0.09	0.08	0.05	0.05	<0.0001	0.005	0.04	0.001	0.011	0.04	0.030	0.01
	RR	0.08	0.38	0.09	0.04	0.009	0.01	0.03	0.06	0.01	0.02	0.366	0.02

4. Discussion

4.1. Soil K appraisal by landscape and soil color

There was more evidence for identification of the soils characterized by low K-level using soil color across the surveyed area: deepest horizon, the FS position of landscape, and the soil color of 10YR (very pale brown-yellow–dark yellowish brown) were fund to be more relevant to this finding. The release of K by organic matters as surface K-source combined with a relative poor (vs. Na) mobility of this nutrient (Hem, 1992) and the downward gradient of bed rock weathering as inner source of K (Wedpohl, 1978) may account for this.

In turn, high and medium levels of soil K were often observed at the SUM and UP positions and characterized by 2.5YR (pink–red) as soil color hues, especially in the topsoil. However, there is chance to observed a variance of this result according to the landscape variability in the studied area somewhere including inselbergs and plateau with outcrop bedrocks (Eschenbrenner and Badarello, 1978; Poss, 1982): of course, encountered young soil (e.g. Regosols) with shallow depth (<40 cm) may have high level of K though, the color hue may range between 7.5 and 10YR. In fact, these yellowish colors are characterized by goethite (α-FeOOH) preceding the reddish soil matrix of lepidocrocite (γ-FeOOH) and hematite (α-Fe$_2$O$_3$) in the course of soil development (Buxbaum and Printzen, 1998; Schwertmann, 1985). As young soil, enriched-K parental material (e.g. feldspar and mica) may have influence in soil composition (Nahon, 1991). When excluding similar cases, reddish soil color hue of 2.5YR at the hill slope (SUM and US) is somewhat a consistent environmental indicator of enriched soil in exchangeable K. However, the occurrence of 2.5YR as soil color was found to be limited in topsoil and FS because of vertical and lateral gradients of tropical soil color (Koné, Yao-Kouamé, et al., 2009). Definitively, enriched soils in K are more limited in the UP position than the SUM on the basis of the highest cumulative frequency of 2.5YR in H and M levels of soil K and the lateral gradient of soil color.

Furthermore, the red pigmentation (e.g. Hematite) of 2.5YR is deriving from ferrihydrate (Schwertmann, 1985) requiring edaphic conditions which may include high temperature, low water activity, low organic matter content, nearly neutral pH, and high contents of Ca and Mg contrasting with the optimum conditions of goethite formation for yellowish pigmentation (Torrent & Barron, 2002). Almost these parameters involved in soil pigmentation are also accounting for major chemical modifiers of fertility capability soil classifications (FCC) defined by Sanchez et al. (2003) emphasizing an opportunity to make easier the adoption this classification system.

Indeed, degraded Ferralsols are richer in coarse particle with dominance of yellow pigmentation coupled with low content of soil exchangeable K even for enriched bedrock K-primary mineral (Koné, Amadji, Touré, Togola, et al., 2013; Koné, Diatta, et al., 2009; Koné, Touré, Amadji, Yao-Kouamé, et al., 2013).

In light of current analysis, soil color use in field is a potential friendly method for soil fertility classification gathering indigenous and scientific methods as additional reliable tool for participatory land use planning in tropical environment.

4.2. Predictability of exchangeable soil K

The linear regressions observed for soil content of exchangeable K according to soil RR and RF, respectively, were exceptionally significant in the same manner characterizing the relation between soil-exchangeable K and water-soluble K in kaolinitic and mixed mineralogy soils Sharpley (1989). In fact, kaolinit is dominant in Ferralsol, but change in mineralogy can occurred throughout soil profile and along a toposequence (Diatta, 1996) though, still remaining 1:1 clay mineral (kaolinit and illit) in topsoil, while smectite can be observed in deep horizons and down slope. Consequently, the lack of fit observed for the linear models of K prediction in 80–150 cm soil depth of middle and foot slopes (Table 5) could account for such neoformations due to soil moisture and geochemistry (Azizi et al., 2011). However, soil content of K can be predicted by soil redness within 0–80 cm depth indifferently to the topographic positions. This result can be considered as a significant advance in methodology of soil survey, especially for annual crops. Actually, up to date, rapid assessment of soil content of K was only referring to laboratory data including complex method as near-infrared measurement (He and Song, 2006).

Furthermore, the current investigation may have implication in the prediction of soil Ca using the model of Pal (1998) for kaolinitic soil toward a readily estimation of soil cation exchangeable capacity (Bigorre, 1999; Larson & Pierce, 1994). Hence, the use of soil redness may increase the adoption of existing method of soil CEC estimation though; the prediction models (RR and RF) of Mg and Na are still required. Nevertheless, the current study may have significant contribution in pedometry, especially when using pedotransfert functions for estimation of soil cation saturation ratio, hence for easily evaluation of soil fertility in field.

The sensitivity of soil color to wetness (Poss et al., 1991) and the change of soil physic and chemical characteristics (Vizier, 1971) as observed for soil content of K (Koné, Diatta, et al., 2009), are further supporting current finding and relevant assertions.

However, there was scant evidence of correlation between soil RR and RF with soil content of exchangeable K, respectively, throughout soil profile as much as observed for Ca and Mg (Tables 3 and 4) asserting that Ca and Mg may be better predicted by soil redness indices than soil K. In fact, divalent cations are more relevant to soil redness (Koné, Diatta, et al., 2009; Segalen, 1969), while colloids dispersion induced by monovalent cations (Grolimund et al., 1998; Kaplan et al., 1996) may alter the red pigmentation. Well, beside of hematite, the red pigmentation of soil is also inducing by amorphous component (Mauricio & Ildeu, 2005; Segalen, 1969) accounting for soil colloidal phase.

In light of these analyses, there is also chance to predict soil contents of Ca and Mg using soil redness indices in away to estimate soil CEC *in situ* during agricultural soil survey. Of such investigation in future, agricultural land assessment may be strengthening, especially for Ferralsols.

5. Conclusion

Reddish topsoil of 2.5YR in color observed at hill slope positions (SUM and US) is identified as most enriched soil in K when referring to a cumulative frequency in high and medium K-levels. In turn, low soil K-level was accounting for deep soil horizon, FS position, and yellowish (7.5 and 10YR) soil while characterizing 50% of 5YR as soil color. Furthermore, it was revealed a predictability of soil K content within 0–80 cm depth using soil redness indices also advocated for *in situ* prediction of soil contents of Ca and Mg respectively toward the estimation of soil cation exchangeable capacity. Hence, further challenge was outlined concerning the prediction of soil contents of Ca and Mg using Munsell chart in Ferralsol environment. Overall, soil color was deemed as promising tool for improvement of soil fertility management.

Acknowledgment
We are thankfull for DCGTx (BNETD) and for Sitapha Diatta for their respective contribution to this study.

Funding
This work was supported by DCGtx [grant number 1990].

Author details
Brahima Koné[1]
E-mail: kbrahima@hotmail.com
Traoré Lassane[2]
E-mail: tlassane@hotmail.com
Sehi Zokagon Sylvain[1]
E-mail: sehisylvain_nung@yahoo.fr
Kouassi Kouassi Jacques[1]
E-mail: kouassi.kouassijacques@yahoo.fr

[1] Earth Science Unit, Soil Science Department, Felix Houphouet-Boigny University, 22 BP 582, Abidjan, Côte d'Ivoire.

[2] Department of Economic Sciences, Biology Sciences, Peleforo Gon Coulibaly University, BP 1328 Korhogo, Côte d'Ivoire.

References
Azizi, P., Mahmood, S., Torabi, H., Masihabadi, M. H., & Homaee, M. (2011). Morphological, physic-chemical and clay mineralogy investigation on gypsiferous soils in southern of Tehran, Iran. *Middle-East Journal of Scientific Research, 7*, 153–161.

Berryman, C., Brower, R., Charteres, C., Davis, H., Davison, R., Eavis, B., ... Yates, R. A. (1984). *Booker tropical soil manual: A handbook for soil survey and agricultural land evaluation in the tropics and subtropics*. London: Longman.

Bertrand, R., Kilian, J., Raunet, M., Guillobez, S., & Bourgeon, G. (1985). Characterizing landscape systems, a prerequisite for environment protection. Methodological approach. *Recherche Agronomie Gembloux, 20*, 545–559.

Bigorre, F. (1999). Contribution of clays and organic matters to soil water holding. Mean and fondamental implication to exchangeable cation capacity. *CR Academic Science, 330*, 245–250.

Böhn, W. (1976). In situ estimation of root length at natural soil profiles. *The Journal of Agricultural Science, 87*, 365–368.

Burger, J. A. (1996). Limitations of bioassays for monitoring forest soil productivity: Rationale and example. *Soil Science Society of America Journal, 60*, 1674–1678. http://dx.doi.org/10.2136/sssaj1996.036159950060000 60010x

Buxbaum, G., & Printzen, H. (1998). Colored poigments: Iron oxide pigments. In G. Buxbaum (Ed.), *Industrial inorganic pigments* (pp. 85–107). Weiheim: VCH. http://dx.doi.org/10.1002/9783527612116

Chopart, J. L. (1985). Root development of some annual crops in west Africa and resistance to drought in intertropical zone. In *For integrated drought management*. Paris: CILF edition.

Diatta, S. (1996). *Grew soils of foot slope on granit-gneiss rock in the centre region of Côte d'Ivoire: Toposequential and spatial structures, hydrologic regime. Consequence for rice cropping* (Doctorate thesis). Henri Point Carré University, Nancy I.

Eschenbrenner, V., & Badarello, L. (1978). *Pedological study of the region of Odiénné (Côte d'Ivoire). Morpho-pedological mape* (p. 123). Paris: Manual N°74. ORSTOM .

Gale, M. R., Grigal, D. F., & Harding, R. B. (1991). Soil productivity index: Predictions of site quality for white spruce plantations. *Soil Science Society of America Journal, 55*, 1701–1709. http://dx.doi.org/10.2136/sssaj1991.03615995005500 060033x

Gretagmacъeth. (2000). *Munsell soil color charts: Year 2000 revised washable edition*. New York, NY: Author.

Grolimund, D., Elimelech, M., Borkovee, M., Barmettle, K., Kretzschmar, R., & Sticher, H. (1998). Transport of in situ mobilized colloidal particles in packed soil columns. *Environmental Science & Technology, 32*, 3562–3569.

He, Y., & Song, H. (2006). Prediction of soil content using near-infrared spectroscopy. Hangzhou: SPIE Newsroom (The International Society of Optical Engineering). doi: http://dx.doi.org/10.1117/2.1200604.0164

Hem, J. D. (1992). *Study and interpretation of the chemical characteristics of natural water* (3rd ed., p. 263). Alexandria, VA: United States Geological Survey Water Supply Paper 2254.

Kaplan, D. I., Sumner, M. E., Bertsch, P. M., & Adriano, D. C. (1996). Chemical conditions conducive to the release of mobile colloids from ultisol profiles. *Soil Science Society of America Journal, 60*, 269–274. http://dx.doi.org/10.2136/sssaj1996.03615995006000010041x

Kiniry, L. M., Scrivener, C. L. & Keener, M. E. (1983). *A soil productivity index based on water depletion and root growth*. Res. Bull. 105 Colombia University of Missouri. (p. 89).

Koné, B. (2007). *Color as indicator of the soils' fertility: Data use for assessing inherent fertility of ferralsols over the latitude 7°N of Côte d'Ivoire* (Doctorate thesis). Cocody University. 146p+annexe.

Koné B., Amadji, G. L., Toure A., Togola A., Mariko M., & Huat, J. (2013). A Case of *Cyperus* spp. and *Imperata cylindrica* occurrences on acrisol of the dahomey gap in South Benin as affected by soil characteristics: A strategy for soil and weed Management. *Applied and Environmental Soil Science, 2103*, Article ID 601058, 7 p.

Koné, B., Bongoua-Devisme, A. J., Kouadio, K. H., Kouadio, K. F., & Traoré, M. J. (2014). Potassium supplying capacity as indicated by soil colour in Ferralsol environment. *Basic Research Journal of Soil and Environmental Science, 2*, 46–55.

Koné, B., Diatta, S., Sylvester, O., Yoro, G., Camara, M., Dohm, D. D., & Assa, A. (2009). Assessment of ferralsol potential fertility by color: Color use in morpho pedology. *Canadian Journal of Soil Science, 89*, 331–342. http://dx.doi.org/10.4141/CJSS07119

Koné, B., Fatogoma, S., & Chérif, M. (2013). Diagnostic of mineral deficiencies and interactions in upland rice yield declining on foot slope soil in a humid forest zone. *International Journal of Agronomy and Agricultural Research, 3*, 11–20.

Koné, B., Touré, A., Amadji, G. L., Yao-Kouamé, A., Angui, P. T., & Huat, J. (2013). Soil characteristics and *Cyperus* spp. occurrence along a toposequence. *African Journal of Ecology, 51*, 402–408. http://dx.doi.org/10.1111/aje.2013.51.issue-3

Koné, B., Yao-Kouamé, A., Ettien, J. B., Oikeh, S., Yoro, G., & Diatta, S. (2009). Modelling the relationship between soil color and particle size for soil survey in Ferralsol environments. *Soil and Environment, 28*, 93–105.

Krasilnikov, P. V., & Tabor, J. A. (2003). Perspectives on utilitarian ethnopedology. *Geoderma, 111*, 197–215. http://dx.doi.org/10.1016/S0016-7061(02)00264-1

Larson, W. E., & Pierce, F. J. (1991). Conservation and enhancement of soil quality. In *Evaluation for sustainable land management in the developing world*. International Board for Soil Research and Management (IBSRAM) (Proceedings 12) (Vol. 2, pp. 175–203). Bangkok.

Larson, W. E., & Pierce, F. J. (1994). The dynamics of soil quality as a measure of sustainable management. In Doran, J. W., et al. (Eds.), *Defining soil quality for a sustainable environment* (pp 37–52). Madison, WI: SSSA. Pub. 35.

Loukili, M., Bock, L., Engles, P., & Mathieu, L. (2000). Geomorphological approach and Geographic Information System (GIS) for land management in Moroco. *Etude et Gestion des Sols, 7*, 37–52.

Mauricio, P., & Ildeu, A. (2005). Color attributes and mineralogical characteristics, evaluated by radiometry of highly weathered tropical soils. *Soil Science Society of America Journal, 69*, 1162–1172.

Milner, K. S., Running, S. W., & Coble, D. W. (1996). Biopsical soil—site model for estimating potential productivity of forested landscape. *Canadian Journal of Soil Sciences, 55*, 228–234.

Nahon, D. B. (1991). *Introduction to the petrology of soils and chemical weathering*. New York, NY: John Wiley and Sons.

Neill, L. L. (1979). *An evaluation of soil productivity based on root growth and water depletion* (M.Sc. Thesis). University of Missouri, Columbia, MO.

Page, A. L. (1982). *Methods of soil analysis. Part 2. Chemical and microbiological properties* (2nd ed.). Madison, WI: SSA, ASA.

Pal, S. K. (1998). Prediction of plant available potassium in kaolinitic soils of India. *Agropedology, 8*, 94–100.

Pearson, C. J., Norman, D. W., & Dixon, J. (1995). *Sustainable dryland cropping in relation to soil productivity - FAO soils bulletin 72*. Rome: FAO.

Pieri, C. J. M. G. (1992). *Fertility of soils. A future for farming in the West African savannah*. Berlin: Springer-Verlag.

Pierce, F. J., Larson, W. E., Dowdy, R. H., & Graham, W. A. P. (1983). Productivity of soils: Assessing long term changes due to erosion. *Journal of Soil and Water Conservation, 38*, 39–44.

Poss, R. (1982). *Etude morphopédologique de la région de Katiola (Côte d'Ivoire)* (p. 142). Paris: ORSTOM. Note explicative N°94.

Poss, R., Fardeau, J. C., Saragonit, H., & Quantin, P. (1991). Potassium release and fixation in Ferralsols (Oxisols) from Southern Togo. *Journal of Soil Science, 42*, 649–660. http://dx.doi.org/10.1111/ejs.1991.42.issue-4

Riquier, J., Cornet, J. P., & Braniao, D. L. (1970). A new system of soil appraisal in terms of actual and potential productivity (1st Approx). *World Soil Res* (p. 44). Rome: FAO.

Torrent, J., Schwertmann, U., Fechter, H., & Alferez, F. (1983). Quantitative relationships between soil color and hematite content. *Soil Science, 136*, 354–358. http://dx.doi.org/10.1097/00010694-198312000-00004

Torrent, J., Schwertmann, U., & Schulze, D. G. (1980). Iron oxide mineralogy of some soils of two river terrace sequences in Spain. *Geoderma, 23*, 191–208. http://dx.doi.org/10.1016/0016-7061(80)90002-6

Sanchez, P. A., Palm, C. A., & Buol, S. W. (2003). Fertility capability soil classification: a tool to help assess soil quality in the tropics. *Geoderma, 114*, 157–185. http://dx.doi.org/10.1016/S0016-7061(03)00040-5

Santana, D. P. (1984). *Soil formation in a toposequence of oxisols from Patos de Minas region* (PhD. Thesis), Minas Gerais State, Brazil, Purdi Univ, West Lafayette, IN.

Schwertmann, U. (1985). The effect of pedogenic environments on iron oxide minerals. *Advances in soil science, 1*, 171–200.

Segalen, P. (1969). Contribution to knowledge of sesquioxyd soil colors in intertropical zone: Yellow soils and red soils. *Cah ORSTOM, Ser Pedol, 7*, 225–236.

Sharpley, A. N. (1989). Relationship between soil potassium forms and mineralogy. *Soil Science Society of America Journal, 52*, 1023–1028.

Stoner, E. R., Baumgardner, M. F., Weismiller, R. A., Beilh, L. L., & Robbinson, B. F. (1980). *Atlas of soil reflectance properties*. Purdue Uni., West Lafayette, IN. Res. Bull. 962 Agric. Exp. Stn.

Sys, C., & Frankart, R. (1971). Soils' assessment in humid tropical zones. *African Soils, 15*, 177–199.

Torrent, J., & Barrón, V. (2002). *Iron oxides in relation to the colour of Mediterranean soils. Applied Study of Cultural Heritage and Clays* (pp. 377–386). Madrid: Consejo Superior de Investigaciones Científicas.

Vizier, J. F. (1971). Study of soil oxydo reduction statut and consequences on iron dynamic in hydromorphic soils. *Cah. ORSTOM. Ser. Pedol., 4*, 373–398.

Webster, R., & Oliver, M. O. (1990). *Statistical methods in soil and land resources survey*. New York, NY: Oxford University Press.

Wedpohl, K. H. (1978). *Handbook of Geochemistry*. Berlin: Springer Verlag.

5

A feasible way to increase carbon sequestration by adding dolomite and K-feldspar to soil

Leilei Xiao[1,2§], Qibiao Sun[1§], Huatao Yuan[1], Xiaoxiao Li[1], Yue Chu[1], Yulong Ruan[3], Changmei Lu[1] and Bin Lian[1]*

*Corresponding author: Bin Lian, Jiangsu Key Laboratory for Microbes and Functional Genomics, Jiangsu Engineering and Technology Research Center for Microbiology, College of Life Sciences, Nanjing Normal University, Nanjing 210023, China
E-mail: bin2368@vip.163.com
Reviewing editor:
Craig O'Neill, Macquarie University, Australia

Abstract: In recent years, many researchers have explored various possible ways to slow down the increase in atmospheric CO_2 concentration as this process poses a serious threat to mankind's survival. Mineral weathering is one possible way. Silicate weathering, for example, causes net carbon sequestration and carbonate weathering occurs relatively rapidly. In this study, dolomite and K-feldspar were added to soil to investigate if these minerals can increase carbon sequestration and also improve the available potassium content. The carbon content of amaranth, the organic and inorganic carbon content of the soil, two kinds of enzymes (polyphenol oxidase and urease), and the available potassium content were all tested. The experimental results show that the minerals accelerate the fixation of organic and inorganic carbon in the soil and also promote amaranth growth. Moreover, the available potassium content was increased when K-feldspar was added. Taken together, adding moderate amounts of carbonate and silicate minerals into the soil is found to be an attemptable way of accelerating CO_2 fixation and improving the potassium content of soil.

Subjects: Agriculture & Environmental Sciences; Environmental Sciences; Soil Sciences

Keywords: soil; mixed mineral; mineral weathering; carbon fixation

1. Introduction

It is widely accepted that human activity has caused the atmospheric CO_2 concentration to rise continually. The average global atmospheric CO_2 concentration reached 395.31 ± 0.10 ppm in 2014 (Le Quéré et al., 2014). Methods of reducing the rate of atmospheric CO_2 enrichment, and investigating what kind of practices are feasible and effective in blocking the trend of increasing atmosphere greenhouse gas (GHG) concentrations, have been attracting more and more attention (Liping &

ABOUT THE AUTHORS

The research of my group is concentrating in the following aspects: effects of microbes on the carbonate weathering and carbon sequestration; microbe–carbonate–silicate mineral interactions and their effects on the carbon migration and conversion; process and mechanisms in the microbe–mineral interaction. The research reported in this paper provides a possibility to slow down CO_2 release and possibly fix atmospheric CO_2 through adding silicate minerals in agricultural practice, which relates to the issues of how to deal with the global change.

PUBLIC INTEREST STATEMENT

Atmospheric CO_2 concentration increases continually. Seeking utilizable methods for blocking the trend attracts more and more attention. China is an agricultural country possessing rich mineral resources. Agricultural operations affect the carbon cycle through uptake, fixation, emission, and transfer of carbon among different pools. Based on these, we added dolomite and K-feldspar into soil to investigate if these minerals can increase carbon sequestration and also improve the available potassium content. Our research results are positive. It provides a possibility to slow down the momentum of elevated atmospheric CO_2 concentration through the practice of agriculture.

Erda, 2001). Notwithstanding the temporary nature of the release of anthropogenic CO_2, it should be noted that large quantities of carbon are also exuded through the roots of plants in the form of organic matter which degrade to gaseous form and ultimately return to the atmosphere (Ryan, Delhaize, & Jones, 2001). This process acts as a dominant conveyor in the global carbon cycle, accounting for ~120 Gt Ca^{-1} which clearly overshadows the 6 Gt Ca^{-1} produced by anthropogenic activities (Renforth, Manning, & Lopez-Capel, 2009). Many GHGs are sequestered by the agricultural and terrestrial ecosystems, and plant biomass and soil, for example, are the major sinks of atmospheric CO_2 (Liping & Erda, 2001). Therefore, carbon sequestration in soil is something that we should value (Lal, 2004).

Schlesinger and Andrews (2000) showed that the CO_2 released from the soil should be considered to be one of the largest sources of flux in the global carbon cycle, and can thus have a large effect on the atmospheric CO_2 concentration. Plants and the soil are the principal parts of the agricultural and terrestrial ecosystems. Thus, it is highly appropriate to consider the importance of the coupled plant–soil system in carbon capture, and to develop ways to enhance these natural processes (Renforth et al., 2009). For example, adopting restorative land use and using recommended management practices with agricultural soils can mitigate the negative impact of elevated atmospheric CO_2 concentrations. In temperate climates, soil organic carbon (SOC) is always a major carbon sink in agricultural soil (Smith, 2004). By adopting reasonable practices, the global SOC sequestration potential is about 0.9 ± 0.3 Pg Ca^{-1}, which offsets one-fourth to one-third of the annual increase in atmospheric CO_2 (Lal, 2004). In arid climates, the role of soil as a carbon sink is often associated with the accumulation of soil *inorganic* carbon (SIC). The maximum capacity of SIC capture technology has been shown to be limited by the availability of Ca-rich minerals (Renforth et al., 2009). Seeking cost-effective means of increasing SOC and SIC is therefore very significant, although this is a long and arduous process.

It has been demonstrated that weathering occupies an important position in present and future carbon cycling, and involvement of microbes can accelerate the weathering process of carbonate (Burford, Fomina, & Gadd, 2003; Xiao et al., 2014) and silicate (Xiao, Lian, Dong, & Liu, 2016; Xiao, Lian, Hao, Liu, & Wang, 2015). However, many researchers ignore the accumulation of weathering products in the soil and thus underestimate their contribution to the global climate (Goudie & Viles, 2012). It is also worth considering methods that can reduce atmospheric CO_2 by reacting silicate minerals to form carbonate minerals (Lackner, 2003; Manning & Renforth, 2013; Seifritz, 1990). This typically involves dissolution of the silicate minerals and subsequent precipitation of stable carbonate minerals (Power, Harrison, Dipple, & Southam, 2013). Manning and Renforth (2013) showed that the pedogenic carbonate should now be considered as a consequence of reactions between plant root exudates and calcium liberated by the silicate dissolution. However, it should be noted that the rate of silicate weathering is limited due to the slow kinetics of the CO_2–silicate reaction process (Oelkers, Gislason, & Matter, 2008). In general, the process is over an order of magnitude slower than that of the carbonate process (Mortatti & Probst, 2003; Wu, Xu, Yang, & Yin, 2008). According to calculations, it will take more than one million years to stabilize the atmospheric CO_2 level through silicate weathering (Goudie & Viles, 2012). Adding 1–2 tons of crushed olivine (grain size < 300 micron) to one hectare of soil will last approximately 30 years in a temperate climate (Schuiling & Krijgsman, 2006). Therefore, it appears to be difficult to ease the increase in atmospheric CO_2 by only employing silicate weathering in the short term. The uptake of CO_2 by atmospheric or soil respiration by carbonate rock dissolution has an important effect on the global carbon cycle and serves as one of the most important sinks (Cao et al., 2012). Liu, Dreybrodt, and Wang (2010) showed that dissolved inorganic carbon is an important but previously underestimated sink for atmospheric CO_2. This contribution to carbon sequestration reaches up to 0.8242 Pg Ca^{-1}, which amounts to 10.4% of the total anthropogenic CO_2 emission (Liu et al., 2010). Dissolution of calcite and dolomite can transform soil-generated CO_2 into alkaline form (HCO_3^-, CO_3^{2-}) (Macpherson et al., 2008).

Agricultural operations affect the carbon cycle through uptake, fixation, emission, and transfer of carbon among different pools (Lal, 2004). It should be feasible to change the carbon content distribution among these pools by anthropogenic manipulation. Recently, several studies (Fan et al., 2014; Mahmoodabadi & Heydarpour, 2014) have shown that application of manure is generally considered to increase carbon sequestration, although Schlesinger (1999) pointed out that net carbon sequestration does occur. Schuiling and Krijgsman (2006) mentioned that spreading finely powdered olivine on farmland could be extensively used to fix CO_2. In this study, we use soil as a substrate to explore if carbonate and silicate minerals, when added to the soil, can cause an increase in net carbon sequestration.

2. Materials and methods

2.1. Minerals

Dolomite, $CaMg(CO_3)_2$, for the study was provided by the Institute of Geochemistry, Chinese Academy of Sciences (Guiyang, China). Analysis using X-ray diffraction (XRD) showed that the samples were doped with small amounts of calcite and sanidine. Analysis of the K-feldspar ($KAlSi_3O_8$) used using XRD showed that quartz, muscovite, and clinochlore were present as impurities. The two kinds of mineral were both crushed and specific-sized particles (100–200 mesh) used in the study.

2.2. Plants

The study is mainly aimed at investigating whether carbon sequestration can be increased by adding carbonate and silicate to the soil. At the same time, we also explored if using K-feldspar was able to provide potassium for plant growth. Therefore, amaranth is used in this study because it is a common vegetable that is planted widely in china and has a strong ability to become enriched in K ions.

2.3. Summary of the experimental method

The experimental pot employed is illustrated in Figure S1. XRD analysis of the soil showed that it is composed of quartz, muscovite, albite ($Na(Si_3Al)O_8$), orthoclase ($KAlSi_3O_8$), and kaolinite-1A ($Al_2Si_2O_5(OH)_4$). Each pot contains 1,500 g of soil and 500 g of mineral powder. Our primary purpose is to detect the effectiveness of mineral, so the addition was relative large. The different amounts added to each pot are shown in Table1. Every treatment has three replicates. To increase the permeability of the soil, 450 g glass beads were added to the pots. The amaranth was watered timely according to the soil moisture. Water (500 ml) was regularly added to the pot every six days, and the soil infiltration water (SIW) collected the next day. The experiment was continued about two months.

2.3.1. Determination of the organic carbon content of amaranth

Whole amaranth plants were collected and dried overnight at 105°C. The dry weight was then measured. The percentage of carbon present was measured using an elemental analyzer (Elementar Vario MACRO, Germany).

2.3.2. Determination of soil parameters

At the end of the experiment, the soil moisture, pH, and activity of polyphenol oxidase and urease were determined. Portions of the naturally air-dried soil samples were used to determine the soil's organic carbon and microbial biomass.

Number	Composition			Abbreviation
	Soil (g)	Dolomite (g)	K-feldspar (g)	
1	2,000	–	–	all-s
2	1,500	500	–	d-s
3	1,500	375	125	d-p-s
4	1,500	125	375	p-d-s
5	1,500	–	500	p-s

Table 1. The different amounts of the two types of mineral added to each device

To measure the soil's moisture content, an aliquot of moist soil (about 5 g) was dried at 105°C. After 5 h, the sample was placed in a desiccator to cool for 30 min. Samples were dried and weighed repeatedly until the weight no longer decreased. Soil pH was determined using CO_2-free deionized water and a 1:1 (w/v) soil-to-water ratio. Samples were shaken for 15 min, left to settle for 30 min, and then the pH measured as in previous reports (Fierer & Jackson, 2006).

Polyphenol oxidase is known to play an important role in carbon cycling in the soil (Sinsabaugh, 1994). The analysis mirrored the work of Carney, Hungate, Drake, and Megonigal (2007). Briefly, (1) litterbags were removed as possible and approximately 2 g of wet weight placed in a blender mini-jar; (2) acetate buffer (60 ml, 50 mM, pH 5) was added and the mixture blended on "whip" for 1 min; (3) homogenate (0.750 ml) was mixed with an equal volume of substrate in a 2-ml tube—tubes were placed in a shaker and incubated for 2 h; (4) the reaction mixture was centrifuged for 2 min at 10,000 g and the absorbance of the supernatant at 460 nm immediately measured on a microplate reader.

Previous work has shown that urease activity has a positive correlation with organic carbon and total nitrogen (Zantua, Dumenil, & Bremner, 1977). For this test, toluene (1 ml) was mixed with 5 g of natural air-dried soil sample for 15 min. Then, urea solution (10 ml, 5%) and citrate buffer (20 ml, 0.96 M, pH 6.7) were added. Meanwhile, as a control, a repeat experiment was performed using an equal volume of distilled water instead of the urea solution. After incubation for 24 h at 37°C, the solution was centrifuged (4,000 g, 10 min). Supernatant (1 ml) was mixed with sodium phenoxide solution (4 ml, 2.7 M) and sodium hypochlorite solution (3 ml, 0.9%) in a 50-ml volumetric flask. After 20 min, the reaction solution was diluted to 50 ml and the absorbance at 460 nm measured. Urease activity was expressed according to the number of milligrams of NH_3–N in 1 g of soil.

For SOC determination, dry soil (2 g) was added to HCl solution (40 ml, 5%), and the mixture blended about 10 min until gas is no longer generated. The tubes were then spun at 8,000 g for 5 min. After this, the samples were rinsed three times with ultrapure water. After drying at 105°C, the residual solids were weighed and the carbon content tested using the elemental analyzer (Elementar Vario MACRO, Germany).

The microbial biomass in the soil was measured using the method outlined by Vance, Brookes, and Jenkinson (1987) involving chloroform fumigation and extraction. Moist soil was fumigated in a sealed desiccator using ethanol-free chloroform for 24 h at 25°C. Water (20 ml) and the same amount of NaOH (1 M) were placed in the bottom to trap any evolved CO_2. Non-fumigated soil was used as a control. Fumigated and non-fumigated soils were subjected to extraction using 0.5 M K_2SO_4 solution for 30 min using an "end-over-end" shaker at 350 rpm and centrifuged. The supernatant was filtered through a 0.45-μm membrane. The filtrate was tested using a total organic carbon analyzer (Shimadzu TOC-VCSN, Japan). The soil microbial biomass carbon (B_C) was estimated using $B_C = E_c/k_{EC}$, where E_c = [organic carbon extracted by K_2SO_4 from fumigated soil–carbon extracted by K_2SO_4 from non-fumigated soil] (Wu, Joergensen, Pommerening, Chaussod, & Brookes, 1990). The value of the k_{EC} parameter (the proportion of the extracted microbial biomass carbon evolved as organic carbon) was taken to be 0.45, following the work of Wu et al. (1990).

The available potassium content of the soil was measured. Briefly, a portion (0.5 g) of crushed dry sample (50–80 mesh) and ammonium acetate solution (50 ml, 1 M) were mixed in an extraction bottle (flask) following the method of Zhu, Lian, Yang, Liu, and Zhu (2013). Then, the bottle was stoppered and oscillated for 30 min. The mixture was filtered using a filter paper, and the filtrate collected for testing using a full-spectrum, direct-reading plasma emission spectrometer (Thermo Fisher Scientific, UK).

2.3.3. Determination of SIW parameters

The pH and temperature of the SIW measured using a pH meter (S20 SevenEasy, Mettler-Toledo). The concentrations of certain cations (K^+, Na^+, Ca^{2+}, and Mg^{2+}) were determined using an atomic absorption analyzer (AA900F, PerkinElmer, US). Anion concentrations (Cl^- and SO_4^{2-}) were determined using ion chromatography (DIONEX ICS-90, US). An acid–base titration method was used to measure the content of the bicarbonate in the aqueous solution according to the published literature with a little modification (Verma, 2004; Zangen, 1962). The SIW was filtered using a 0.45-μm microporous membrane and 20 ml titrated with a standardized HCl solution. The above parameters (pH, water temperature, ion concentrations) were imported into appropriate software (MINTEQ) to calculate the saturation index of the calcite, and the disordered-dolomite and ordered-dolomite.

2.4. Statistical analysis

StatSoft's STATISTICA 6.0 software was used to analyze the data. The significance of the differences between the treatments was tested separately using one-way ANOVA tests followed by Fisher LSD tests for mean comparisons. All analyses were performed in triplicate. The data shown correspond to the means (along with the standard deviation) of at least three independent experiments.

3. Results

3.1. Amaranth carbon content

For the average carbon content, dolomite or/and K-feldspar is conducive to plant growth (see Figure 1). The minerals thus improve organic carbon fixation. From a statistical point of view, only adding dolomite (d-s) or increasing the K-feldspar content (p-d-s and p-s) were beneficial to the formation of organic carbon sinks. However, there was no difference between the all-s and d-p-s treatments.

3.2. Soil parameters

Soil moisture content was not significantly different (statistically) among the five different kinds of treatment (ranging from 11.6% to 16.8%). Dolomite significantly increased the soil pH, but K-feldspar did not (Figure 2(a)). The amount of dolomite added and the elevation of the soil pH were positively correlated. As can be seen from Figure 2(b) (compared with all-s treatment), the increase in SOC per gram of moist soil was significantly boosted after adding a large amount of dolomite (p-s and d-p-s). However, adding a large amount of K-feldspar did not have this effect (p-d-s and p-s). This may be due to the dissolution of the dolomite, which can consume gaseous or liquid CO_2. In contrast, K-feldspar does not have this ability over a relatively short period of time.

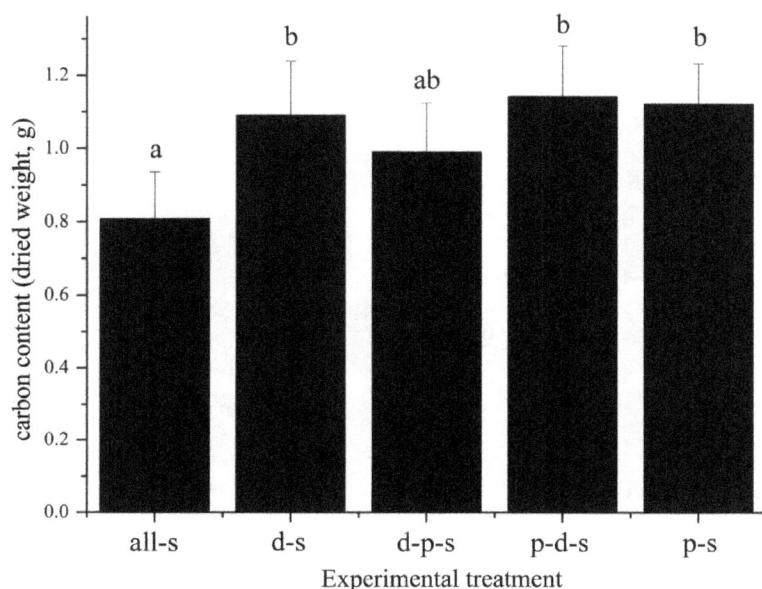

Figure 1. Amaranth carbon content after different minerals are added.

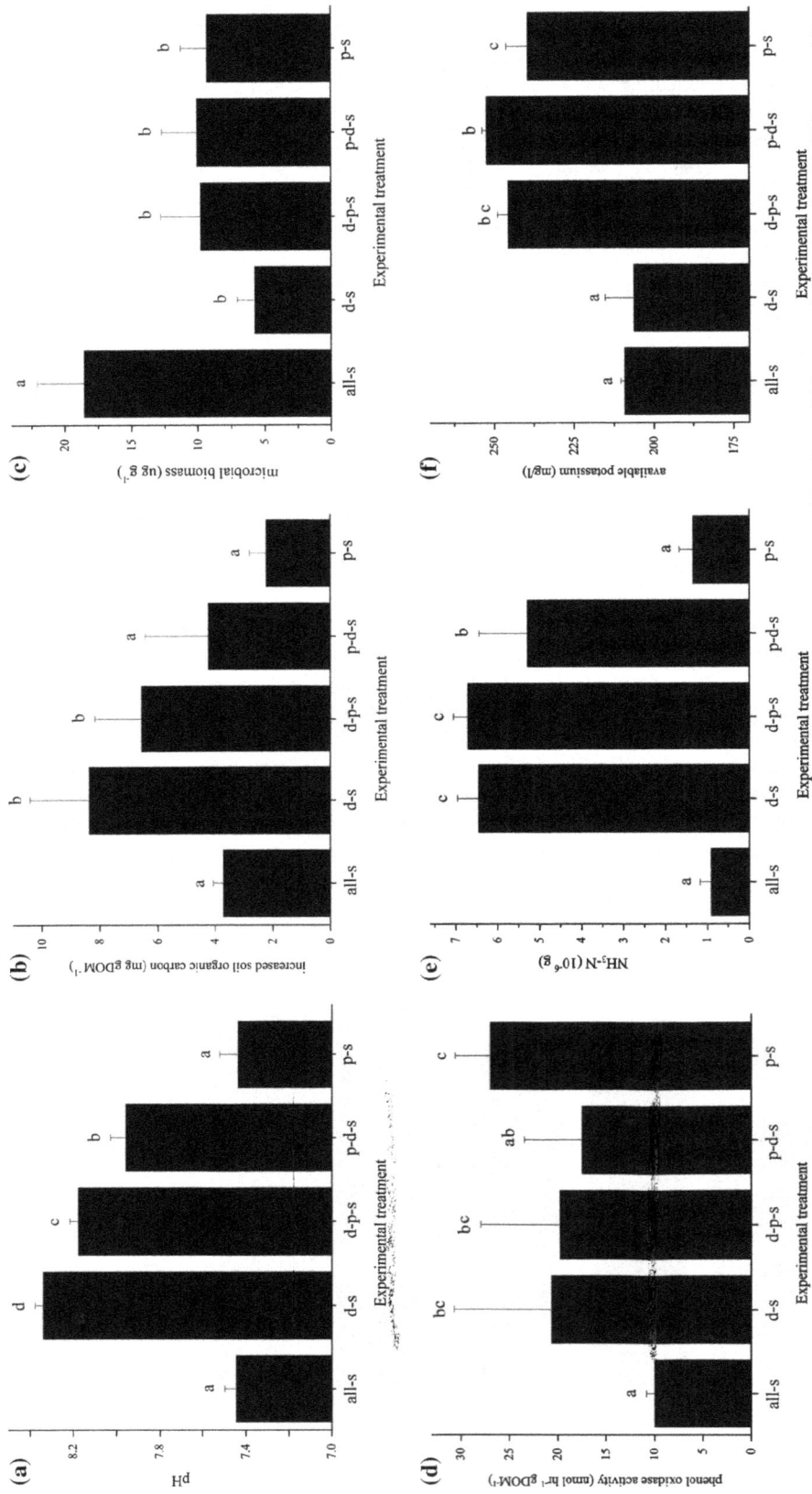

Figure 2. The effect of adding mixed minerals (dolomite and feldspar) to the soil on several soil parameters: (a) soil pH, (b) increased SOC, (c) microbial biomass, (d) polyphenol oxidase, (e) urease, and (f) available potassium.

To our surprise, the mineral (dolomite or feldspar) caused the soil microbial biomass to decline a certain amount over the experimental period. The effect of the added minerals on microbial survival in the microenvironment may be key here. Compare Figure 2(b) and (c). The microbial biomass and increase in organic carbon show some negative correlation. However, different minerals had no significant effect on soil microbial biomass. From Figure 2(a)–(c), the increase in SOC may be mainly pH dependent and little to do with the total number of micro-organisms. The minerals made the activity of the polyphenol oxidase (except for all-p) and urease (except for p-d-s) increase (Figure 2(d) and (e)). There was no significant difference between dolomite and K-feldspar with respect to polyphenol oxidase activity. Nevertheless, dolomite alone induced an increase in soil urease activity. As was expected, the amount of available potassium rose to a certain extent after K-feldspar was added (see Figure 2(f)). However, excessive addition did not significantly increase the available potassium content.

3.3. SIW parameters

3.3.1. pH
Both dolomite and K-feldspar promoted an increase in the pH of the SIW (Figure 3(a)). The impact of these two minerals on the pH of the SIW was little different. In the initial stages of the trial, the average pH in the sample with dolomite added was the highest. As the experiment progressed, a mixture of the minerals was most propitious to enhancing the SIW's pH.

3.3.2. The concentration of HCO_3^-
The addition of minerals had a great effect on HCO_3^- concentration (Figure 3(b)). No matter which mineral (dolomite or feldspar) was added, there was a significant difference at each of the eight sampling times (compared with the all-s treatment). Overall, the change in HCO_3^- concentration showed a sudden increase, followed by a rapid decrease, and then it gently changed. Dolomite alone was more influential with respect to HCO_3^- concentration compared to K-feldspar. However, adding

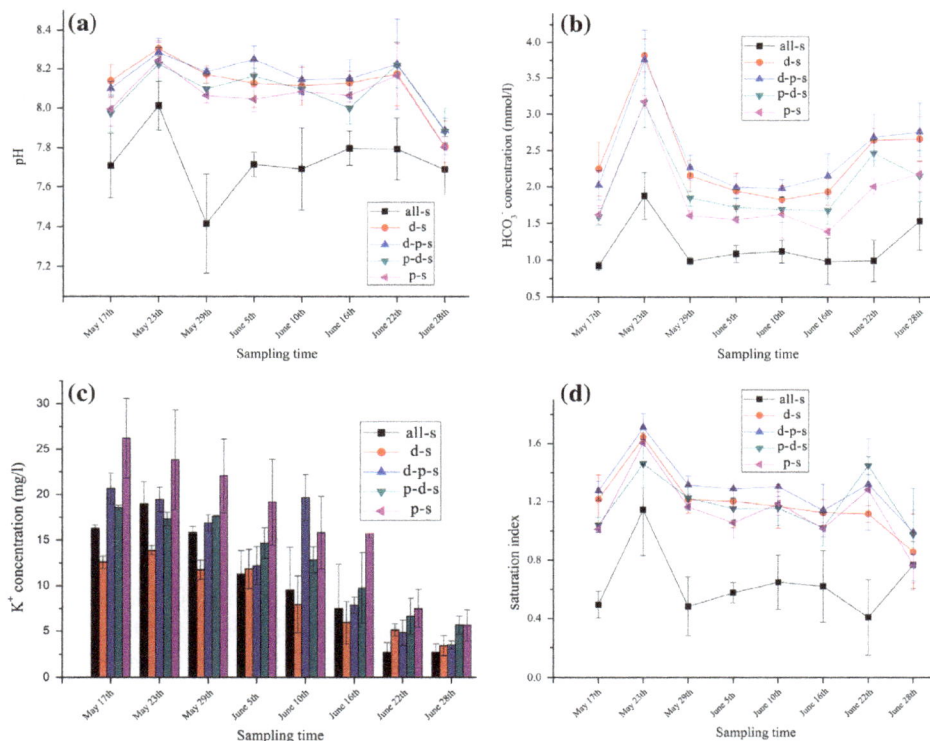

Figure 3. The effect of adding mixed minerals (dolomite and feldspar) to the soil on several of the parameters of the soil filtrate: (a) pH, (b) concentration, (c) K⁺ concentration, and (d) calcite saturation index.

more dolomite did not cause more HCO_3^- to be produced. If standard deviation is ignored, then the HCO_3^- produced was the most in the d-s treatment at the first two sampling times, yet mixed minerals are more conducive to generating HCO_3^- in the long term.

3.3.3. K+ concentration

Overall, the K^+ concentration gradually decreased (see Figure 3(c)). Only the average K^+ concentration in the p-s sample always exceeded that in all-s. Using mixed minerals or only K-feldspar did not significantly increase the concentration of K^+ in soil in the beginning of this experiment. The ability of K-feldspar to release K^+ can be seen in the latter part of the experiment (22 June and 28 June). Thus, it is feasible that, for long-term farming, fertilizers containing K-feldspar could be used to continue to provide K^+.

3.3.4. Saturation index calculation

MINTEQ software was used to calculate the saturation index of the carbonate using the following parameters: water temperature, pH, and the concentrations of K^+, Na^+, Ca^{2+}, Mg^{2+}, Cl^-, SO_4^{2-}, and HCO_3^- ions. Three kinds of carbonate, calcite, ordered-dolomite, and disordered-dolomite, were selected. Interestingly, the saturation indices of these three carbonates were almost the same (Figures 3(d) and S2). In the case of the calcite saturation index (Figure 3(d)), no matter which minerals were added, the calcite saturation index increased significantly. Comparing the effects of the different minerals on the saturation index, it seems mixed minerals (dolomite and K-feldspar) are more conducive to increasing the saturation index.

4. Discussion

The atmospheric CO_2 concentration continues to rise; thus researchers are constantly looking at a variety of solutions. Some of the research has shown that only a minor increase in the natural uptake is required to compensate for the extra anthropogenic CO_2 emission (Oelkers et al., 2008; Salek, Kleerebezem, Jonkers, Witkamp, & van Loosdrecht, 2013). Up until the last decade, mineral weathering has been a subject of some concern. However, this is the most important way in which nature keeps the CO_2–levels stable (Schuiling & Krijgsman, 2006). This oversight may be because people have not found a practical way to accelerate weathering (Schuiling & Krijgsman, 2006) or they have ignored the natural regulation effect. In this study, we have investigated the impact on CO_2 fixation of adding mixed dolomite–K-feldspar power to the soil. The results suggest that artificially adding suitable amount of dolomite and K-feldspar to the soil is likely to increase the amount of SOC. This is similar to the finding of previous reports investigating the potential of artificial soils (i.e. made by adding demolition waste or basic slag to soil). These were used to capture some of the transferred carbon as geologically stable $CaCO_3$ (Renforth et al., 2009). Artificial soils were also prepared by blending compost with dolerite and basalt quarry fines to be used for the purpose of CO_2 capture (Manning, Renforth, Lopez-Capel, Robertson, & Ghazireh, 2013).

The significantly increased pH value of the soil implies that adding dolomite to the soil is possible to mitigate the effects of acid rain (Allen & Brent, 2010; Teir, Eloneva, Fogelholm, & Zevenhoven, 2006) and/or alleviates soil acidification caused by agricultural fertilizer. Although the addition of the minerals made the total microbial biomass decrease, the increased polyphenol oxidase and urease activity suggests that adding minerals may promote the reproduction and activity of the microbes in the soil which are associated with the carbon and nitrogen cycles.

Manning and Renforth (2013) studied sequestration of atmospheric CO_2 through coupled plant–mineral reactions in urban soils. They showed that the rate-limiting factor seems to be the availability of Ca, not carbon (Manning & Renforth, 2013). Our results showed that the increase in SOC was doubled by dolomite addition, while K-feldspar addition did not do this. Both dolomite (d-s) and K-feldspar (p-f-s and p-s) improved the total carbon content of the amaranth. This means that the mineral may act as a regulatory component of the soil. After adding K-feldspar, the availability of K was significantly increased. Potassium is the most abundant cation in plants, comprising up to 10% of a plant's dry weight (Leigh & Wynjones, 1984). In China, soluble potassium is a scarce resource—there is only a 35% self-supply, so most is sourced via imports (Sun et al., 2013). Adding K-bearing minerals to soil is an attemptable way to improve the potassium content of the soil.

The following reaction occurs when dolomite is added to the soil:

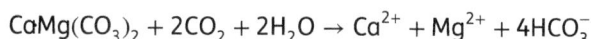

$$CaMg(CO_3)_2 + 2CO_2 + 2H_2O \rightarrow Ca^{2+} + Mg^{2+} + 4HCO_3^-$$

The experimental results showed that the amount of HCO_3^- generated was in some way related to the amount of dolomite added. In other words, the more dolomite added meant that more inorganic carbon produced (to some extent). However, the amount of Ca^{2+} in the d-s sample was lower than in the d-p-s-treated sample in the latter part of the experiment. Precipitation of carbonate mineral is controlled by the saturation state of the soil solution which itself depends on the activities of the dissolved species (cation and bicarbonate). A higher saturation index is beneficial to the formation of carbonate precipitation. From the saturation index of calcite or dolomite (Figures 3(d) and S2), we see that the saturation index of the minerals added is always higher than the all-s sample. This indicated that the addition of minerals can accelerate the fixation of gaseous carbon into the form of relatively stable inorganic carbon, e.g. HCO_3^- and CO_3^{2-} as well. Moreover, others minerals, such as anorthite, can be tested as well. Further researches are still needed to explore if the long-term minerals addition can cause soil desertification. Anyway, these results show some prospects for mineral application in the farming.

5. Conclusions

The addition of minerals was likely to accelerate the fixation of organic and inorganic carbon in the soil. Moreover, the available K content in the soil was increased when K-feldspar was added. Taken together, adding moderate amounts of carbonate and silicate minerals into soil is an attemptable way to accelerate CO_2 fixation and improve the soil's K content. However, further research is needed to improve the application of mixed minerals so as to increase soil quality and carbon sequestration.

Funding

This work was jointly supported by National Natural Science Foundation of China [Grant number: 41373078].

Author details

Leilei Xiao[1,2]
E-mail: ai-yanzi@163.com
Qibiao Sun[1]
E-mail: sunqibiao001@163.com
Huatao Yuan[1]
E-mail: 1164650987@qq.com
Xiaoxiao Li[1]
E-mail: lanlin717@foxmail.com
Yue Chu[1]
E-mail: 466393948@qq.com
Yulong Ruan[3]
E-mail: ryl880035@163.com
Changmei Lu[1]
E-mail: luchangmei@njnu.edu.cn
Bin Lian[1]
E-mail: bin2368@vip.163.com

[1] Jiangsu Key Laboratory for Microbes and Functional Genomics, Jiangsu Engineering and Technology Research Center for Microbiology, College of Life Sciences, Nanjing Normal University, Nanjing 210023, China.

[2] Key Laboratory of Coastal Biology and Utilization, Yantai Institute of Coastal Zone Research, Chinese Academy of Sciences, Yantai 264003, China.

[3] Key Laboratory of Karst Environment and Geological Hazard Prevention, Ministry of Education, Guizhou University, Guiyang 550003, China.

[§] Co-first author.

Authors' contributions

BL, CL, and LX designed the experiments; BL and LX wrote the main manuscript text; HY, LX, QS, XL, YC, and YR carried out the experiments, and LX prepared all figures. All authors reviewed the manuscript.

References

Allen, D. J., & Brent, G. F. (2010). Sequestering CO_2 by mineral carbonation: Stability against acid rain exposure. *Environmental Science & Technology, 44*, 2735–2739.

Burford, E., Fomina, M., & Gadd, G. (2003). Fungal involvement in bioweathering and biotransformation of rocks and minerals. *Mineralogical Magazine, 67*, 1127–1155. http://dx.doi.org/10.1180/0026461036760154

Cao, J. H., Yuan, D. X., Chris, G., Huang, F., Yang, H., & Lu, Q. (2012). Carbon fluxes and sinks: The consumption of atmospheric and soil CO_2 by carbonate rock dissolution. *Acta Geologica Sinica-English Edition, 86*, 963–972.

Carney, K. M., Hungate, B. A., Drake, B. G., & Megonigal, J. P. (2007). Altered soil microbial community at elevated CO_2 leads to loss of soil carbon. *Proceedings of the National Academy of Sciences, 104*, 4990–4995. http://dx.doi.org/10.1073/pnas.0610045104

Fan, J., Ding, W., Xiang, J., Qin, S., Zhang, J., & Ziadi, N. (2014). Carbon sequestration in an intensively cultivated sandy loam soil in the North China Plain as affected by compost and inorganic fertilizer application. *Geoderma, 230*, 22–28. http://dx.doi.org/10.1016/j.geoderma.2014.03.027

Fierer, N., & Jackson, R. B. (2006). The diversity and biogeography of soil bacterial communities. *Proceedings of the National Academy of Sciences, 103*, 626–631. http://dx.doi.org/10.1073/pnas.0507535103

Goudie, A. S., & Viles, H. A. (2012). Weathering and the global carbon cycle: Geomorphological perspectives. *Earth-Science Reviews, 113*, 59–71. http://dx.doi.org/10.1016/j.earscirev.2012.03.005

Lackner, K. S. (2003). Climate change: A guide to CO2 Sequestration. *Science, 300*, 1677–1678. http://dx.doi.org/10.1126/science.1079033

Lal, R. (2004). Soil carbon sequestration to mitigate climate change. *Geoderma, 123*, 1–22. http://dx.doi.org/10.1016/j.geoderma.2004.01.032

Le Quéré, C., Moriarty, R., Andrew, R. M., Peters, G. P., Ciais, P., Friedlingstein, P., ... Zeng, N. (2014). Global carbon budget 2014. *Earth system science data discussions, 7*, 521–610. http://dx.doi.org/10.5194/essdd-7-521-2014

Leigh, R. A., & Wynjones, R. G. (1984). A hypothesis relating critical potassium concentrations for growth to the distribution and functions of this ion in the plant cell. *New Phytologist, 97*, 1–13. http://dx.doi.org/10.1111/nph.1984.97.issue-1

Liping, G., & Erda, L. (2001). Carbon sink in cropland soils and the emission of greenhouse gases from paddy soils: A review of work in China. *Chemosphere-Global Change Science, 3*, 413–418. http://dx.doi.org/10.1016/S1465-9972(01)00019-8

Liu, Z. H., Dreybrodt, W., & Wang, H. J. (2010). A new direction in effective accounting for the atmospheric CO_2 budget: Considering the combined action of carbonate dissolution, the global water cycle and photosynthetic uptake of DIC by aquatic organisms. *Earth-Science Reviews, 99*, 162–172. http://dx.doi.org/10.1016/j.earscirev.2010.03.001

Macpherson, G. L., Roberts, J. A., Blair, J. M., Townsend, M. A., Fowle, D. A., & Beisner, K. R. (2008). Increasing shallow groundwater CO_2 and limestone weathering, Konza Prairie, USA. *Geochimica et Cosmochimica Acta, 72*, 5581–5599. http://dx.doi.org/10.1016/j.gca.2008.09.004

Mahmoodabadi, M., & Heydarpour, E. (2014). Sequestration of organic carbon influenced by the application of straw residue and farmyard manure in two different soils. *International Agrophysics, 28*, 169–176.

Manning, D. A. C., & Renforth, P. (2013). Passive sequestration of atmospheric CO_2 through coupled plant-mineral reactions in urban soils. *Environmental Science & Technology, 47*, 135–141.

Manning, D. A. C., Renforth, P., Lopez-Capel, E., Robertson, S., & Ghazireh, N. (2013). Carbonate precipitation in artificial soils produced from basaltic quarry fines and composts: An opportunity for passive carbon sequestration. *International Journal of Greenhouse Gas Control, 17*, 309–317. http://dx.doi.org/10.1016/j.ijggc.2013.05.012

Mortatti, J., & Probst, J.-L. (2003). Silicate rock weathering and atmospheric/soil CO_2 uptake in the Amazon basin estimated from river water geochemistry: Seasonal and spatial variations. *Chemical Geology, 197*, 177–196. http://dx.doi.org/10.1016/S0009-2541(02)00349-2

Oelkers, E. H., Gislason, S. R., & Matter, J. (2008). Mineral carbonation of CO_2. *Elements, 4*, 333–337. http://dx.doi.org/10.2113/gselements.4.5.333

Power, I. M., Harrison, A. L., Dipple, G. M., & Southam, G. (2013). Carbon sequestration via carbonic anhydrase facilitated magnesium carbonate precipitation. *International Journal of Greenhouse Gas Control, 16*, 145–155. http://dx.doi.org/10.1016/j.ijggc.2013.03.011

Renforth, P., Manning, D. A. C., & Lopez-Capel, E. (2009). Carbonate precipitation in artificial soils as a sink for atmospheric carbon dioxide. *Applied Geochemistry, 24*, 1757–1764. http://dx.doi.org/10.1016/j.apgeochem.2009.05.005

Ryan, P. R., Delhaize, E., & Jones, D. L. (2001). Function and mechanism of organic anion exudation from plant roots. *Annual Review of Plant Physiology and Plant Molecular Biology, 52*, 527–560. http://dx.doi.org/10.1146/annurev.arplant.52.1.527

Salek, S. S., Kleerebezem, R., Jonkers, H. M., Witkamp, G.-J., & van Loosdrecht, M. C. M. (2013). Mineral CO_2 sequestration by environmental biotechnological processes. *Trends in Biotechnology, 31*, 139–146. http://dx.doi.org/10.1016/j.tibtech.2013.01.005

Schlesinger, W. H. (1999). Carbon and agriculture: Carbon sequestration in soils. *Science, 284*, 2095. http://dx.doi.org/10.1126/science.284.5423.2095

Schlesinger, W. H., & Andrews, J. A. (2000). Soil respiration and the global carbon cycle. *Biogeochemistry, 48*, 7–20. http://dx.doi.org/10.1023/A:1006247623877

Schuiling, R. D., & Krijgsman, P. (2006). Enhanced weathering: An effective and cheap tool to sequester CO_2. *Climatic Change, 74*, 349–354. http://dx.doi.org/10.1007/s10584-005-3485-y

Seifritz, W. (1990). CO_2 disposal by means of silicates. *Nature, 345*, 486. http://dx.doi.org/10.1038/345486b0

Sinsabaugh, R. S. (1994). Enzymic analysis of microbial pattern and process. *Biology and Fertility of Soils, 17*, 69–74. http://dx.doi.org/10.1007/BF00418675

Smith, P. (2004). Soils as carbon sinks: the global context. *Soil Use and Management, 20*, 212–218. http://dx.doi.org/10.1079/SUM2004233

Sun, L., Xiao, L., Xiao, B., Wang, W., Pan, C., Wang, S., & Lian, B. (2013). Differences in the gene expressive quantities of carbonic anhydrase and cysteine synthase in the weathering of potassium-bearing minerals by *Aspergillus niger*. *Science China Earth Sciences, 56*, 2135–2140. http://dx.doi.org/10.1007/s11430-013-4704-4

Teir, S., Eloneva, S., Fogelholm, C. J., & Zevenhoven, R. (2006). Stability of calcium carbonate and magnesium carbonate in rainwater and nitric acid solutions. *Energy Conversion and Management, 47*, 3059–3068. http://dx.doi.org/10.1016/j.enconman.2006.03.021

Vance, E., Brookes, P., & Jenkinson, D. (1987). An extraction method for measuring soil microbial biomass C. *Soil Biology and Biochemistry, 19*, 703–707. http://dx.doi.org/10.1016/0038-0717(87)90052-6

Verma, M. P. (2004). A revised analytical method for HCO_3^- and CO_3^{2-} determinations in geothermal waters: An assessment of IAGC and IAEA interlaboratory comparisons. *Geostandards and Geoanalytical Research, 28*, 391–409. http://dx.doi.org/10.1111/ggr.2004.28.issue-3

Wu, J., Joergensen, R., Pommerening, B., Chaussod, R., & Brookes, P. (1990). Measurement of soil microbial biomass C by fumigation-extraction—an automated procedure.

Soil Biology and Biochemistry, 22, 1167–1169. http://dx.doi.org/10.1016/0038-0717(90)90046-3

Wu, W., Xu, S., Yang, J., & Yin, H. (2008). Silicate weathering and CO_2 consumption deduced from the seven Chinese rivers originating in the Qinghai-Tibet Plateau. Chemical Geology, 249, 307–320. http://dx.doi.org/10.1016/j.chemgeo.2008.01.025

Xiao, L., Hao, J., Wang, W., Lian, B., Shang, G., Yang, Y., ... Wang, S. (2014). The up-regulation of carbonic anhydrase genes of bacillus mucilaginosus under Soluble Ca^{2+} deficiency and the heterologously expressed enzyme promotes calcite dissolution. Geomicrobiology Journal, 31, 632–641. http://dx.doi.org/10.1080/01490451.2014.884195

Xiao, L., Lian, B., Hao, J., Liu, C., & Wang, S. (2015). Effect of carbonic anhydrase on silicate weathering and carbonate formation at present day CO_2 concentrations compared to primordial values. Scientific Reports, 5, 7733. http://dx.doi.org/10.1038/srep07733

Xiao, L., Lian, B., Dong, C., & Liu, F. (2016). The selective expression of carbonic anhydrase genes of Aspergillus nidulans in response to changes in mineral nutrition and CO_2 concentration. MicrobiologyOpen, 5, 60–69. http://dx.doi.org/10.1002/mbo3.2016.5.issue-1

Zangen, M. (1962). Titration of carbonate-bicarbonate leach solutions. Journal of Applied Chemistry, 12, 92–96.

Zantua, M. I., Dumenil, L. C., & Bremner, J. M. (1977). Relationships between soil urease activity and other soil properties1. Soil Science Society of America Journal, 41, 350–352. http://dx.doi.org/10.2136/sssaj1977.03615995004100020036x

Zhu, X., Lian, B., Yang, X., Liu, C., & Zhu, L. (2013). Biotransformation of earthworm activity on potassium-bearing mineral powder. Journal of Earth Science, 24, 65–74. http://dx.doi.org/10.1007/s12583-013-0313-6

6

Dynamics of soil organic carbon stocks in the Guinea savanna and transition agro-ecology under different land-use systems in Ghana

Enoch Bessah[1]*, Abdullahi Bala[2], Sampson K. Agodzo[3] and Appollonia A. Okhimamhe[1]
*Corresponding author: Enoch Bessah, Department of WASCAL, Federal University of Technology, Bosso Campus, Minna, PMB 65, Niger State, Nigeria
E-mail: enoch.bessah@gmail.com
Reviewing editor:
Nir Krakauer, City College of New York, USA
This paper is one of the outcomes of master's thesis submitted to the Federal University of Technology, Minna, Nigeria.

Abstract: This study was to assess and predict soil organic carbons stocks (SOCS) under the major land use/cover types in Kintampo North Municipal located in the Guinea savanna through the transition agro-ecological zone of Ghana. Random field sampling was done on 34 plots and 24 sample points at depths 0–10 cm, 10–20 cm and 20–30 cm per plot with soil corer. Soil bulk density, pH, particle size distribution and SOC were determined using standard laboratory procedures and computations. Results were subjected to both statistical and Geo-statistical analyses. The SOCS in each land use decreased with depth. The mean SOC for the five land-use systems studied were 11.33 t/ha, 7.95 t/ha and 6.08 t/ha at 0–10 cm, 10–20 cm and 20–30 cm, respectively. The vertical variability in SOC distribution across the considered land use/cover types was statistically significant ($p < 0.05$) but the statistical difference amongst land use/cover types was insignificant. The determined mean SOCS were 30.02 (±13.20) t/ha for savanna woodland as the highest and 22.01 (±8.92) t/ha for cashew plantation as the lowest at total depth (0–30 cm). The spatial distribution of SOC stocks ranged between 12 t/ha to about 33 t/ha.

Subjects: Agriculture; Environmental Management; Environmental Sciences; Forestry; Soil Sciences

Keywords: soil organic carbon stocks; land use/cover types; geostatistical kriging; Kintampo North Municipal

ABOUT THE AUTHORS

Research discipline of authors includes climate change assessment and adaptation, soil and land use engineering, irrigation and water resources management, integrated agriculture, conservation agriculture, postharvest engineering, irrigation and drainage, agricultural water management, soil mechanics applications, rural water supply, hydrological modelling, soil macro and micro nutrient assessment, remote sensing applications, carbon dioxide assessment, performance evaluation of agriculture machinery amongst others. This research is a part of the climate change adaptation and mitigation via adapted land use for carbon dioxide control under the West African Science Service Center on Climate Change and Adapted Land Use (WASCAL) Project. Other aspect of this project is the relation of climate change to water resources, energy, human security, biodiversity, agriculture and west African climate systems.

PUBLIC INTEREST STATEMENT

This study was to determine how savanna woodland, cropland and tree plantations affected soil organic carbon stocks (SOCS) in Kintampo North Municipal located in the Guinea savanna through the transition agro-ecological zone of Ghana. Soil samples were picked on 34 plots at depths 0–10 cm, 10–20 cm and 20–30 cm to see how soil organic carbon varies with depth. The types of tree plantations considered in the study were cashew, teak and mango. savanna woodland had the highest SOCS as a land cover and cashew plantation had the lowest amount of SOCS. Running spatial distribution of SOCS in ArcGIS gave values between 12 t/ha to about 33 t/ha all over the municipal. It implies that soil in Kintampo North Municipal stores moderate amount of organic carbons which is vital for climate change mitigation if land use change and soil preparation by ploughing are controlled.

1. Introduction

The largest emission of CO_2 from soils result from land use change and especially drainage of organic soils (Houghton & Goodale, 2004) however, drainage plays an important role to improve soil quality in agriculture (Callesen et al., 2003; Valipour, 2014). Land use change in controlling CO_2 emissions in sub-Saharan Africa may be more critical than in other regions and have been discovered to be an uncertain component in global carbon cycle for the continent (Grieco, Chiti, & Valentini, 2012). The most typical example of incomplete estimates in sub-Saharan Africa is as a result of inadequate and reliable data for various carbon pools. Adu-Bredu, Abekoe, Tachie-Obeng, and Tschakert (2010) studied carbon stocks under four land-use systems in three different ecological systems in Ghana. They reported that soils have the highest carbon stocks in Bawku, which is in the Sudan savanna zone with mean values of 34.05 MgCha^{-1}, 32.02 MgCha^{-1}, 32.14 MgCha^{-1} and 23.64 MgCha^{-1} for fallow, cultivated land, natural forest and teak plantation, respectively. The results from the study show that soils in the savanna zone and forest zone in Ghana have the highest soil carbon content.

Carbon exists as inseparable components of biomass and soil organic matter. Its storage in soil organic matter is important in mitigating global climate change and improves the livelihood of resource-poor farmers. It increases land productivity through improved soil properties such as nutrient supply and moisture retention (van Keulen, 2001). Degradation and deforestation have impacted negatively on both vegetation and soil carbon stock. Soils in Africa have been reported to lose 136 gigatonnes of carbon between 1850 and the late 1990s (United Nations Environment Programme, 2012). About 33% of carbon lost in Africa soils within this period was attributed to land degradation and soil erosion. Soil organic carbon (SOC) is also an energy source for organism decomposition (Melillo et al., 2002) and can be lost through increased soil respiration (Conant et al., 2008). Soil Carbon sequestration alone is surely not the only way to fight climate change but it is realistic to link climate change with soil carbon conservation, as soil carbon sequestration is cost competitive, of immediate availability, does not require the development of new and unproven technologies, and provides comparable mitigation potential to that available in other sectors (Schils et al., 2008).

SOC is one of the largest and active carbon pools. Globally, the estimates of SOC storage range from 1,200 to 1,600 pg in the top 1-m soil depth, inorganic component amounts to 695–930 pg down to the same depth (Batjes, 1996; Sombroek, Nachtergaeke, & Hebel, 1993), which is mostly stored in arid and semi-arid regions (Díaz-Hernández, Fernández, & González, 2003). To mitigate global warming, carbon sequestration, which redistribute carbon from the air to other pools would help to reduce the rate of atmospheric CO_2 release (Soil Science Society of America, 2001; Tieszen, 2000). Other parameters affecting organic matter behaviour in soils include moisture status, temperature, oxygen supply (drainage), acidity, nutrient supply, clay content and mineralogy. Erosion, decomposition and leaching are important soil processes causing carbon concentrations to decrease in the soil (Lal, 2003). Land-use systems conserve, sequestrate, or release carbon into the atmosphere directly or indirectly. United Nations Framework Convention on Climate Change (UNFCCC) reported in 2000 that agricultural soils could be made into a net sink of carbon dioxide. As much as 0.9 ± 0.3 Pg of carbon could be absorbed by agricultural soils annually through improved management practices designed to increase agricultural productivity (Lal, 2004). Proven and documented management practices that will sequester carbon in the transition zone will contribute immensely to the fight against global carbon release from soils (Hedlund, 2015; Woomer, Touré, & Sall, 2004). Bationo, Kihara, Vanlauwe, Waswa, and Kimetu (2007) reported that surface disturbance from cultivated systems decreases soil carbon contents because tree cover is reduced and mineralisation takes place faster. Soil organic matter (SOM) increases soil structural stability and resistance to rainfall impact; rate of infiltration and fauna activities (Bationo et al., 2007) and 58% of soil organic matter is made up of SOC (Natural Land & Water Resources Audit, 2008).

It has been reported that temperatures in the tropics, including the savanna zone of Ghana are increasing (Nutsukpo, Jalloh, Zougmore, Nelson, & Thomas, 2012). This has led to instability in precipitation and increased food insecurity as well. Increased temperatures in the tropics leads to increased decomposition rates reducing carbon stored as organic matter in the soil. Increased soil respiration as a result of increasing temperature would release a higher concentration of CO_2 into the atmosphere due to carbon loss from the soil (Melillo et al., 2002). Therefore, this study was to assess and predict soil organic carbon stocks (SOCS) under the major land use/cover types (cropland, savanna woodland, teak plantation, mango plantation and cashew plantation) in the Guinea savanna and transitional agro-ecological zone of Ghana.

2. Materials and methods

2.1. Study area

Kintampo North Municipal where the study was done is located between latitudes 8°45' N and 7°45' N and longitudes 1°20' W and 0°1' W. It is surrounded by five districts in the country namely Central Gonja District to the north; Bole District to the west; East Gonja District to the north-east (all in the northern region); Kintampo South District to the south and Pru District to the south-east. The municipality has a surface area of about 5,108 km², and occupies about 12.9% of the total land area of Brong Ahafo Region (Strategic Environmental Assessment, 2010). In terms of location and size, the municipal is strategically positioned at the centre of the country as shown in Figure 1. It is the transit point between the northern and southern sectors of Ghana. The municipal comes under the interior wooded savanna or tree savanna and transitional agro-ecology. Every transitional zone is believed to have once existed as forest. Its current transformation is attributed to prevailing savanna conditions resulting from man's activities. This is evident by the existence of riparian forest where anthropogenic activities are limited.

The municipality experiences the tropical continental or interior savanna type of climate (Strategic Environmental Assessment, 2010). The mean annual rainfall is between 1,400 mm and 1,800 mm. It occurs in two seasons from May to July as minor and from September to October as major (Strategic Environmental Assessment, 2010). The mean monthly temperature ranges from 30°C in March to 24°C in August with mean annual temperatures between 26.5°C and 27.2°C (Strategic Environmental Assessment, 2010). Kintampo North Municipal is found within the Voltain Basin and the Southern Plateau physiographic regions and is elevated between 60 and 150 m above sea level. Voltain plateau occupies the southern part of the municipal with series of escarpments. About 40% of the surface area is on Voltarian formation rocks and the municipal is covered with 80% of it. Voltarian formation rocks are mainly sedimentary and exhibit horizontal alignments. Examples are sand stone, shale, mudstone and limestone. Two main groups of soils are in the municipal. The first

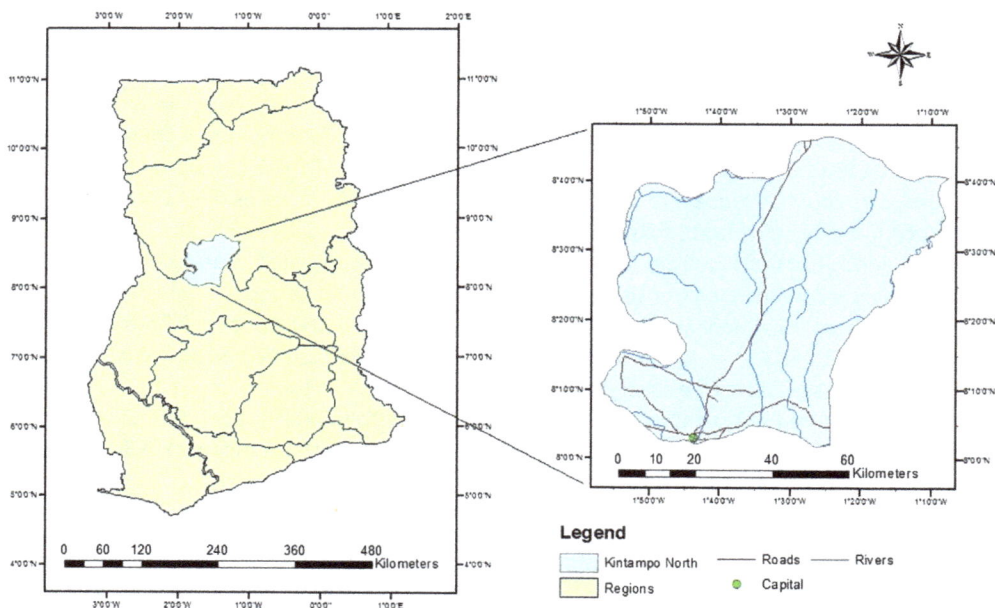

Figure 1. Map of Ghana showing study area.

which is groundwater laterite soils or Plinthosols (FAO) or Plinthaquox (USA, Soil Taxonomy) covers nearly 60% of the municipal particularly in the interior wooded savanna zone. Savanna ochrosols or Alfisols (USDA classification) is the other soil group mostly found in the south and south-western parts. The municipal has 71% of her population in the agriculture sector and land is a very important resource (Strategic Environmental Assessment, 2010).

2.2. Study methods

The municipal is covered by different types of land cover and land use types. However, the land use types considered for sampling were savanna woodland, teak plantation, mango plantation, cashew plantation and crop land (mixed cropping). A measuring tape was used to delineate 50 × 50 m plot (N) and Garmin eTrex 10 GPS device was used to pick location coordinates of sample sites in the middle of the plot. Soil core sampler (5 cm diameter × 5 cm length) was used to collect samples at three (3) points diagonally on each plot at each depth 0–10 cm, 10–20 cm and 20–30 cm (Intergovernmental Panel on Climate Change, 2007; United Nations Framework Convention on Climate Change, 2000), for bulk density determination. This amounted to nine samples per plot. Samples were taken from all six plots for the selected land use/cover types except crop land where sampling was done in 10 plots. Another set of soil samples were obtained by sampling at five points (5 on a die method) on each plot at the three depths as done for the previous sampling. The samples obtained from the five points, for each depth, were thoroughly mixed together to get composite samples for the determination of SOC concentrations, pH and particle size analysis as recommended by World Bank Electronic Institute (WBI) for Clean Development Mechanism (CDM) project (World Bank Electronic Institute, 2014). This amounted to 15 samples per plot. Samples were taken from six plots for all land use/cover types except crop land where sampling was done in 10 plots. On each plot, therefore, a total of 24 samples were collected for the two sets of laboratory analyses for all three depths. A total of 816 samples were collected from the 34 plots across all considered land use/cover types in the study area. The limit of the top 30 cm of the soil was chosen because it often represents the limit of the visible humic horizons.

Bulk densities were determined by oven drying at 105°C till constant weight was attained, Walkley–Black wet oxidation method was used for organic carbon concentration determination. The hygrometer method for the particle size analysis and soil–water ratio method for pH determination. Pluske, Murphy, and Sheppard (2014) formula was used for the SOC stocks calculation as follows:

$$SOC = 100\rho_d ZC \hspace{4cm} (1)$$

where SOC is the soil organic carbon stocks (t/ha); ρ_d is the soil bulk density (Mg/m³); Z is the soil depth (m); C is the carbon concentration (%).

StatistiXL 2007, Excel 2007, IBM SPSS 20 and ArcGIS 10.1 were used for all statistical and geo-statistical (kriging) analysis, respectively.

3. Results

3.1. Dry bulk density

The dry bulk density for all collected samples is shown in Table 1. The mean bulk densities for total depth 0–30 cm across the five land use/cover types increased from 1.34 g/cm³ under mango plantation to 1.46 g/cm³ in cropland. Dry bulk density increased with increasing soil depth for all land use/ cover types; this result is consistent with the observation of other works in Ghana (Agboadoh, 2011; Dawoe, 2009; Dowuona & Adjetey, 2010). The decrease in bulk density with increasing soil depth may be due to the decrease in SOC with depth. At depth 0–10 cm, mango Plantation recorded the least density of 1.27 g/cm³ and teak and cropland all recording the highest at 1.42 g/cm³. High bulk density means the soil has a high percentage of sand as confirmed by the particle size analysis (Table 2) and therefore is expected to have less nutrient (SOC). Mango plantation had the least dry bulk density 1.35 g/cm³ at depth 10–20 and cashew plantation at depths 20–30 cm recorded 1.40

Table 1. Mean values of dry bulk density in g/cm³ across the depths and land uses						
Land use/cover types	N	0–10 cm	10–20 cm	20–30 cm	0–30 cm	Area (ha)
		Mean (Std. Dev.)				
Savanna woodland	18	1.35 (±0.12)	1.47 (±0.07)	1.47 (±0.08)	1.43 (±0.08)	303,519.17
Mango plantation	18	1.27 (±0.05)	1.35 (±0.07)	1.41 (±0.09)	1.34 (±0.05)	8,548.56 (planta-tion)
Cashew plantation	18	1.40 (±0.07)	1.42 (±0.08)	1.40 (±0.08)	1.40 (±0.07)	
Teak plantation	18	1.42 (±0.10)	1.41 (±0.06)	1.45 (±0.04)	1.43 (±0.06)	
Crop land	30	1.42 (±0.08)	1.47 (± 0.07)	1.49 (±0.08)	1.46 (±0.07)	129,859.40

Table 2. Mean particle size distribution and textural class of sampled plots						
Land use types	N	Sand (%)	Silt (%)	Clay (%)	Textural class	Area (ha)
		Mean (Std. Dev.)				
Savanna woodland	6	70.14 (±6.24)	24.73 (±5.71)	5.13 (±2.70)	Sandy Loam	303,519.17
Mango plantation	6	75.50 (±4.51)	20.70 (±5.11)	3.80 (±0.74)	Loamy Sand	8,548.56 (planta-tion)
Cashew plantation	6	70.97 (±6.45)	22.95 (±3.65)	6.08 (±3.09)	Sandy Loam	
Teak plantation	6	73.36 (±5.79)	21.81 (±5.67)	4.83 (±2.33)	Sandy Loam	
Crop land	10	74.08 (±7.35)	21.18 (±5.74)	4.74 (±3.12)	Sandy Loam	129,859.40

Notes: N—Number of plots sampled. Each composite sample is made up of 5 samples for each depth.

g/cm³. The largest deviation in values was observed under teak plantation, cashew and mango at ±0.10, ±0.08 and ±0.09 respectively.

3.2. Soil texture

A summary of the soil texture characteristics and class under the land use/cover types is shown in Table 2. The clay content for all soils was generally low. Clay content was higher under cashew plantation and least in mango with a variation from 3.80% to 6.08%. Residual cover from the leaves of cashew was a major cover for its plantation compared with other land use and therefore has the potential to reduce erosion and eluviation. This might be a major reason for the variation in clay content amongst land use. Sand fractions ranged from 70.14% in savanna woodland to 75.50% in mango plantation. The textural class was Loamy sand for only mango plantation and the remaining land use types were sandy loam. Sand correlated negatively ($R = -0.583$, $p = 0.00$ for 0–30 cm) with SOC at all depths implying that, where sand is high, clay is expected to decrease resulting in decreased SOCS. Therefore, cashew plantation with the highest clay content was expected to record more SOC content (Figure 2) but was rather the least which might be due to the intercropping of tubers crops with cashew for over 10 years till their canopy cover the spaces between them.

3.3. Soil pH

The mean determined pH in the study at the three considered depths is summarised in Table 3. Averagely, cropland measured the lowest mean pH value of 6.69 whilst savanna woodland had the highest mean pH value of 7.10. Intermediate values were recorded by teak, cashew and mango at 6.87, 6.87 and 6.90, respectively. Soil pH increased with soil depth in all land use/cover types. The soil pH correlated positively with clay content ($R = 0.049$, $p = 0.937$) implying that lower clay content soils should record lower pH. Crop yields are normally high in soils with pH values between 6.0 and 7.5 but most nutrient elements are usually available in the pH range of 5.5–6.5 (Motsara & Roy, 2008). This confirms the study area as a good agricultural land. The range of pH at total depth 0–30 cm at 6.69–7.10 means the soils are neutral (pH approx. 7) and do not need any chemical treatment.

Table 3. Mean pH values in selected land use/cover in relation to soil depth

Land use types	N	0–10 cm	10–20 cm	20–30 cm	0–30 cm	Area (ha)
		Mean (Std. Dev.)				
Savanna woodland	6	7.01 (±0.11)	7.14 (±0.18)	7.15 (±0.31)	7.10 (±0.19)	303,519.17
Mango plantation	6	6.86 (±0.39)	6.91 (±0.25)	6.94 (±0.21)	6.90 (±0.26)	8,548.56 (plantation)
Cashew plantation	6	6.82 (±0.44)	6.90 (±0.45)	6.91 (±0.27)	6.87 (±0.38)	
Teak plantation	6	6.89 (±0.22)	6.90 (±0.38)	6.82 (±0.47)	6.87 (±0.34)	
Crop land	10	6.64 (±0.20)	6.72 (±0.20)	6.71 (±0.26)	6.69 (±0.03)	129,859.40

Notes: N—Number of plots sampled. Each composite sample is made up of 5 samples for each depth.

3.4. SOC concentration

The mean SOC concentrations are reported in Table 4 for all three depths by land use/cover type. SOC concentrations decreased from 1.03, 0.64 and 0.48% in savanna woodland to 0.72, 0.47 and 0.38% in cashew plantation at depth 0–10 cm, 10–20 cm and 20–30 cm, respectively. It was only at depth 20–30 cm that cashew plantation and teak plantation recorded the same amount of organic carbon concentration. SOC concentration decreased with increasing depth across all land use/cover types which is consistent with the followings studies (Agboadoh, 2011; Dawoe, 2009; Hairiah & van Noordwijk, 2000; Jiao et al., 2010). Savanna woodland and crop land had the widest coverage in the municipal. Therefore, the SOC concentration of these two land uses indicate on average an appreciable stored content which is not protected and can be lost via land use change.

3.5. Soil organic carbon stocks

The SOC stocks distribution for all the land use/cover types ranged between 4.95 and 21.18 t/ha, 3.42 and 16.07 t/ha, 1.38 and 12.74 t/ha and 9.80 and 49.63 t/ha for depths 0–10 cm, 10–20 cm,

Table 4. Mean values of soil organic carbon concentration (%)

Land use types	N	0–10 cm	10–20 cm	20–30 cm	0–30 cm	Area (ha)
		Mean (Std. Dev.)				
Savanna woodland	6	1.03 (±0.44)	0.64 (±0.33)	0.48 (±0.32)	0.72 (±0.35)	303,519.17
Mango plantation	6	0.85 (±0.32)	0.60 (±0.20)	0.46 (±0.23)	0.64 (±0.23)	8,548.56 (plantation)
Cashew plantation	6	0.72 (±0.30)	0.47 (±0.16)	0.38 (±0.20)	0.52 (±0.22)	
teak plantation	6	0.77 (±0.24)	0.57 (±0.26)	0.38 (±0.18)	0.57 (±0.20)	
Crop land	10	0.82 (±0.29)	0.55 (± 0.23)	0.41(±0.21)	0.59 (±0.22)	129,859.40

Notes: N—Number of plots sampled. Each composite sample is made up of 5 samples for each depth.

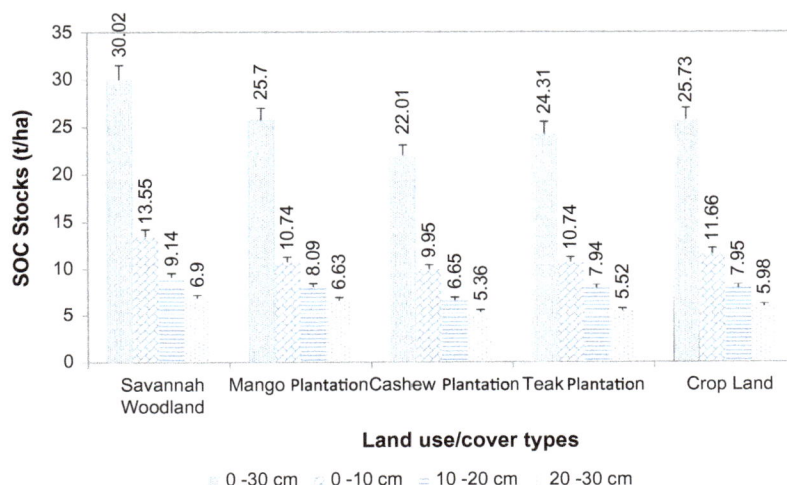

Figure 2. Mean SOC stocks under different land use/cover types.

20–30 cm and 0–30 cm, respectively (Figure 2). All the land use/cover types recorded a decreasing SOC stocks with increasing depth. The mean SOC stocks for all land use/cover types were 11.3, 7.95 and 6.08 t/ha for depths 0–10 cm, 10–20 cm and 20–30 cm, respectively. Savanna woodland measured the highest SOC stocks at all depths and cashew plantation the least. Observations during sampling revealed that soils under plantations were less disturbed after crops are fully grown compared to croplands. Fully grown mango and cashew plantations had the highest cover or litter on top soil and less direct heating of soil from sunshine. It was not so for teak plantation due to the regular bush fires in the municipal. Deviations ranged between ±2.76 and ±4.73, ±2.32 and ±4.33, ±2.48 and ±4.33 and ±7.74 and ±13.20 for depths 0–10 cm, 10–20 cm, 20–30 cm and 0–30 cm, respectively. Savanna woodland recorded the highest variability or deviations at all depths.

Analysis of variance (ANOVA) at 95% confidence indicated no significant difference in SOC stocks amongst the land use types for all the depths ($p = 0.542$ for 0–10 cm, $p = 0.831$ for 10–20 cm, $p = 0.988$ for 20–30 cm and $p = 0.785$ for 0–30 cm) but there was a significant difference ($p < 0.05$) across the three depths under all land use/cover types sampled. SOC stocks correlated positively under the Pearson correlation (Table 5) with clay and finest particles (clay + silt) which indicate that as clay and the finest particles in the soil increases, SOC stocks also increases as confirmed by other studies (Agboadoh, 2011; Hairiah & van Noordwijk, 2000; Jiao, Xu, Zhao, & Yang, 2012). Bulk density and sand content correlated negatively with SOC stocks at all depths showing the inverse proportionality between % sand, bulk density and SOC stocks. High % sand content gives soil higher weight and less % clay therefore bulk density is expected to be higher in all soils with high sand content as SOC will be low in those soils. pH had a positive correlation at all depths except 20–30 cm implying less nutrient at that depth.

3.6 Spatial distribution of SOC stocks

The SOC stocks at 0–30 cm depths were spatially distributed by ordinary kriging in Figure 3. The SOC stocks range majorly between 12 t/ha and about 33 t/ha. The northern part had the least estimates which can be attributed to the limited data collected in that area and also the less SOC stocks measured across agro-ecological zones from the south to the north in the Guinea savanna. The study area is averagely storing about 25 t/ha SOCS. The district has a potential to increase the amount of SOC stocks stored in its soils if sustainable management practices are adopted to increase the sequestration capacity and decrease release from such activities.

Table 5. Pearson correlation (R) result of SOC stocks with other parameters				
	Soil organic carbon stocks; R (significance)			
	0–10 cm	**10–20 cm**	**20–30 cm**	**0–30 cm**
Bulk Density	−0.416 (0.014)	−0.248 (0.157)	−0.114 (0.529)	−0.345 (0.045)
pH	0.076 (0.669)	0.031 (0.860)	−0.130 (0.471)	0.018 (0.919)
% OC	0.980 (0.000)	0.993 (0.000)	0.690 (0.000)	0.993 (0.000)
% Sand	−0.282 (0.106)	−0.081 (0.649)	−0.297 (0.093)	−0.276 (0.114)
% Silt	0.215 (0.222)	−0.069 (0.698)	0.098 (0.587)	0.108 (0.545)
% Clay	0.268 (0.125)	0.321 (0.065)	0.446 (0.009)	0.454 (0.007)
% Silt + Clay	0.282 (0.106)	0.081 (0.649)	0.297 (0.093)	0.233 (0.205)

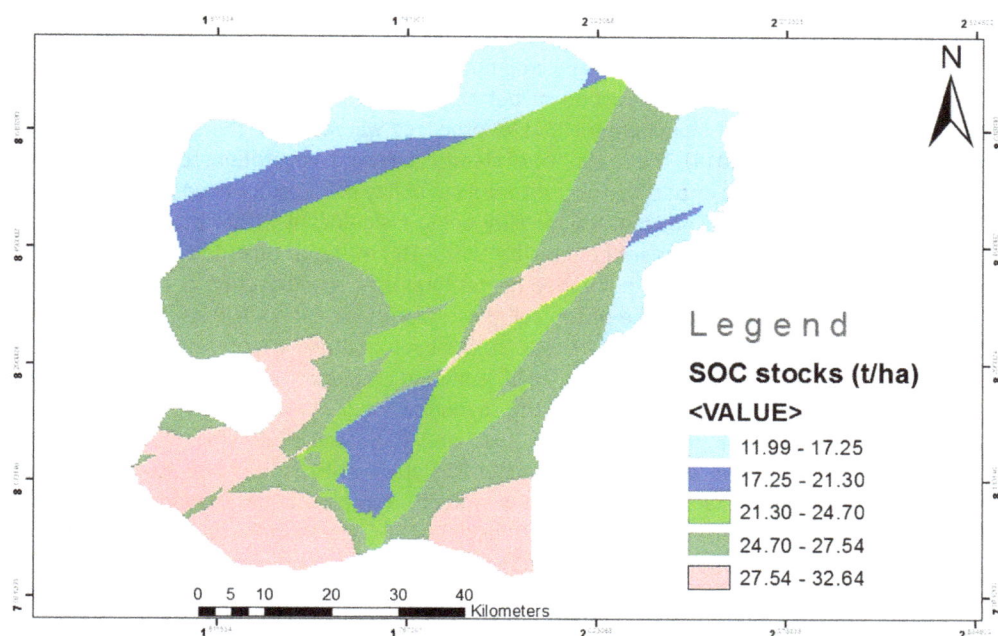

Figure 3. Spatial distribution of SOC stocks by kriging.

4. Discussion

The determined values of the dry bulk density suggest that the soil is within the sandy and loam textural classes (Table 2). Bulk density was higher in croplands compared to the report of Agboadoh (2011) which was averagely 1.34 g/cm³ for the top 0–20 cm at Bechem District in the forest zones and lower than 1.65 g/cm³ for cultivated field in Upper East Region of Ghana (Dawidson & Nilsson, 2000). This shows that dry bulk density increases from the southern to the northern part of Ghana. These higher values under cropland over savanna woodland might be as a result soil compaction due to the use of heavy agricultural machinery.

Yao et al. (2010) reported that land use types affect soil texture characteristics and the nutrient availability which confirms the result in this study. Cashew plantation with the highest clay content also recorded the least SOC stocks which is contrary to most findings (Agboadoh, 2011; Hairiah & van Noordwijk, 2000; Su, Xiong, Zhu, Ye, & Ye, 2006). This was due to the continuous intercropping of tuber crops with cashew until they are fully grown and developed closed canopies. Crop yields are normally high in soils with pH values between 6.0 and 7.5 but most nutrient elements are usually available in the pH range of 5.5–6.5 (Motsara & Roy, 2008). This confirms the study area with mean pH range of 6.64–7.15 at depths 0–10 cm, 10–20 cm and 20–30 cm as a good agricultural land. The range of pH at total depth 0–30 cm was 6.69–7.10 implying the soils in Kintampo North Municipal are neutral (pH approx. 7) and do not need any chemical treatment.

The insignificant difference in horizontal distribution of SOC stocks amongst land use/cover types is contrary to the report of Anderson-Teixeira, Davis, Masters, and Delucia (2009) that conversion of uncultivated land to agriculture purposes result in significant SOC stock loss but Yao et al. (2010) however had no significant differences for SOC stocks under 10–year-old teak plantation, cocoa plantation and recurrent fallow. Dawoe (2009) also recorded insignificant difference amongst SOC stocks under forest 3, 15 and 30 years' cocoa plantation in moist semi-deciduous zone of Ghana. The SOC stocks decreased with increasing depth across all land use/cover types as reported in several findings (Agboadoh, 2011; Dawoe, 2009; Follett, Kimble, Pruessner, Samson-Liebig, & Waltman, 2009; Hairiah & van Noordwijk, 2000; Jiao et al., 2010, 2012; Morisada, Ono, & Kanomata, 2004; Sheikh, Kumar, & Bussmann, 2009; Su et al., 2006). Savanna woodland recorded the highest mean of SOC stocks at 30.02 t/ha with a very high variability of 43.97% and cropland followed as the highest amongst the cultivated lands or systems. This can be attributed to the use of fertilizers as part of the land use management practices by crop farmers as reported by Wang, Zhang, Song, Liu, and Ren (2010). Therefore, the introduction of better land use management practices such as Sustainable Agricultural Land Management (SALM) practices will increase the stored SOC stocks (Verified Carbon Standards, 2014). The relevance of climate, soil type, vegetation, terrain and topography in the study area has no impact on the horizontal variability in SOC stocks due to it homogeneity (Su et al., 2006). Therefore, horizontal variability being insignificant in this study can also be attributed to land use and land cover change. The top 0–10 cm depth recorded the highest SOC stocks under all land use/cover types but varied across land use types because land use management practices have a higher influence at top soil (Post, Izaurralde, Mann, & Bliss, 2001; Su et al., 2006). Savanna woodland recorded 14.30 % SOC stocks over cropland which was second highest and cashew having the least SOC stocks was 14.46 % behind cropland. The SOC stocks is similar to the findings of Dowuona and Adjetey (2010) in the savanna zone of Ghana, ranging between 16.23 and 33.03 t/ha in 2005, 18.08 and 44.89 t/ha in 2007 and 20.85 and 52.54 t/ha in 2009 at depth 0–20 cm under field treatment of maize to sequester SOC stocks. Also, mean SOC stocks values of 30.02 t/ha, 24.31 t/h and 25.73 t/ha for savanna woodland, teak plantation and cropland, respectively, is comparable with Adu-Bredu et al. (2010) results of SOC stocks in Bawku in the Guinea savanna zone of Ghana. Natural forest recorded 32.14 t/ha, cultivated or cropland was 32.02 t/ha and teak plantation was the least at 23.64 t/ha. The decreasing trend was same in both studies. Tan, Tieszen, Tachie-Obeng, Liu, and Dieye (2008) with General Ensemble Biogeochemical Modeling System also estimated SOC stocks in the transition zone to be 21.2 t/ha at the top 20 cm for the year 2000, which is similar to the findings of this study.

5. Conclusion

From this study, SOC ranged between 4.95 and 21.18 t/ha, 3.42 and 16.07 t/ha, 1.38 and 12.74 t/ha and 9.80 and 49.63 t/ha for the depths 0–10 cm, 10–20 cm, 20–30 cm and 0–30 cm, respectively. Generally, SOC stocks decreased with increasing depth. The study identified savanna woodland which mostly suffers from deforestation as having the highest SOC stocks in the municipal. This is a natural and undisturbed ecosystem at the time of the study. Cropland recording high SOC stocks could be attributed to some land use management practices. This implies the adoption of better management practices like SALM practices which include use of cover crops, returning composted crop residuals to the field, manure management and the introduction of trees into landscapes will increase the carbon storage capacity of soils. It is necessary to consider the SALM concept and disseminate to farmers through the Ministry of Agriculture in order to address both food security and climate change issues. Finally, education on climate change, carbon credit, SALM and their relevance in sustaining livelihood should start in the municipal since carbon sequestration has a higher probability on productivity and further investigate SOCS effect on yield under the land use/cover types considered.

Acknowledgements
Special recognition goes to Ministry of Food and Agriculture (MoFA), Kintampo Municipal for their assistance on the field and to the farmers of the Kintampo North Municipal who participated in the study.

Funding
This work has been funded by the German Federal Ministry of Education and Research (BMBF) through West Africa Science Centre of Climate change and Adapted Land use (WASCAL).

Author details
Enoch Bessah[1]
E-mail: enoch.bessah@gmail.com
Abdullahi Bala[2]
E-mail: abdullahi_bala@yahoo.com
Sampson K. Agodzo[3]
E-mail: skagodzo7@usa.net
Appollonia A. Okhimamhe[1]
E-mail: aimiosino@yahoo.com
[1] Department of WASCAL, Federal University of Technology, Bosso Campus, Minna, PMB 65, Niger State, Nigeria.
[2] Department of Soil Science, Federal University of Technology, Bosso Campus, Minna, PMB 65, Niger State, Nigeria.
[3] Department of Agricultural Engineering, Kwame Nkrumah University of Science and Technology, PMB, Kumasi, Ghana.

References

Adu-Bredu, S., Abekoe, M. K., Tachie-Obeng, E., & Tschakert, P. (2010, November). *Carbon stock under four land-use systems in three varied ecological zones in Ghana.* Presented at the Open Science Conference on "Africa and Carbon Cycle: The Carbo Africa Project", Accra.

Agboadoh, D. M. Y. (2011). *Soil organic carbon stocks in croplands of the Bechem Forest District, Ghana* (Unpublished master's thesis). Kwame Nkrumah University of Science and Technology, Kumasi, Ghana.

Anderson-Teixeira, K. J., Davis, S. C., Masters, M. D., & Delucia, E. H. (2009). Changes in soil organic carbon under biofuel crops. *GCB Bioenergy, 1,* 75–96. http://dx.doi.org/10.1111/gcbb.2009.1.issue-1

Bationo, A., Kihara, J., Vanlauwe, B., Waswa, B., & Kimetu, J. (2007). Soil organic carbon dynamics, functions and management in West African agro-ecosystems. *Agricultural Systems, 94,* 13–25. http://dx.doi.org/10.1016/j.agsy.2005.08.011

Batjes, N. H. (1996). Total carbon and nitrogen in the soils of the world. *European Journal of Soil Science, 47,* 151–163. http://dx.doi.org/10.1111/ejs.1996.47.issue-2

Callesen, I., Liski, J., Raulund-Rasmussen, K., Olsson, M. T., Tau-Strand, L., Vesterdal, L., & Westman, C. J. (2003). Soil carbon stores in Nordic well-drained forest soils-relationships with climate and texture class. *Global Change Biology, 9,* 358–370. http://dx.doi.org/10.1046/j.1365-2486.2003.00587.x

Conant, R. T., Drijber, R. A., Haddix, M. L., Parton, W. L., Paul, E. A., Plante, A. F., ... Steinweg, J. G. (2008). Sensitivity of organic matter decomposition to warming varies with its quality. *Global Change Biology, 14,* 868–877. http://dx.doi.org/10.1111/gcb.2008.14.issue-4

Dawidson, E., & Nilsson, C. (2000). *Soil organic carbon in the upper east region, Ghana—Measurement and modelling* (Unpublished master's thesis). Lunds Universitets Naturgeografiska Institution, Sweden.

Dawoe, E. (2009). *Conversion of natural forest to Cocoa Agroforest in Lowland Humid Ghana: Impact on plant biomass production, organic carbon and nutrient dynamics* (Unpublished doctoral dissertation). Kwame Nkrumah University of Science and Technology, Kumasi, Ghana.

Díaz-Hernández, J. L., Fernández, E. B., & González, J. L. (2003). Organic and inorganic carbon in soils of semiarid regions: a case study from the Guadix–Baza basin (Southeast Spain). *Geoderma, 114,* 65–80. http://dx.doi.org/10.1016/S0016-7061(02)00342-7

Dowuona, G. N. N., & Adjetey, E. T. (2010, August). *Assessment of carbon storage in some savanna soils under different land-use systems in Ghana.* Presented at ICID+18 Conference, Fortaleza, Ceará, Brazil.

Follett, R. F., Kimble, J. M., Pruessner, E. G., Samson-Liebig, S., & Waltman, S. (2009). In R. Lal & R. F. Follett (Eds.), *Soil carbon sequestration and the greenhouse effect: Soil organic carbon stocks with depth and land use at various US sites* (2nd ed., Chap. 3, pp. 29–66). Madison, WI: Soil Science Special Publication 57.

Grieco, E., Chiti, T., & Valentini, R. (2012, April 22–27). Land use change and carbon stocks dynamics in sub-saharan Africa—Case study of Western Africa – Ghana. In *EGU General Assembly 2012* (p. 12218). Vienna.

Hairiah, K., & van Noordwijk, M. (2000). In A. Gillison (Ed.), *Soil properties and carbon stocks* (Aboveground Biodiversity Assessment Working Group Summary Report 1996–99: Impact of Different Land Uses on Biodiversity, pp. 143–254). Nairobi.

Hedlund, K. (2015, July). *Best practices on soil quality management.* Paper presented at Expo Milano 2015, Italy.

Houghton, R. A., & Goodale, C. L. (2004). Effect of Land-Use change on the carbon balance of terrestrial ecosystems. *Ecosystems and land use change Geophysical Monograph Series, 153,* 85–98. http://dx.doi.org/10.1029/GM153

Intergovernmental Panel on Climate Change. (2007). In R. K. Pachauri & A. Reisinger (Eds.), *Contribution of Working Group I, II, and III to the Fourth Assess Report of the Intergovernmental Panel on Climate Change: Climate Change (2007) Synthesis Report.* Geneva. Retrieved from http://www.ipcc.ch/

Jiao, J. G., Yang, L. Z., Wu, J. X., Wang, H. Q., Li, H. X., & Ellis, E. C. (2010). Land use and soil organic carbon in China's village landscapes. *Pedosphere, 20,* 1–14. http://dx.doi.org/10.1016/S1002-0160(09)60277-0

Jiao, Y., Xu, Z., Zhao, J., & Yang, W. (2012). Changes in soil carbon stocks and related soil proper-ties along a 50-year grassland-to-cropland conversion chronosequence in an agro-pastoral ecotone of Inner Mongolia, China. *Journal of Arid Land, 4,* 420–430. http://dx.doi.org/10.3724/SP.J.1227.2012.00420

Lal, R. (2003). Soil erosion and the global carbon budget. *Environment International, 29,* 437–450. http://dx.doi.org/10.1016/S0160-4120(02)00192-7

Lal, R. (2004). Soil carbon sequestration to mitigate climate change. *Geoderma, 123,* 1–22. http://dx.doi.org/10.1016/j.geoderma.2004.01.032

Morisada, K., Ono, K., & Kanomata, H. (2004). Organic carbon stocks in forest soils in Japan. *Geoderma, 119,* 23–32.

Melillo, J. M., Steudler, P. A., Aber, J. D., Newkirk, K., Lux, H., Bowles, F. P., & Morrisseau, S. (2002). Soil warming and carbon-cycle feedbacks to the climate system. *Science, 298,* 2173–2176. http://dx.doi.org/10.1126/science.1074153

Motsara, M. R., & Roy, R. N. (2008). *Guide to laboratory establishment for plant nutrient analysis, FAO fertilizer and plant nutrition bulletin 19.* Rome: FAO.

Natural Land & Water Resources Audit. (2008). *Soil conditions—Status of information for reporting against indicators under National Natural Resource Management Monitoring and Evaluation Framework.* National Land & Water Resources Audit (NLWRA). Braddon.

Nutsukpo, D. K., Jalloh, A., Zougmore, R., Nelson, G. C., & Thomas, T. S. (2012). *West African agriculture and climate change* (Chapter 6). Ghana: IFPRI/CGIAR. Retrieved from http://www.ifpri.org/sites/default/files/publications/rr178ch06.pdf

Pluske, W., Murphy, D., & Sheppard, J. (2014). *Fact sheets total organic carbon.* Retrieved April 25, 2014, from http://www.soilquality.org.au/factsheets/organic-carbon

Post, W. M., Izaurralde, R. C., Mann, L. K., & Bliss, N. (2001). Monitoring and verifying changes of organic carbon in soil. *Climatic Change, 51,* 471–495.

Schils, R., Kuikman, P., Liski, J., Van Oijen, M., Smith, P., Webb, J., ... Hiederer, R. (2008). *Review of existing information on the interrelations between soil and climate change* (Technical Report 2008-048). European Communities. ISBN: 978-92-79-20667-2. doi:10.2779/12723

Sheikh, A. M., Kumar, M., & Bussmann, R. (2009). Altitudinal variation in soil organic carbon stocks in coniferous subtropical and boradleaf temperate forests in Garhwal Himalaya. *Carbon Balance and Management, 4,* 1–6. doi:10.1186/1750-0680-4-6

Soil Science Society of America. (2001). *Soil Science Society of America: Carbon sequestration in soils: Position of the Soil Science Society of America.* SSSA Ad Hoc Committee S893 Report: USA.

Sombroek, W. G., Nachtergaeke, F. O., & Hebel, A. (1993). Amounts, dynamics and sequestrations of carbon in tropical and subtropical soils. *AMBIO, 22,* 417–426.

Strategic Environmental Assessment. (2010). *Kintampo Municipal Assembly, Strategic Environmental Assessment (SEA).* Ghana: Environmental Protection Agency (EPA).

Su, Z. Y., Xiong, Y. M., Zhu, J. Y., Ye, Y. C., & Ye, M. (2006). Soil organic carbon content and distribution in a small landscape of Dongguan, South China. *Pedosphere, 16,* 10–17. http://dx.doi.org/10.1016/S1002-0160(06)60020-9

Tan, Z., Tieszen, L. L., Tachie-Obeng, E., Liu, S., & Dieye, A. M. (2008). Historical and simulated ecosystem carbon dynamics in Ghana: Land use, management, and climate. *Biogeosciences Discussions, 5*, 2343–2368. http://dx.doi.org/10.5194/bgd-5-2343-2008

Tieszen, L. L. (2000). *Carbon sequestration in semi-arid and sub-humid Africa*. South Dakota: U.S. Geological Survey. Retrieved from http://edcintl.cr.usgs.gov/ip

United Nations Environment Programme. (2012). *New report underlines Africa's vulnerability to climate change*. Retrieved from http://www.unep.org/Documents.Multilingual/Default. asp?DocumentID=485&ArticleID=5409&l=en

United Nations Framework Convention on Climate Change. (2000). *United Nations convention on climate change*. Bonn. Retrieved from http://www.unfccc.de/resource/ process/components/response/*

Valipour, M. (2014). *Handbook of drainage engineering problems*. Foster City, CA: OMICS Group eBooks.

van Keulen, H. (2001). (Tropical) soil organic matter modelling: problems and prospects. *Nutrient Cycling in Agroecosystems, 61*, 33–39. http://dx.doi.org/10.1023/A:1013372318868

Verified Carbon Standards. (2014). *Adoption of sustainable agricultural land management* (Approved VCS Methodology VM0017, Version 1, Sectoral Scope 14). Retrieved from http://v-c-s.org

Wang, Z. M., Zhang, B., Song, K. S., Liu, D. W., & Ren, C. Y. (2010). Spatial variability of soil organic carbon under maize monoculture in the Song-Nen Plain, Northeast China. *Pedosphere, 20*, 80–89. http://dx.doi.org/10.1016/S1002-0160(09)60285-X

Woomer, P. L., Touré, A., & Sall, M. (2004). Carbon stocks in Senegal's Sahel transition zone. *Journal of Arid Environments, 59*, 499–510. http://dx.doi.org/10.1016/j.jaridenv.2004.03.027

World Bank Electronic Institute. (2014). *Carbon monitoring in CDM afforestation and reforestation projects*. The World Bank Institute. E-course. module 3.

Yao, M. K., Angui, P. K. T., Konate, S., Tondoh, J. E., Tano, Y., Abbadie, L., & Benest, D. (2010). Effects of land use types on soil organic carbon and nitrogen dynamics in mid-west Cote d'Ivoire. *European Journal of Scientific Research, 40*, 211–222.

7

Creation of subtropical greenhouse plan for the Flora Exhibition Grounds using GIS

Zdena Dobesova[1]*

*Corresponding author: Zdena Dobesova, Faculty of Science, Department of Geoinformatics, Palacký University, 17 listopadu 50, 771 46 Olomouc, Czech Republic
E-mail: zdena.dobesova@upol.cz
Reviewing editor:
Louis-Noel Moresi, University of Melbourne, Australia

Abstract: This paper describes a cartographical and Geographic Information System (GIS) work on a poster for the subtropical greenhouse. The subtropical greenhouse is part of a collection of greenhouses in Olomouc that are located in the centre of the city (Czech Republic) near Smetana Park. An overall plan exists for the collection of greenhouses and botanical gardens in the area. We have created a poster regarding the subtropical greenhouse plan. Partial plan for the subtropical greenhouse shows the detailed positions of approximately 120 plants. This plan is central information on the big presented poster (format A0). Plans arose from the cooperation of cartographers and botanists using GIS. All digital maps and plans were created using ArcGIS software after punctual measure in the field. Beside text information on the poster, there are pictures of selected plants. They are accompanied with maps of their native range. The maps of the native range are original cartographical part of the same authors. The poster also contains an old proposal for the subtropical greenhouse, which was created by a leading garden architect, I. Otruba, in 1991 before construction of the greenhouse. The book regarding species in subtropical greenhouses was issued in 2013. It contains descriptions of 33 select species. The exposition represents mainly Mediterranean flora. Each species is described with text and includes illustrations of fruits, leafs, flowers, and habitus.

ABOUT THE AUTHOR

She holds a PhD degree from Technical University of Ostrava, Faculty of Mining and Geology since 2007. Research interests are GIS, digital cartography, the visual programming language in GIS, scripting in Python for ArcGIS, spatial databases. She has several lectures for study branch Geoinformatics and Geography at Palacky University in Olomouc, Czech Republic. She is an author of 8 books and more than 70 articles from journals and conferences.

PUBLIC INTEREST STATEMENT

The design of greenhouse plan is a demonstration that the geoscience field has a wider application that only geography. The basic vector graphic was digitized using the geographical information systems. There are two most important parts of the area of greenhouses: the shapes of buildings and the positions and shapes of plants. Data could be used repetitively for different tasks. The article presents one printed output—big poster with the plan of the subtropical greenhouse. Other utilizations of the same data are also mentioned. One of them is the interactive web map of greenhouses with the detailed position of plants. Next utilization of the same data is on the plans in three botanical books about species in palm, tropical and subtropical greenhouses. The biggest advantage is using one source of digital vector data for several purposes. The results could be accomplished only by intensive cooperation specialists as geoinformatics and botanists.

Maps of the native ranges of each species are an original important part of the book. Greenhouses are open to the public, and students from Palacky University participate in botany and environmental lectures there. The visitors are tended to be intrigued by the astonishing variety of subtropical plants.

Subjects: Botany; Earth Sciences; Geography; Cartography

Keywords: GIS; biogeography; thematic map; native range; plan

1. Introduction

The collection greenhouses on the Exhibition Grounds are situated in Smetana Park, Olomouc. They have been declared to be cultural monuments. The greenhouses, together with the Palacký University botanical garden, are a part of the Union of Botanical Gardens of the Czech Republic (Figure 1).

The collection greenhouses consist of four separate greenhouses. The largest and oldest is the palm conservatory (Figure 2). The range and richness of the collections are among the largest and most interesting in the Czech Republic. The greenhouses occupy an area of approximately 4,100 m², of which 3,040 m² is open to the public. Every season, dozens of the most exotic species bud, bloom and spawn there. The greenhouses are open to visitors nearly year round.

The botanical garden belongs to Palacký University and neighbours the greenhouses. The garden area is approximately 5,600 m². The greenhouses and botanical garden create a unique area for spending time or learning, in addition to their classical function of conserving endangered species. Special botanical expositions are designed there as a part of the annual Flora Fairground.

The maintenance and education programs at the greenhouses and garden require the creation of detailed area plans and plant registries. Staff and students from the Department of Geoinformatics and the Department of Botany at Palacky University have created the BotanGIS information portal. Moreover, printed plans, posters and books have been produced. This article presents a poster that encompasses all of the interesting map documents regarding the subtropical greenhouse.

Figure 1. Location of Olomouc in the Czech Republic.

Figure 2. The entrance to collection greenhouses by architect J. Pelikan.

2. History and exposition of greenhouses

Before the current greenhouses, there was an orangery with beautiful wooden carvings. Its history dates back to 1886 when the wooden orangery structures were relocated from the castle park in Velká Bystřice to Smetana Park in Olomouc. It was built as the first palm greenhouse, with more than 90 types of tropical plants imported from Moravian castle greenhouses (Dančák, Šupová, Škardová, Dobesova, & Vávra, 2013b).

The conservatory expositions are divided into individual greenhouses according to the ecological requirements of particular plants, such as palm trees, cacti, tropical species and subtropical plants. Some species are represented by very old specimens, which have been grown over many years. The cacti and succulents collection is the most important and undoubtedly stands out nationwide. The subtropical crops, bromeliads and orchids collection is also of great importance. Tourists and professionals alike are attracted by the valuable cycad trees and Ceratozamia specimens, as well as many other dominant palm trees. Permanent specimens are tagged with a valid Latin name, family, distribution range and identification number (ID). This identifier can be used find information in the greenhouse plans and the BotanGIS database.

3. History and exposition of the subtropical greenhouse

The subtropical greenhouse is a typical 1980s construction. The original subtropical greenhouse project was created in 1991 by a leading garden architect, Prof. Ing. Ivar Otruba, CSc., who designed the exposition as an Israeli garden. The original plan is part of the presented poster.

The first plants (*Ceratonia siliqua*) were ceremonially planted in 1994 by the former Israeli ambassador in the Czech Republic. The current exhibition is the result of a systematic plan, which is conducted by gardeners who work for the Flora Exhibition Grounds Olomouc, JSC and cooperate with Palacký University, Olomouc. In 2012, the greenhouse received roof renovations (Dančák, Šupová, Škardová, Dobesova, & Vávra, 2013a).

The major plants in the composition come from subtropical zone across the world (e.g. the Mediterranean, Asia Minor and Australia), many of which are useful plants (Figure 3). Citrus fruits, which comprise a large portion of the exhibited plants, are spontaneous cultural hybrids of native species. In addition to commonly known species, such as mandarin, lemon and orange trees, less known citrus trees also exist, including *Citrus medica* whose fruit can grow to two kilograms and are used by the perfumery and confectionery industries.

The entrance is lined with Chinese plant *Poncirus trifoliate*, whose thorny shoots attract attention. This species is used as a rootstock for citruses. *Actinidia chinensis* is an interesting plant, with large,

Figure 3. Exposition in the subtropical greenhouse.

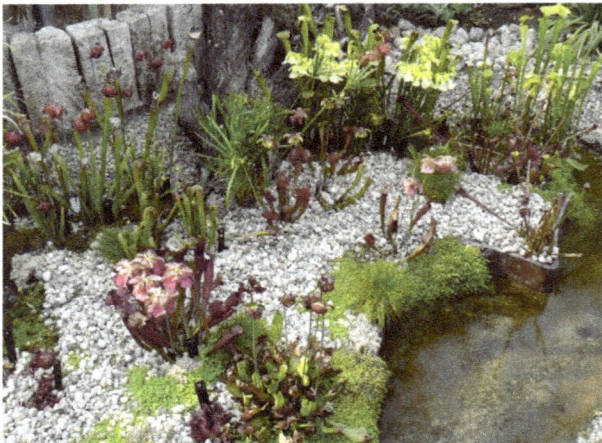

Figure 4. Carnivorous plants exposition.

felt-like leaves. Representatives of Mediterranean flora include fig trees (*Ficus carica*), olive trees (*Olea europaea*) and aromatic rosemary (*Rosmarinus officinalis*). Twigs from myrtle (*Myrtus communis*), a Mediterranean shrub, have white flowers and dark fruits, making them popular choices for wedding decorations. The pomegranate tree (*Punica granatum*) is a scarlet flowering species. In the back of the greenhouse, near the lake, a very popular seasonal exhibit, carnivorous plants, is managed by the Science Faculty and the Department of Botany (Figure 4).

Tomato-like fruits, known as kaki, grow on the Japanese persimmon tree (*Diospyros kaki*). The evergreen Australian tea trees (*Melaleuca*) represent another remarkable species, from which essential oils with unique bactericidal effects are extracted. The remainder of the greenhouse is lined with strawberry trees (*Psidium cattleianum*), which produce red-purple fruits with a strawberry flavour.

4. Methods for plan creation

Some examples of digital plans for parks, and arboretums using Geographic Information System (GIS) exist. GIS technology is helping community garden managers inventory, maintain, and manage their plant collections. Shields (2010) describes the Davis Arboretum of University of California, where GIS software catalogue and map more than 30,000 plants. Arnold Arboretum of Harvard University (2011) has web map application with point positions of plants above aerial photo. Plants are labelled by ID number or by scientific or common name. There is a possibility to switch between point symbols of plants according to family, country or plant size. The crown expression by polygons is not used. Alliance for Public Gardens GIS and firm Esri published Alliance for Public Gardens GIS (2011) as a tool for creating a public garden GIS.

Some map solutions are used only a simple flash-based tree map like Kew Royal Botanical Gardens: Kew Gardens Map (2017). There is only reduced selection of visitor interesting trees. Very seldom is used the GIS for greenhouses or small detail botanical area.

Unfortunately, detailed documentation of the greenhouse collections, including the plant origins and ages, does not exist. It was destroyed by a fire in the Archives of the Exhibition Grounds, which occurred in the early 1990s. The remaining documentation was completely destroyed during the 1997 floods.

The only remaining subtropical greenhouse plan that still exists is the plan proposed by I. Otruba. This plan is presented in the poster. It contains suggestions for pavements, bridges, lakes and major plants. The contours are also mapped in the plan. The elevation difference between the entrance and rear of the greenhouse is more than 1 m. This elevated arrangement creates a more attractive exhibit. The basic arrangement of the greenhouse has remained unchanged, including elevated soils, bridges, and the lakes.

The discussion of geoinformatics-cartographers and botanists started before the creation of plans. Some problems also arose during digital plan creation. These questions were solved. What data to measure and digitize firstly, buildings and plants? What level of detail? How express the plants, by point or polygons? What is a correct shape of the polygon for plants to select, schematic or punctual? How assure the visibility of all plants, especially small plants under big plants in the plan? How many data-sets prepare in various scales for printed plan, poster, and interactive map? Next text describes the final solutions and procedures.

A systematic gathering of detailed greenhouse plans began in 2011. First, the greenhouse and botanical garden buildings were measured and digitized using ArcGIS. This resulted in an orientation plan for the entire area, depicting all greenhouses and botanical gardens without position of plants (Figure 5). Some technical building parts as heating, electricity, etc. were omitted. The main purpose

Figure 5. Collection greenhouses and Palacký University botanical garden overview plan.

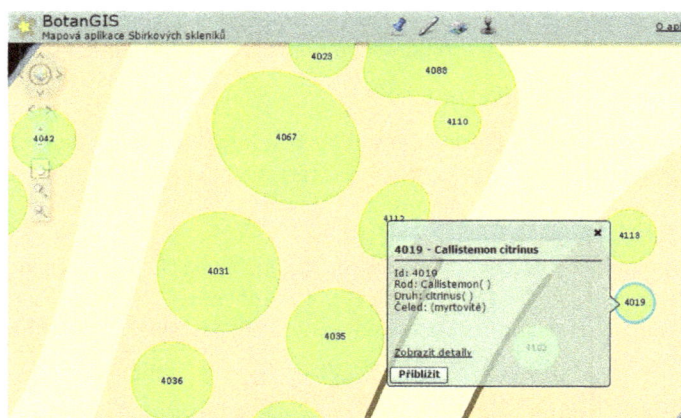

Figure 6. Interactive subtropical greenhouse plan via the portal BotanGIS with a plant callout (www.botangis.upol.cz).

of the mapping is to better imagine expositions and produce a recommended tour route for visitors. The red line denotes the recommended route, which is coloured to attract attention within the plan.

The greenhouse and botanical garden plans have similar map keys. The colour and symbol choices mainly fulfil the cartographic associativity rule. Cartographers recommend green colours for parks and greenhouse expositions. Dark green is used for six rockeries in the botanical garden. Yellow was used for footpaths in both areas. Plant beds were rendered in brown (Dobesova, Vavra, & Netek, 2013).

The format of the plan is A4. This plan is used as overview map in the presented poster. The print version is available for visitors at the entrance to the greenhouses (both in Czech language both English language).

The second part of the project collected data regarding plant position and crown size. The greenhouse plants are drawn as circles or ovals, expressing the actual plant cover of each specimen. The plant list was stored in the database. Each plant received a unique ID number that identifies the plant in the database and plan. The plan and database are both parts of the BotanGIS botanical information system (www.botangis.upol.cz). The plant database is linked to the interactive plans (Figure 6). It is possible to search for plants based on ID, genus, family and other descriptors (Nétek, Dobesova, & Vavra, 2014). The same digital data-sets were used for printed plan, poster (Appendix A) and interactive map. The differences in scales are not so big to generalize data and to create separate data sets for each scale and type of plan output. The reusability of vector data is a big advantage of creation and storing data in GIS.

Three books were issued in 2003 regarding palm, tropical and subtropical greenhouses (Dančák et al., 2013a, 2013b, 2013c). Select species are described in these books. Each book contains a detailed plan of each greenhouse.

The subtropical greenhouse book contains both the original plan from 1991 and the contemporary plan, with a complete list of plants (Dančák et al., 2013a). The book describes only 33 species. These interesting plants are denoted in the plan using a dark green colour, while others are light green. In addition, some species have more than one individual plant in the greenhouse. In the book, each species is described by three paragraphs: distribution and ecology, morphology description and interesting facts about the species. Species are illustrated using photos of fruits, leafs, flowers and habitus (Figure 7).

Moreover, each plant is depicted by a native species range map. The definition of the *native range of the species* term is based on each plant species evolving at a specific place on Earth, where it

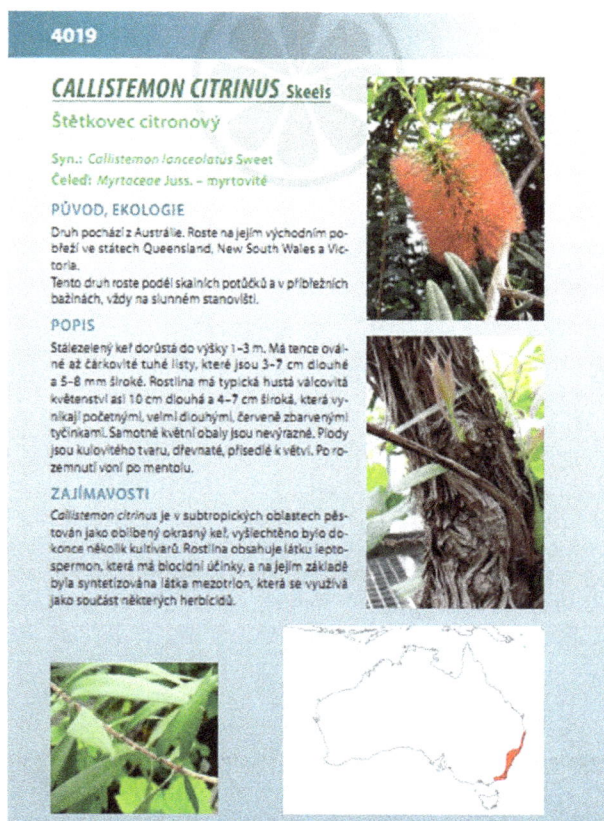

Figure 7. An example list of the Australian *Callistemon citrinus* species (Dančák et al., 2013a).

typically occurs at present. However, a large number of species have expanded into neighbouring and even distant areas. If they did so spontaneously, via natural processes and without human contributions, these areas are called *native ranges* (Pyšek et al., 2004; Smith, 1986; Webb, 1985).

All native range maps were created by cartographers at the Department of Geoinformatics, with help from botanists. A collection of 115 native range maps were created for the three above-mentioned greenhouse books (Dančák et al., 2013a, 2013b, 2013c). Some of them are original works. Native range maps are examples of chorochromatic maps. Chorochromatic maps only illustrate nominal data for specific areas using various colours (Kraak & Ormeling, 2003; Voženílek & Kaňok, 2011).

Experimental testing of visualization methods verified good understanding of chorochromatic maps (Pődör & Kiszely, 2014). Two types of regional delimitations were used for the native range maps. If areas were well known and detailed in the botanical literature, we used a red polygon with a detailed outline (e.g. *Actinidia chinensis* and *Arundo donax* on the poster). At times, the coastline region became discontinuous and comprised islands (Figure 8). Accurately drawing the regions proved difficult, with botanists verifying the final accuracy.

If a native range was approximated, only a red oval outline was used (e.g. *Punica granatum*, *Citrus madurensis*, *Citrus sinesis*, *Citrus limon* and *Lycianthes ranntonei* on the poster—Appendix A). Botanists were uncertain of all of the native ranges, especially for early domesticated crops. In these cases, they assumed or inferred the native ranges. Various visualization methods exist for approximating data (Brus, Vozenilek, & Popelka, 2013). Delimitation using smooth ovals is simple and comprehensive. Therefore, it is an effective generalization method.

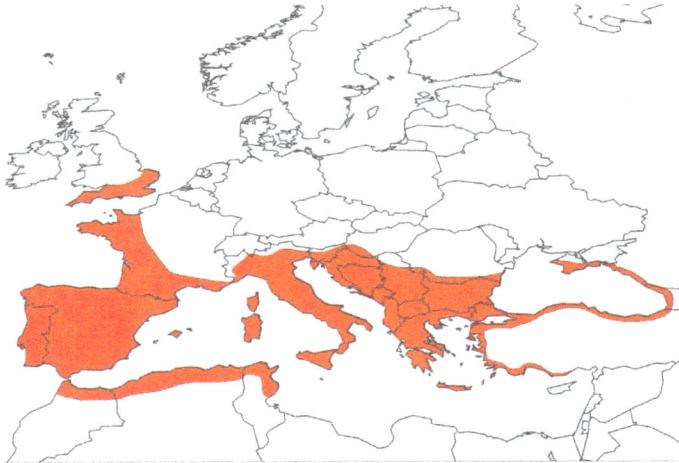

Figure 8. Europe native ranges for *Ruscus aculeatus* (Dančák et al., 2013a).

5. The contribution of cooperation

The presented utilization of GIS is a cross-disciplinary example of cooperation between botanists and geoinformatics. Chain of discussions and partial solutions of problems resulted in the series of useful map results. One of the solved problematic tasks were the recommendations and the decision how to correctly expressed suitable shape of crowns of plants from the botanical point of view. Mainly circles and ellipses were chosen. The jagged snake shapes were used only sporadically in case of three bushes. The smoothed shapes were preferred. Exact detail shapes are not necessary. The shapes could be various due to the pruning of bushes and trees in seasonal maintenance. The form of the crow expression also respects the botanic tradition, that uses only simple circles for crowns. From the cartographical point, the unsmoothed shapes also disturbed reading of plan and have non-aesthetic impression.

We also solved the expressing of cover by plant crowns to be all visible in plans. ArcGIS served very helpfully by the operation *Sort* for polygon shapes. Descending sort according to polygon area moved smaller shapes higher to be visible in the plan under bigger crowns. Without botanists were not possible to decide the correct set of important plants to be inserted to the plans. Some plants are seasonal, and they were not included in plans.

The outcome of GIS research is that data collecting could be used for mapping in very detailed scale, as the position of plants in greenhouses. The distances between plants were sometimes in centimeters in reality. The next outcome is the experience with the multiple utilizations of the same data. One source of digital vector data (shapefiles) was used for several outputs: interactive map on the web, orientation maps in greenhouses as big posters, and maps in printed books about planted species. Both administrators of greenhouses and visitors use all maps in everyday work. The map outputs are helpful also for botanical education that takes place in the area of greenhouses. All these useful experiences increase the skills and knowledge in using GIS in the fields of mapping botanical objects.

5.1. Software

The ArcGIS for Desktop 10.2 software, which is produced by Esri, was used to create the following plans: the subtropical greenhouse plan, the collection greenhouse area orientation plan and the native range maps. The AutoCAD Map 3D Raster Design 2015 software, which is produced by Autodesk, was used to clean and de-skew the scanned, original plan from 1991. This old plan was necessary to prepare before presentation on the poster. Some speckles and pencil notes were there. All these unwanted scraps were cleaned. The poster was designed and printed using AutoCAD Map 2015. This software was powerful to collect and maintained all partial plans, maps, photos and texts to one big informational poster.

5.2. Data

The source data were taken from the "Data and Maps for ArcGIS" data-set, which is available in ArcGIS (Esri, 2015). Topographic data for the continents and political boundaries were used as base layers for native range maps. The data about position and crowns of plants were measured manually in the field.

6. Conclusions

The creation of the BotanGIS information portal provided a digital data-set for the collection greenhouses and botanical garden in Olomouc. The set was used to create a series of maps with several purposes and scales. The overview map contains all of the greenhouses, buildings and botanical gardens, as well as the main tour route. The detailed plans for each greenhouse depict plant locations and crown shapes. The use of ArcGIS to store and illustrate the data allows for the publication of plans in different forms. The first form is the interactive BotanGIS portal. The second form is the plans printed in the three books. Finally, plans were used as important poster features (format A0).

The content of poster for subtropical greenhouse is depicted in this article and presented as Appendix A. Posters were placed near the greenhouse entrances. ArcGIS was also used to produce native range maps.

Funding
This work was supported by the European Social Fund (project CZ.1.07/2.3.00/20.0170) of the Ministry of Education, Youth, and Sports of the Czech Republic.

Author details
Zdena Dobesova[1]
E-mail: zdena.dobesova@upol.cz
ORCID ID: http://orcid.org/0000-0002-3989-5951
[1] Faculty of Science, Department of Geoinformatics, Palacký University, 17 listopadu 50, 771 46 Olomouc, Czech Republic.

References
Alliance for Public Gardens GIS. (2011). The ArcGIS public garden data model. Retrieved from http://publicgardensgis.ucdavis.edu/downloads/data-model/

Arnold Arboretum of Harvard University. (2011). Retrieved from http://arboretum.harvard.edu/explorer/

Brus, J., Vozenilek, V., & Popelka, S. (2013). An assessment of quantitative uncertainty visualization methods for interpolated meteorological data. In *Computational science and its applications – ICCSA, Lecture Notes in Computer Science* (vol. 7974, pp. 166–178). Ho Chi Minh City: Springer Berlin Heidelberg.

Dančák, M., Šupová, H., Škardová, P., Dobesova, Z., & Vávra, A. (2013a). *Zajímavé rostliny subtropického skleníku Výstaviště Flora Olomouc* [Interesting species in the subtropical greenhouse of flora Olomouc exhibition grounds] (48 p.). Olomouc: Palacký University. ISBN 978-80-244-3548-0.

Dančák, M., Šupová, H., Škardová, P., Dobesova, Z., & Vávra, A. (2013b). *Zajímavé rostliny palmového skleníku Výstaviště Flora Olomouc* [Interesting species in the palm greenhouse of flora Olomouc exhibition grounds] (65 p.). Olomouc: Palacký University. ISBN 978-80-244-3672-2.

Dančák, M., Šupová, H., Škardová, P., Dobesova, Z., & Vávra, A. (2013c). Zajímavé rostliny tropického skleníku Výstaviště Flora Olomouc [Interesting species in the tropical greenhouse of flora Olomouc exhibition grounds] (52 p.). Olomouc: Palacký University. ISBN 978-80-244-3885-6.

Dobesova, Z., Vavra, A., & Netek, R. (2013). Cartographic aspects of creation of plans for botanical garden and conservatories. In *SGEM 2013 13th International Multidisciplinary Scientific Geo Conference, Proceedings* (vol. I, pp.653–660). Sofia: STEF92 Technology Ltd. ISBN 978-954-91818-9-0. doi:10.5593/SGEM2013/BB2.V1/S11.006

Esri. (2015, June 6). Data and maps for ArcGIS. Retrieved from http://www.esri.com/data/data-maps/data-and-maps-dvd

Kew Royal Botanical Gardens: Kew Gardens Map. (2017). Retrieved from http://www.kew.org/visit-kew-gardens/plan-your-visit/map

Kraak, M. J., & Ormeling, F. (2003). *Cartography: Visualization of geospatial data* (2nd ed., 129 p.). Harlow: Prentice Hall.

Nétek, R., Dobesova, Z., & Vavra, A. (2014). Innovation of botany education by cloud-based geoinformatics system. In Q. Wang (Ed.), *Innovative use of online platforms for learning support and management, International Journal of Information Technology and Management* (Vol. 13, No. 1, pp. 15–31). Geneva: Inderscience Enterprises Ltd.. ISSN online: 1741-5179, ISSN print: 1461-4111. doi:10.1504/IJITM.2014.059149

Pödör, A., & Kiszely, M. (2014). Experimental investigation of visualization methods of earthquake catalogue maps. *Geodesy and Cartography, 40*, 156–162. doi:10.3846/20296991.2014.987451

Pyšek, P., Richardson, D. M., Rejmánek, M., Webster, G. L., Williamson, M., & Kirschner, J. (2004). Alien plants in checklists and floras: Towards better communication between taxonomists and ecologists. *Taxon, 53*, 131–143.

Shields, B. (2010). *Public gardens grow research capability with GIS, 2010.* Retrieved from http://www.esri.com/news/arcwatch/0810/uc-davis.html

Smith, P. M. (1986). Native or introduced? Problems in the taxonomy and plant geography of some widely introduced annual brome-grasses. *Proceedings of the Royal Society of Edinburgh. Section B. Biological Sciences, 89B*, 273–281.

Voženílek, V., & Kaňok, J. (2011). *Metody tematické kartografie: vizualizace prostorových jevů* [Methods of thematic Cartography: Visualisation of Spatial Phenomena] (216 p.). Olomouc: Palacky University. ISBN 978-80-244-2790-4.

Webb, D. A. (1985). What are the criteria for presuming native status? *Watsonia, 15*, 231–231.

8

Geospatial approach to study the spatial distribution of major soil nutrients in the Northern region of Ghana

Mary Antwi[1]*, Alfred Allan Duker[2], Mathias Fosu[3] and Robert Clement Abaidoo[4,5]

*Corresponding author: Mary Antwi, Department of Crop and Soil Sciences, Kwame Nkrumah University of Science and Technology, Kumasi, Ghana

E-mails: martwi2007@yahoo.com, maryamoah1982@gmail.com

Reviewing editor:
Saied Pirasteh, University of Waterloo, Canada

Abstract: Spatial distribution of soil nutrients is not normally considered for smallholder farms in Ghana resulting in blanket fertilizer application which leads to low efficiencies of some applied nutrients. This study focuses on applying geospatial analyses to map 120 maize farms in 16 districts of the Northern region of Ghana to identify nutrient distribution. Soil samples were taken from these 120 locations and analysed for contents of nitrogen (N), phosphorus (P) and potassium (K). Spatial models of the contents were generated through geostatistical analysis to map the status of N, P and K nutrients across the locations. Study results indicated that proportion of area deficient in N is 97%, P is 72% and K is 12%. Distribution pattern for N and K nutrients were clusters of low or high contents at specific locations; and that of P was random. Outcome of this study could enhance site-specific nutrient recommendation in Ghana.

Subjects: Agriculture; Soil Sciences; Statistics; Technology

Keywords: smallholder farmers; geospatial analysis; spatial distribution; major soil nutrients; soil fertility

1. Introduction

Agriculture is the main economic activity in the Northern region of Ghana. According to statistics provided by FAO (AQUASTAT, 2005), majority of the population in the region are smallholder farmers

ABOUT THE AUTHORS

The author's key research activities are focused on the use of Geographical Information Systems to study the spatial pattern within natural resources. These activities include modelling and mapping underlying spatial features and factors that contribute to the continuous degradation of these natural resources in order to make decisions that could control and monitor the resources. The author's research interest in the presented paper, therefore, falls in line with these activities and presents the models and maps of major soil nutrients that will aid appropriate fertilizer application at different locations according to their initial contents. The paper relates to the broader call on research to improve soil fertility and control the continuous soil degradation in Ghana and Sub-Saharan Africa due to the inappropriate fertilizer application by smallholder farmers resulting in low crop yields which also contributes to food insecurity in the region.

PUBLIC INTEREST STATEMENT

The article describes the pattern of distribution of soil nitrogen (N), phosphorus (P) and potassium (K) nutrients contents in 16 districts of the Northern region of Ghana. Although the region is noted for producing most of the major food crops in Ghana, this study showed that the soils in most of the districts have low levels of N, P and K nutrients. The article, therefore, provides maps of distribution of these major soil nutrients contents at different locations after assessing the extent of their levels in the study districts. The generated maps showed locations where the nutrients levels are very low, low or adequate to support the production of maize crops in the study area. This study is to aid decision-making during application of N, P and K fertilizers to the soil so that high efficiencies of applied fertilizers would be attained to give desired crop yields.

and about 80% of the land area is used for cropping. The soils, however, in these land areas of Sub-Saharan Africa are poor, especially in nutrient levels (Wairegi, 2011). Improper fertilizer rates application (Martey et al., 2014) and poor land management (Nkonya, 2004) contribute to the low contents of nutrients in these soils. This had resulted to persistently low crop yields. To increase the productivity of these soils for enhanced crop yields and increase the income of smallholder farmers, spatial distribution of the major soil nutrients needs to be mapped and based on the results of the distribution, appropriate fertilizer needs could be recommended for localised intervention. Since soil is not a renewable resource (Haghdar, Malakouti, Bybordi, & Ali, 2012), the concept of soil nutrient content assessment becomes very important for higher agricultural productivity and the economic development of every country.

In addition, adoption and implementation of soil fertility management concepts may vary, it is, therefore, necessary to assess the soil nutrient levels within locations where smallholder farmers are cropping. The mapped results of the soil nutrient assessment could then be used for effective monitoring of changes that might occur between cropping systems' and seasons' overtime. Monitoring of the nutrient levels will enable stakeholders to assess soil fertility improvement or otherwise in such localities. The tedious and costly conventional methods needed to obtain soil nutrient information will also be reduced when nutrient levels are mapped because those conventional methods are no more affordable (Behrens & Scholten, 2006). Accordingly, mapping of the nutrient levels will provide spatial soil nutrients information that can be used as a decision support tool. Hence, developing spatial distribution maps of soil nutrients is important in the "breadbasket" regions of which Northern region of Ghana is one (Adesina, 2009), since it will help refine agricultural management practices, improve sustainable resource use as well as provide a base against which future soil nutrients can be recommended at site-specific locations (Fairhurst, 2012; Reetz & Rund, 2004). Mapping of soil nutrient levels, especially nitrogen (N), phosphorus (P) and potassium (K) would also facilitate proper monitoring and review of recommended farming technologies at locations from time to time. This might also help in the evaluation of the impact of a particular technology at a particular time (e.g. every 10 years) in a particular location depending on assessment of the soil quality (Wang & Gong, 1998). In addition, due to the growing knowledge in precision agriculture (Buick, 1997; Ping, Wang, & Jin, 2009), researchers and decision-makers in soil science would be in a better position to implement location-based technologies, if approximate levels of soil nutrients in specific locations in the region are mapped. This approach will improve soil fertility management results and increase the interest of smallholder farmers to invest more in the agricultural sector.

The aims of this study were therefore (i) to generate appropriate nutrient models and parameters in order to produce a spatial distribution map of major soil nutrient contents across the Northern region of Ghana and (ii) to evaluate their pattern of distribution through spatial modelling of the N, P and K contents. The results of the study would, therefore, reveal the spatial variation and pattern of distribution of N, P and K nutrient contents across the study area and their evaluation would help to make appropriate site-specific nutrient analysis.

2. Materials and methods

2.1. Study area

The study was carried out in 16 out of 22 districts in the Northern region of Ghana (Figure 1), which is one of the regions classified as the "breadbasket" area of Ghana (Adesina, 2009). This is because most of the major food crops in the country are cultivated in this region and they include maize, rice, cowpea and yam. The region covers an area of about 70,384 km² and is the largest of the 10 regions in Ghana. The study districts, however, covered an area of approximately 40,000 km². It lies in a geographical location of latitudes N9° 30' and N10° 00'and longitudes W0° 51'and W1° 00' with a mean elevation of 149 m above sea level (Getamap, 2006). The mean annual rainfall of the area ranges from 750 to 1,050 mm and the mean temperature is 28°C which can fall as low as 14°C in the night of December/January and rise as high as 40°C during the day in February/March. The region is located in the Guinea savanna agro-ecological zone. Some of the major soils found in the region

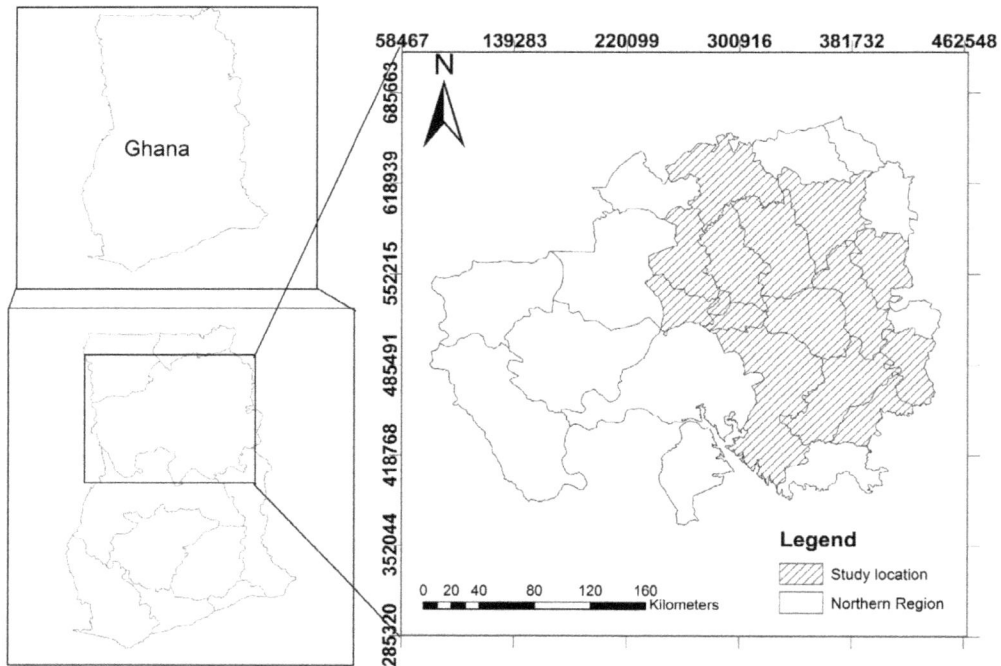

Figure 1. Map of study area.

include Lixisols, Luvisols, Acrisols and Gleysols (Dedzoe, Senayah, & Asiamah, 2001). The study, however, considered maize-cultivated fields since maize crop is one of the most cultivated cereals in the region and is regarded as one of the fundamental crops in the "food security equation" (Sauer, Hardwick, & Wobst, 2006).

2.2. Methods

The fields of eight maize smallholder farmers were chosen from each district that has data on particulars of farmers regarding their farming activities that have been recorded in the database provided by Savanna Agricultural Research Institute (SARI), as well as farmers who do not have their particulars included in the database. Each field was selected from different communities within a district to ensure uniformity of dispersion in the distribution of selected fields. Locations of selected maize fields, based on the objective of this study and for mapping purposes, were obtained with the Garmin GPS. Fifteen districts were found to have complete smallholder farmers' data on maize production from 2012 to 2014 cropping seasons in the database provided by SARI and theses districts were selected to be used in this study. A total of 120 locations were therefore obtained. A map showing the locations of the farms within the districts was produced (Figure 2).

2.2.1. Soil sampling and analysis

Soil samples were taken from each of these 120 locations previously cropped to maize, between May 2013 and March 2014 with the soil auger. Twenty cores of 0–20 cm depth of soil were taken and hand-mixed thoroughly in a bucket to homogenise the sample. A composite sample was then taken from the bulk to represent that location as shown in Figure 2. The soils were analysed for total N, available P and exchangeable K contents; total N was determined by Kjedahl's method, available P by Bray I method and exchangeable K by ammonium acetate extraction (Matula, 2009).

2.2.2. Statistical and geostatistical analysis of nutrient contents

The soil nutrient contents obtained through the laboratory analysis were subjected to Genstat (twelfth edition) statistical descriptive analysis. Some statistical parameters that were observed for the purpose of this study were the mean, standard deviation (SD), coefficient of variation (CV), skewness and kurtosis of the data distribution (Table 1).

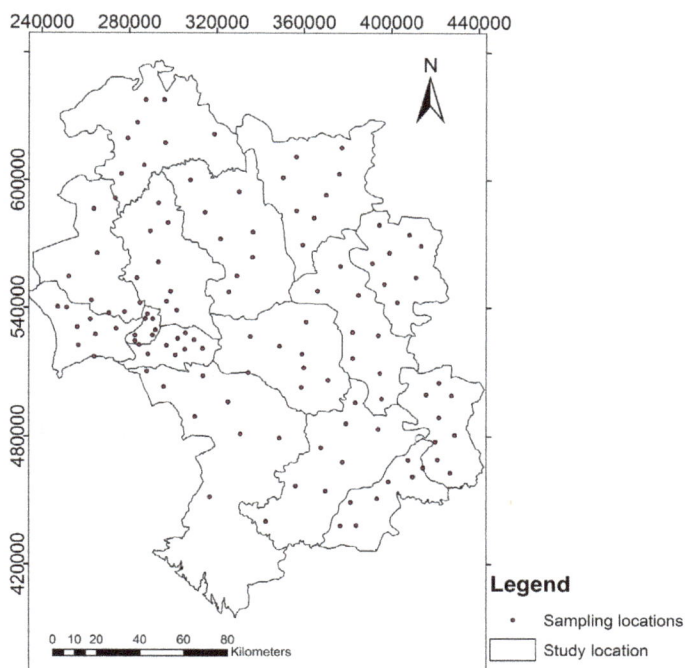

Figure 2. Map of soil sampling locations.

Geostatistical analysis that employs the use of simple point kriging and simulations (ESRI, 2010) was used to model the total N, available P and exchangeable K contents to produce the spatial distribution maps. The simple kriging model is expressed as follows:

$$Z(s) = \mu + \varepsilon(s) \tag{1}$$

where $Z(s)$ = the predicted value at the prediction location

μ = a known constant

$\varepsilon(s)$ = estimated error

Simple kriging assumes normality within the data before modelling. However, in this study, data exploration of the soil nutrient contents revealed that the levels were not normally distributed as shown by the elements of test of normality, skewness and kurtosis values generated by the data statistics (Table 1).

Table 1. Statistical parameters of major soil nutrient contents (*n = 120*)

Variable	Total N (%)	P (mg kg⁻¹)	K (cmol_c kg⁻¹)
S.D.	0.02	3.61	0.11
C.V. (%)	29.85	70.56	67.30
Mean	0.07	5.12	0.16
Minimum	0.03	1.24	0.06
Maximum	0.13	15.57	0.73
Skewness	0.11	0.99	2.53
Kurtosis	2.43	2.96	7.71

A data transformation was therefore applied to the original nutrient levels to render the contents normalised before modelling. A logarithmic transformation was applied to the total N and exchangeable K contents and a normal score transformation (ESRI, 2010; Harter, 1961; Royston, 1982) was applied to the available P contents.

The normalised soil nutrient contents were then analysed geostatistically by fitting different semi-variogram models iteratively, to measure the spatial variation within the soil nutrient contents (Liu et al., 2006; Matheron, 1963). In addition, the semi-variogram provided the necessary input parameters for spatial interpolation of kriging (Krige, 1951). ESRI (2010) ArcGIS defines the semi-variogram as follows:

$$\gamma(s_i, s_j) = 1/2 \, var(Z(s_i) - Z(s_j)) \tag{2}$$

where var is the variance, s_i and s_j are two different locations and; Z is the difference in their values.

The semi-variogram model that fitted the soil nutrients phenomena more accurately and provided the least root mean square error (RMSE) compared to others was selected for each of the nutrient levels. The RMSE was evaluated using the formula by Chai and Draxler (2014) as follows:

$$RMSE = \sqrt{\frac{1}{N} \sum_{i=1}^{N} (x_i - \overline{x_i})^2} \tag{3}$$

where N is the sample size, x_i is the observed value and $\bar{x}_i$ is the mean value for the observed sample values.

The parameters that were obtained from the semi-variogram were the sill, *nugget* and *range*. The sill represented the amount of variation defined by the spatial correlation structure and it is the value of the semi-variogram at which the model first levels out (given as partial sill plus the nugget). The range is the lag distance from where the model levels off and the nugget is the variability error (measurement error) obtained at shorter distances than the typical sampling interval (Bohling, 2005). The nugget-to-sill ratio was then used to classify the spatial dependence within the nutrient contents.

The prediction model for the soil nutrients was finally validated to measure the accuracy of the prediction map generated showing the distribution of the nutrients. The prediction model that gave the minimum average standard error (i.e. which fits the distribution more accurately) and the least RMSE was considered to be simulated as a measure of uncertainty within the predictions.

The simple kriged maps of the major soil nutrients across the study area were then simulated (generated from 10 realisations from different statistical parameters, i.e. mean, median, SD, upper value, lower value, first and second quartiles, minimum and maximum values and the percentile) to generate stochastic models of the surfaces since simple kriged maps produced smooth surfaces.

Spatial autocorrelation, using the Moran's index (Moran, 1948) was then calculated to assess the significance of the pattern of distribution within the nutrient contents. The calculation was done as follows:

$$I_{Nu} = \frac{n_{(loc)}}{S_0} \frac{\sum_{i=1}^{n} \sum_{j=1}^{n} (Nu_{(content)i} - \overline{Nu}_{(content)}) \times loc_{ij} \times (Nu_{(content)j} - \overline{Nu}_{(content)})}{\sum_{i=1}^{n} (Nu_{(content)i} - \overline{Nu}_{(content)})^2} \tag{4}$$

where $n_{(loc)}$ is the number of farm locations where soil samples were taken, loc_{ij} is the element in the spatial weights matrix corresponding to the pairs of locations i, j and $Nu_{(content)i}$ and $Nu_{(content)j}$ are nutrient contents in location i, j, respectively; and $\overline{Nu}_{(content)}$ are the mean nutrient content values.

The spatial weight matrix was generated for each of the nutrient contents and denoted by

$$S_0 = \sum_{i=1}^{n} \sum_{j=1}^{n} loc_{ij} \tag{5}$$

where loc_{ij} collectively defined the neighbourhood structure over the entire study location.

The z-score value for each of the nutrient contents and the Moran's index, I obtained from the spatial autocorrelation analysis were then used to specify the pattern of distribution that existed within the soil N, P and K contents. The probability value obtained was then used to assess the significance of the distribution, whether dispersed, clustered or of a random nature.

3. Results and discussion

3.1. Statistical description of major soil nutrient contents in the study area

Description of the untransformed N, P and K contents in the study area showed that their statistical distributions were all positively skewed (with skewness values of N = 0.11, P = 0.99 and K = 2.53; Figure 3). K contents had higher peak (Kurtosis = 7.71) with N and P contents showing near normal peaks (Kurtosis for N = 2.43; P = 2.96). For a standard normal distribution, skewness should be equal to zero and kurtosis equals to three as reported by Jondeau and Rockinger (2003). After data transformations, the N, P and K contents followed a near normal distribution which rendered the data values appropriate for modelling.

N and K contents followed log-normal distribution with positive skewness (see Table 1), an indication that large proportions of the study location have low to moderate concentrations (found within the range of 0.05–0.1% for N and 0.1 and 0.25 $cmol_c$ kg^{-1} for K; Table 1) whether clustered or random.

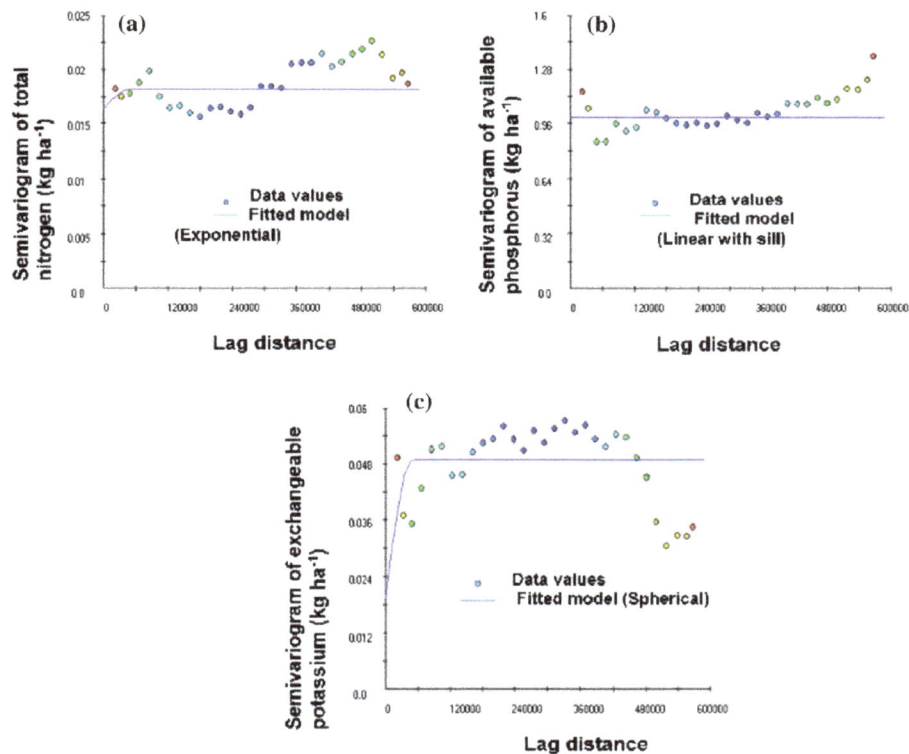

Figure 3. Fitted semi-variograms illustrating the strength of statistical correlation between major soil nutrients; (a) Nitrogen, (b) Phosphorus, and (c) Potassium in selected districts of the Northern region of Ghana.

Few locations recorded relatively high contents (i.e. N being more than 0.1% and K more than 0.25 $cmol_c$ kg^{-1}) as reported by Wopereis, Defoer, Idinoba, Diack, and Dugué (2009) (see Table 2).

The CV determined for the nutrient concentrations showed that N contents had high variations within its distribution (CV value 29.85%; see Table 1), which signifies relatively low dispersion across the districts; those for P and K concentrations were of very high values (CV values of 70.56 and 67.30%, respectively) indicating more dispersion in their distributions. High CV value implies that the data distribution is more variable (dispersed) and, hence, less stable and less uniform (nCalculators, 2013). However, the contents normalised after applying the data transformations.

The near normal distribution of K after log-transformation might be due to the fact that the difference between the minimum and maximum contents were not so large (0.06 and 0.73 $cmol_c$ kg^{-1}, respectively); as compared to that of P contents (Table 1). The fact that some locations recorded P contents as low as 1.24 mg kg^{-1} and others, as high as 15.57 mg kg^{-1} might account for the reason why P contents followed a normal score distribution.

The locations that recorded relatively high p-values were those locations where the smallholder farmers were incorporating about 3 t ha^{-1} of only cattle manure or including NPK (15:15:15) fertilizers plus sulphate of ammonia to the soil (according to the data provided by SARI and interviews with farmers). According to Zhang, Johnson, and Fram (2002), available P builds up in soil when animal manure is applied at a high rate to meet nitrogen requirements and that could have contributed to the high P contents in those locations. The differences in the spatial distribution of the soil nutrient concentrations across the region may thus be attributed to differences in nutrient management practices (Tsirulev, 2010), differences in soil forming processes, inherent heterogeneity in parent material at the different locations, as well as land use pattern and amount of fertilizer used (Liu et al., 2006) by the smallholder farmers. The distribution of the major nutrients confirms the assertion that spatial variation of soil nutrients exist even in neighbouring fields as has been previously reported by Goovaerts (1998), van der Zaag (2010) and Voortman, Brouwer, and Albersen (2002).

When the nutrient contents in the study area were compared with the soil fertility status table (Table 4) produced by Wopereis et al. (2009), only 3% of the locations had N contents within average range for maize production leaving about 97% of the locations having N contents in the soil below recommended average. 28% of the locations studied had levels of P within average contents, whilst 72% had P contents below average. Twelve per cent of the located areas had good K concentrations, 61% were within average, whilst 27% had K contents below average. The low to moderate contents might have resulted from continuous cropping of maize on the same piece of farmland (as confirmed by the farmers to be their practice), low rates of nutrient fertilizers applied at such locations (Mwangi, 1996), soil nutrient loss through soil erosion (Barrows & Kilmer, 1963) and export of nutrients through harvested produce (including straw and stover collection) from the farm (Doran, Wilhelm, & Power, 1984).

3.1. Models of the spatial dependence of major soil nutrients
The semi-variogram models that were used to derive parameters needed to explain the spatial dependence of the soil nutrient contents are presented in Figure 3. These models were obtained based on the semi-variogram model that presented the least RMSE as a measure of uncertainty as shown in Table 3.

Table 2. Nutrient concentration levels indicating soil fertility status for maize production at a depth of 0–20 cm (Wopereis et al., 2009)			
Nutrient level	**N (%)**	**P (mg kg^{-1})**	**K ($cmol_c$ kg^{-1})**
Good	>0.1	>25	>0.25
Adequate	−	6–25	0.10–0.25
Low	0.05–0.1	3–6	0.05–0.10
Very low	<0.05	<3	<0.05

In general, variables of nutrient contents that have a nugget-to-sill ratio less than 0.25 are regarded to have strong spatial dependence within them (i.e. spatial relationship that exists in variable pattern) (Cambardella et al., 1994; Liu et al., 2006). The spatial dependence is considered moderate if the ratio is between 0.25 and 0.75 and weak if it more than 0.75. The nugget-to-sill ratio obtained from the semi-variogram model (Figure 3) for N, P and K contents in the study were 0.64, 0.39 and 0.62, respectively, an indication that their spatial dependencies were moderate. The moderate spatial dependencies within the nutrient contents imply that the degree of association between the variables at different locations may increase as the distances become close to each. This suggests that there could be a possible continuity of the N, P and K variables exhibiting similarities in their values at shorter distances (less than 50 km as shown by the range distance in this study; Table 3). Smallholder farmers located within shorter distances are likely to adopt similar fertilizer management strategies regardless of their soil nutrient variations, which might affect N, P and K contents in the soils in a similar pattern (Adler, Raff, & Lauenroth, 2001). As distances increase, fertilizer management strategies may differ and the dependencies could become weaker or stronger depending on impact of the management (Jonsson & Moen, 1998). Therefore, as previously reported by Luo, Ding, Mi, Yu, and Wu (2009), Pringle, Doak, Brody, Jocqué, and Palmer (2010) soil fertility management should be consistent within patterns of spatial distribution of nutrient contents in the soil in order to manage the considerable variation of the nutrient contents in the study area.

Nitrogen and K nutrient contents recorded in the study area showed a positive low nugget, an indication that sampling error, random and other inherent variations that existed in the variables (Bohling, 2005; Clark, 2010; Liu et al., 2006) were minimal. Phosphorus contents, however, showed a high nugget effect indicating random and inherent variations within the variables. The considerable range of variations within the N, P and K contents might be caused by effects of variable farm level soil fertility management (Tittonell, Vanlauwe, Leffelaar, Rowe, & Giller, 2005; Trangmar, Yost, & Uehara, 1985) across considerable distances from the locations.

The prediction uncertainty generated by the cross-validation of the model were 0.02 kg ha^{-1} for N, 0.98 kg ha^{-1} for P and 0.11 kg ha^{-1} for K (Table 4). These values were less than 1 and hence considered appropriate for the model. The obtained root mean square standardised were also close to 1 for the N, P and K content variables suggesting that none of the variables were under-estimating or over-estimating the predictions as reported by (Hawkins & Sutton, 2011).

Table 3. Parameters for variogram model for major soil nutrients (NPK)					
Major soil nutrient	Model	RMSE*	Nugget	Partial sill	Range (m)
Total N (kg ha^{-1})	Spherical	0.00213	–[a]	–	–
	Exponential	0.00211	0.0159	0.009	50000
	Gaussian	0.00212	–	–	–
	Linear with sill	0.00212	–	–	–
Available P (kg ha^{-1})	Spherical	0.1120	–	–	–
	Exponential	0.1186	–	–	–
	Gaussian	0.1179	–	–	–
	Linear with sill	0.1120	0.3666	0.582	50000
Exchangeable K (kg ha^{-1})	Spherical	0.0083	0.0316	0.019	50000
	Exponential	0.0084	–	–	–
	Gaussian	0.0084	–	–	–
	Linear with sill	0.0084	–	–	–

*RMSE (root mean square error).

[a]Values that were not considered in the model.

Table 4. Measure of uncertainties in the prediction estimates of N, P and K contents in 15 districts of the Northern region of Ghana		
Transformed major soil nutrient	Average standard error	Root mean square standardised
Total nitrogen (kg ha⁻¹)	0.02	0.97
Available phosphorus (kg ha⁻¹)	0.98	0.97
Exchangeable potassium (kg ha⁻¹)	0.11	0.99

3.2. Spatial distribution and autocorrelation of major soil nutrients

The simulated maps from the mean values of the nutrient contents (generated from 10 realisations from different statistical parameters) are presented in Figures 4a–4c. The means were presented because according to simulation concepts by ESRI (2010), the means do not change over the spatial domain of the data. In addition, the mean has a Gaussian distribution around the true value, as stated by the central limit theorem (Engblom, Ferm, Hellander, & Lötstedt, 2009) and will therefore provide a better representation of the distribution.

The nutrients contents ranged from very low to adequate levels for maize cultivation in the study area. The differences in the variation within the distribution might be attributed to factors such the differences in elevation topography of the study area (McKenzie, 2013), soil pH that might influence nutrient levels (Wang, Bai, Huang, Deng, & Xiao, 2011) as well as different fertilizer application strategies (Bationo, Waswa, Okeyo, Maina, & Kihara, 2011; Zingore, Murwira, Delve, & Giller, 2007) as practiced by smallholder farmers in the different districts.

The spatial autocorrelation test that was done to test the significance of the distribution of the soil major nutrient contents are presented in Table 5. The hypothesis for the pattern analysis was that the nutrients levels across the study area were randomly distributed. In the theory of random patterns described by ESRI (2010), when p-value is very small (in this study $p < 0.05$) and z-value is either

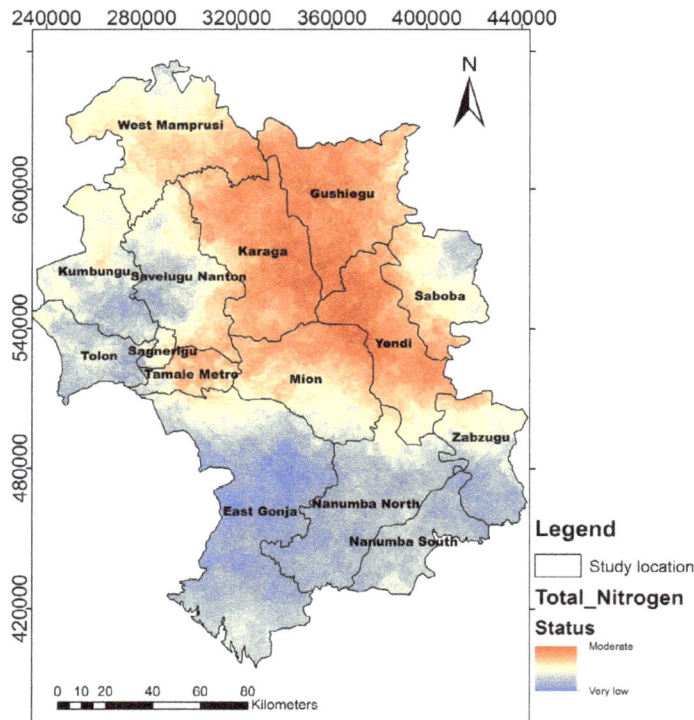

Figure 4a. Spatial distribution of total soil nitrogen contents in 16 districts within the Northern region of Ghana.

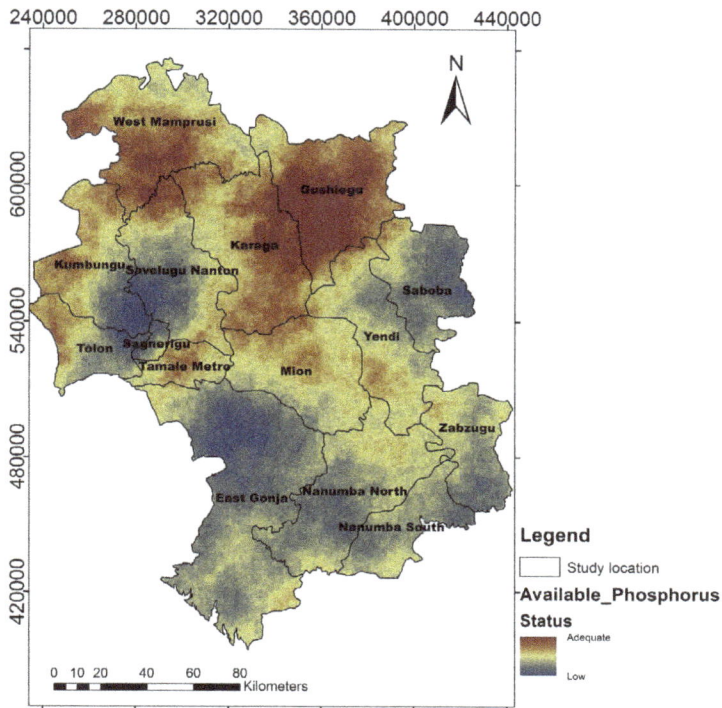

Figure 4b. Spatial distribution of available soil phosphorus contents in 16 districts within the Northern region of Ghana.

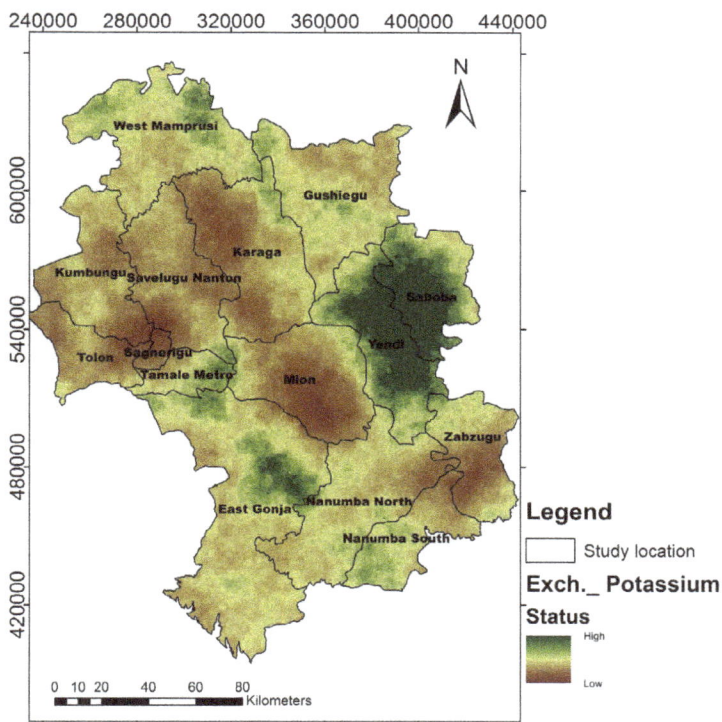

Figure 4c. Spatial distribution of exchangeable soil potassium contents in 16 districts within the Northern region of Ghana.

Table 5. Test of significance of pattern analysis for soil nutrient concentration; ($p < 0.05$) and ($1.96 < z < -1.96$)

	Nitrogen (N)	Phosphorus (P)	Potassium (K)
Moran's Index	0.28	0.04	0.27
Expected Index	−0.01	−0.01	−0.01
Variance	0.002	0.002	0.002
z-score	6.73	1.18	6.70
p-value	0.0003	0.24	0.0001

very high or very low ($1.96 < z < -1.96$), the spatial pattern is not likely to reflect a random form of distribution. In addition, a negative Moran's I index value indicates that the data are dispersed and a positive value indicates a tendency of clustering (clusters of high values only or low values only) at particular locations (Anselin, 1996). Test of significance for values returned by the analysis of the major soil nutrients indicated that N and K have clustered distributions in the study area (Table 5); with low levels clustered at one location and high levels at the other. On the other hand, the pattern of distribution of P did not appear to be significantly different from a random distribution.

Management strategies towards soil N, P and K nutrients enhancement could be implemented in the districts using the spatial distribution maps (Figures 4a–4c) as guide. Soil spatial distribution maps, therefore, provide a quick reference and reliable means by which variability within soil nutrients can be assessed to make decisions on fertilizer allocation at specific locations (Schnug, Panten, & Haneklaus, 1998).

4. Conclusion

Geospatial analysis of soil nutrient contents in the study area has proved to be essential in identifying locations in the Northern region of Ghana, where N, P and K levels are relatively low, moderate and high, respectively. Large proportions of the area recorded nutrient levels below average (N = 97%, P = 72% and K = 12%) which indicated that the study area has low nutrient levels. Models of the distribution maps suggest that N and K nutrients levels were clustered spatially and the distribution pattern of P in the study area was a random one. The soil nutrient information on low levels of N, P and K contents in the study area could be improved using the spatial distribution maps as a guide for fertilizer allocation and management taking into account the pattern within the distribution. The spatial distribution maps generated through this study therefore provided foreknowledge of the N, P and K nutrients status in the districts which could be used by research scientists as bases for fertilizer recommendations. When these considerations are made, proper site-specific nutrient recommendations could be promoted in order to increase soil nutrient fertility in the region.

Acknowledgement
Support: Secondary data were obtained from the Savanna Agricultural research institute (SARI).

Funding
The work was supported by the Alliance for Green Revolution in Africa (AGRA) project.

Author details
Mary Antwi[1]
E-mails: martwi2007@yahoo.com, maryamoah1982@gmail.com
ORCID ID: http://orcid.org/0000-0002-6226-5464
Alfred Allan Duker[2]
E-mails: duker@itc.nl, duker@alumni.itc.nl
Mathias Fosu[3]
E-mail: mathiasfosu@yahoo.co.uk
Robert Clement Abaidoo[4,5]
E-mail: abaidoorc@yahoo.com

[1] Department of Crop and Soil Sciences, Kwame Nkrumah University of Science and Technology, Kumasi, Ghana.
[2] Department of Geomatic Engineering, Kwame Nkrumah University of Science and Technology, Kumasi, Ghana.
[3] Council for Scientific and Industrial Research, Savanna Agriculture Research Institute, Tamale, Ghana.
[4] College of Agriculture and Natural Resources, Kwame Nkrumah University of Science and Technology, Kumasi, Ghana.
[5] International Institute of Tropical Agriculture, Ibadan, Nigeria.

References

Adesina, A. (2009). *Taking advantage of science and partnerships to unlock growth in Africa's breadbaskets*. (A. Adesina, Vice President, Speaker). Wageningen, The Netherlands: Alliance for a Green Revolution in Africa (AGRA) at the Science Forum. Retrieved November 10, 2015, from http://www.scienceforum2009.nl/Portals/11/Adesina-pres.pdf

Adler, P., Raff, D., & Lauenroth, W. (2001). The effect of grazing on the spatial heterogeneity of vegetation. *Oecologia, 128*, 465–479.

Anselin, L. (1996). The Moran scatterplot as an ESDA tool to assess local instability in spatial association. *Spatial Analytical Perspectives on GIS, 111*, 111–125.

AQUASTAT. (2005). *Food and agriculture organization's information system on water and agriculture* Retrieved June 7, 2013, from http://www.fao.org/nr/water/aquastat/countries_regions/GHA/index.stm

Barrows, H. L., & Kilmer, V. J. (1963). Plant nutrient losses from soils by water erosion. *Advanced Agronomy, 15*, 303–316. http://dx.doi.org/10.1016/S0065-2113(08)60401-0

Bationo, A., Waswa, B., Okeyo, J. M., Maina, F., & Kihara, J. M. (2011). *Innovations as key to the green revolution in Africa: Exploring the scientific facts*. Berlin: Springer Science & Business Media. http://dx.doi.org/10.1007/978-90-481-2543-2

Behrens, T., & Scholten, T. (2006). Digital soil mapping in Germany—A review. *Journal of Plant Nutrition and Soil Science, 169*, 434–443. http://dx.doi.org/10.1002/(ISSN)1522-2624

Bohling, G. (2005). Introduction to geostatistics and variogram analysis. *Kansas Geological Survey*, 1–20. Retrieved from http://people.ku.edu/~gbohling/cpe940

Buick, R. D. (1997). *Precision Agriculture: An integration of information technologies with farming*. Paper presented at the 50th N.Z. Plant Protection Conference, New Zealand.

Cambardella, C. A., Moorman, T. B., Novak, J. M., Parkin, T. B., Karlen, D. L., Turco, R. F., ... Konopka, A. E. (1994). Field-scale variability of soil properties in Central Iowa soils. *Soil Science Society of America Journal, 58*, 1501–1511. http://dx.doi.org/10.2136/sssaj1994.03615995005800050033x

Chai, T., & Draxler, R. R. (2014). Root mean square error (RMSE) or mean absolute error (MAE)?–Arguments against avoiding RMSE in the literature. *Geoscientific Model Development, 7*, 1247–1250. http://dx.doi.org/10.5194/gmd-7-1247-2014

Clark, I. (2010). Statistics or geostatistics? Sampling error or nugget effect? *The Journal of the South African Institute of Mining and Metallurgy, 110*, 307–312.

Dedzoe, C., Senayah, J., & Asiamah, R. (2001). Suitable agro-ecologies for cashew (Anacardium occidetale L) production in Ghana. *West African Journal of Applied Ecology, 2*, 103–115.

Doran, J. W., Wilhelm, W. W., & Power, J. F. (1984). Crop residue removal and soil productivity with no-till corn, sorghum, and soybean1. *Soil Science Society of America Journal, 48*, 640–645. doi:10.2136/sssaj1984.03615995004800030034x

Engblom, S., Ferm, L., Hellander, A., & Lötstedt, P. (2009). Simulation of stochastic reaction-diffusion processes on unstructured meshes. *SIAM Journal on Scientific Computing, 31*, 1774–1797. http://dx.doi.org/10.1137/080721388

ESRI. (2010). *The Principles of Geostatistical Analysis (3)*. Retrieved January 20, 2015, from http://maps.unomaha.edu/Peterson/gisII/ESRImanuals/Ch3_Principles.pdf

Fairhurst, T. (2012). *Handbook for integrated soil fertility management*. Nairobi: Africa Soil Health Consortium.

Getamap. (2006). *Northern Region/Ghana*. Retrieved October 22, 2013, from www.getamap.net

Goovaerts, P. (1998). Geostatistical tools for characterizing the spatial variability of microbiological and physico-chemical soil properties. *Biology and Fertility of Soils, 27*, 315–334. doi:10.1007/s003740050439

Haghdar, A., Malakouti, J. Mohammad, Bybordi, A., & Ali, K. (2012). Study on spatial variation of some chemical characteristics of dominant soil series using geostatistics in Iran: Case study of Heris region. *Journal of Food, Agriculture & Environment, 10*, 977–982.

Harter, H. L. (1961). Expected values of normal order statistics. *Biometrika, 48*, 151–165. doi:10.2307/2333139

Hawkins, E., & Sutton, R. (2011). The potential to narrow uncertainty in projections of regional precipitation change. *Climate Dynamics, 37*, 407–418. doi:10.1007/s00382-010-0810-6

Jondeau, E., & Rockinger, M. (2003). Conditional volatility, skewness, and kurtosis: Existence, persistence, and comovements. *Journal of Economic Dynamics and Control, 27*, 1699–1737. doi:10.1016/S0165-1889(02)00079-9

Jonsson, B. G., & Moen, J. (1998). Patterns in species associations in plant communities: The importance of scale. *Journal of Vegetation Science*, 327–332. http://dx.doi.org/10.2307/3237097

Krige, D. G. (1951). A statistical approach to some basic mine valuation problems on the Witwatersrand. *Journal of Chemical and Metallurgical Mining Society of South Africa, 52*, 119–139.

Liu, D., Wang, Z., Zhang, B., Song, K., Li, X., Li, J., ... Duan, H. (2006). Spatial distribution of soil organic carbon and analysis of related factors in croplands of the black soil region, Northeast China. *Agriculture, Ecosystems and Environment, 113*, 73–81. http://dx.doi.org/10.1016/j.agee.2005.09.006

Luo, Z., Ding, B., Mi, X., Yu, J., & Wu, Y. (2009). Distribution patterns of tree species in an evergreen broadleaved forest in eastern China. *Frontiers of Biology in China, 4*, 531–538. http://dx.doi.org/10.1007/s11515-009-0043-4

Martey, E., Wiredu, A. N., Etwire, P. M., Fosu, M., Buah, S. S. J., Bidzakin, J., ... Kusi, F. (2014). Fertilizer adoption and use intensity among smallholder farmers in Northern Ghana: A case study of the AGRA soil health project. *Sustainable Agriculture Research, 3*, 24–36. doi:0.5539/sar.v3n1p24.

Matheron, G. (1963). Principles of geostatistics. *Economics Geology, 58*, 1246–1266.

Matula, J. (2009). A relationship between multi-nutrient soil tests (Mehlich 3, ammonium acetate, and water extraction) and bioavailability of nutrients from soils for barley. *Plant Soil Environment, 55*, 173–180.

McKenzie, R. H. (2013). Understanding soil variability to utilize variable rate fertilizer technology. *Government of Alberta, Agriculture and Rural Development*. Retrieved from http://www.farmingsmarter.com/understanding-soil-variability/

Moran, P. A. (1948). The interpretation of statistical maps. *Journal of the Royal Statistical Society. Series B (Methodological), 10*, 243–251.

Mwangi, W. M. (1996). Low use of fertilizers and low productivity in sub-Saharan Africa. *Nutrient Cycling in Agroecosystems, 47*, 135–147. http://dx.doi.org/10.1007/BF01991545

nCalculators. (2013). *Coefficient of variation*. Retrieved Novembr 07, 2013, from http://ncalculators.com/math-worksheets/coefficient-variation-example.htm

Nkonya, E. (2004). *Strategies for sustainable land management and poverty reduction in Uganda* (Vol. 133, pp. 1–3). Washington, DC: International Food Policy Research Institute.

Ping, H. E., Wang, H., & Jin, J. (2009, November 5–7). *GIS based soil fertility mapping for SSNM at village level in China*. This presentation was made at the IPI-OUAT-IPNI International Symposium, OUAT, Bhubaneswar. *The role and benefits of potassium in improving nutrient management for food production, quality and reduced environmental damage*. Beijing: International Plant Nutrition Institute.

Pringle, R. M., Doak, D. F., Brody, A. K., Jocqué, R., & Palmer, T. M. (2010). Spatial pattern enhances ecosystem functioning in an African Savanna. *PLoS Biology, 8*, e1000377. doi:10.1371/journal.pbio.1000377

Reetz, H. F., & Rund, Q. B. (2004). *GIS in nutrient management– A 21st century paradigm shift.* Paper presented at the Proceedings of ESRI International User Conference, 24th, San Diego, CA 1328.

Royston, J. P. (1982). Algorithm AS 177: Expected normal order statistics (exact and approximate). *Journal of the Royal Statistical Society. Series C (Applied Statistics), 31*, 161–165. doi:10.2307/2347982

Sauer, J., Hardwick, T., & Wobst, P. (2006). Alternate soil fertility management options in Malawi: An economic analysis. *Journal of Sustainable Agriculture, 29*, 29–53.

Schnug, E., Panten, K., & Haneklaus, S. (1998). Sampling and nutrient recommendations-the future. *Communications in Soil Science & Plant Analysis, 29*, 1455–1462.

Tittonell, P., Vanlauwe, B., Leffelaar, P., Rowe, E. C., & Giller, K. E. (2005). Exploring diversity in soil fertility management of smallholder farms in western Kenya: I. Heterogeneity at region and farm scale. *Agriculture, Ecosystems & Environment, 110*, 149–165.

Trangmar, B. B., Yost, R. S., & Uehara, G. (1985). Application of geostatistics to spatial studies of soil properties. *Advances in agronomy, 38*, 45–94.

Tsirulev, A. (2010). Spatial variability of soil fertility parameters and efficiency of variable rate fertilizer application in the Trans-Volga Samara region. *Better Crops, 94*, 26–28.

van der Zaag, P. (2010). Viewpoint—Water variability, soil nutrient heterogeneity and market volatility—Why Sub-Saharan Africa's green revolution will be location-specific and knowledge-intensive. *Water Alternatives, 3*, 154–160.

Voortman, R. L., Brouwer, J., & Albersen, P. J. (2002). *Characterization of spatial soil variability and its effect on Millet Yield on Sudano-Sahelian coversands in SW Niger* (S. W. P. Centre for World Food Studies). *Stichting Onderzoek Wereldvoedselvoorziening van de Vrije Universiteit.*

Wairegi, L. (2011, May 25–26). *Framework for decision support tools for integrated soil fertility management in sub-saharan Africa (Draft).* Paper presented at the African Soil Health Consortium inaugral workshop.

Wang, Q., Bai, J., Huang, L., Deng, W., & Xiao, R. (2011). Soil nutrient distribution in two typical paddy terrace wetlands along an elevation gradient during the fallow period. *Journal of Mountain Science, 8*, 476–483. http://dx.doi.org/10.1007/s11629-011-1122-y

Wang, X., & Gong, Z. (1998). Assessment and analysis of soil quality changes after eleven years of reclamation in subtropical China. *Geoderma, 81*, 339–355. doi:10.1016/S0016-7061(97)00109-2

Wopereis, M. C. S., Defoer, T., Idinoba, P., Diack, S., & Dugué, M. J. (2009). Integrated soil fertility management. *Participatory learning and action research (PLAR) for integrated rice management (IRM) in Inland Valleys of Sub-Saharan Africa:Technical manual.* Cotonou, Benin: Africa Rice Centre.

Zhang, H., Johnson, G. V., & Fram, M. (2002). *Managing phosphorus from animal manure.* Stillwater: Division of Agricultural Sciences and Natural Resources, Oklahoma State University.

Zingore, S., Murwira, H. K., Delve, R. J., & Giller, K. E. (2007). Influence of nutrient management strategies on variability of soil fertility, crop yields and nutrient balances on smallholder farms in Zimbabwe. *Agriculture, Ecosystems & Environment, 119*, 112–126. doi:10.1016/j.agee.2006.06.019

9

Remote sensing of vegetation cover changes in the humid tropical rainforests of Southeastern Nigeria

Friday Uchenna Ochege[1,2]* and Chukwunonyelum Okpala-Okaka[2]
*Corresponding author: Friday Uchenna Ochege, Laboratory for Cartography and GIS, Department of Geography & Environmental Management, University of Port Harcourt, Choba, Rivers State, Nigeria; Faculty of Environmental Studies, Department of Surveying and Geoinformatics, University of Nigeria, Enugu, Nigeria
E-mail: uchenna.ochege@uniport.edu.ng
Reviewing editor:
Louis-Noel Moresi, University of Melbourne, Australia

Abstract: This study demonstrates a 30-year multi-temporal variations in vegetation cover changes as a means of filling the vegetation knowledge gap in the humid tropical forests of southeastern Nigeria. Landsats 4TM, 5TM and 7ETM+ data-sets were accessed and analysed using the Maximum Likelihood Classification algorithm to discriminate and geovisualize the spatiotemporal variations in the general vegetation and other land cover types, from 1984 to 2014. This was supported with detailed field surveys in dry and rainy seasons of 2011 and 2014 to ascertain the status of wide-ranging vegetation cover stands. A 44% vegetation decline was recorded given the reduction in dense vegetation spatial extent from 330.63 km^2 in 1984 to 170.87 km^2 in 2014. Sparse vegetation equally increased in spatial extent by 25% given the variations registered from 6.86 km^2 in 1984 to 97.16 km^2 in 2014. The reduction in vegetation cover was found to have been replaced by increase in other land cover types—residential (18.97 km^2) and industrial areas (39.87 km^2). Suggesting that, heterogeneity in the spatial distribution of land resources, in addition to weak concerns towards preserving the accruing benefits of vegetation resources attracted anthropogenic phenomenon (e.g. urbanization) to vegetated

ABOUT THE AUTHORS

Friday Uchenna Ochege is a PhD candidate with special interest in the application of Cartographic, GIS and Remote Sensing techniques to environment related issues. He has been involved in internationally and nationally funded research projects. We are currently working on the application of geospatial techniques to modeling climate change impacts on agriculture and watershed resilience and resilient landscapes in the Humid Tropics of Southeastern Nigeria. We aim to develop frameworks for non-technical policy makers that will guide in sustainable land use.

Chukwunonyelum Okpala-Okaka is a professor in the Department of Surveying and Geoinformatics, Faculty of Environmental studies at the University of Nigeria. His area of research interest is Cartography, Remote Sensing and GIS modelling. He currently teaches both undergraduate and postgraduate courses in Cartography and GIS and its environmental applications, within Nigeria. He has over 29 years of teaching experience with several scientific publications to his credit.

PUBLIC INTEREST STATEMENT

This study provides baseline information on the spatial variations and temporal patterns of vegetation cover and other land-use changes in the humid tropical rainforest of Nigeria using satellite data-sets. Key factors responsible for the changes have been identified and documented. The validated results helped to reveal interclass confusion especially as earlier report assert that the region lacks adequate monitoring framework and information on vegetation cover changes. With this research, appropriate place-specific reinvestigations and or decisions can be made, such as rapid identification of alternatives where necessary, and to incorporate same in the formulation of sound policies required in the sustainable governance and management of natural resources in the humid tropical rainforest.

areas. As such, strengthening institutional monitoring and urban planning frameworks would help to improve sustainable governance of the tropical rainforests.

Subjects: Vegetation; Environmental Studies & Management; GIS, Remote Sensing & Cartography

Keywords: remote sensing; vegetation cover changes; tropical rainforest; Southeastern Nigeria

1. Introduction

Vegetation is the general plant life or the total plant cover forming parts of the biological system and it is the primary producer of any ecosystem (Ochege, 2014b). Local and regional wide-ranging natural vegetations are important component on earth and they govern all forms of life (Millennium Ecosystem Assessment, 2005). They provide food, oxygen, fertility and finally enhance survival of all living beings (Matlack, 1994; Mckey, Waterman, Gartlan, & Struhsaker, 1978; Newbold et al., 2014). For the earth's environment, natural vegetation constitutes the biologically richest ecosystems and play vital roles in regional hydrology, carbon storage and the global climate dynamics (Du et al., 2015; Forkel et al., 2013; Igbawua, Zhang, Chang, & Yao, 2016).

The benefits accruing to humankind from the natural functioning of a healthy and productive vegetation system cannot be over emphasized. Because, the natural vegetation provides not only the basic needs of life (i.e. food, clothing and shelter), but enables purification of water bodies (Friberg et al., 2011), manages diseases (Fisher & Turner, 2008), regulates climate and the functioning of biosphere (Millennium Ecosystem Assessment, 2005) and provides mankind with spiritual fulfilment that contributes to improving quality of life (Food & Agriculture Organisation, 2013). As such, vegetation remains the lifeblood of human societies around the world (United Nations Environment Programme & the International Institute for Sustainable Development, 2004).

Nevertheless, human being has long distinguished himself from other species by shaping ecosystem forms and processes using fire, tools and technologies that are beyond the capacity of other organisms including natural vegetation stands (Smith, 2007). As a result, man is inextricably and unavoidably attached to the environment, which often time gives him the nerve to impact negatively on forests and natural vegetations resulting in other land uses.

Based on the considerations of the Intergovernmental Panel on Climate Change (IPCC), land use and land cover are not technically synonymous. Land cover is the observed physical and biological cover of the Earth's land, such as vegetation or man-made features, whilst the total arrangements, activities and inputs undertaken in a certain land cover type (e.g. a set of human actions) for both social and economic purposes (e.g. grazing, timber extraction, conservation) are referred to as land use (Watson et al., 2000). Since vegetation is continuously changing—alteration in the surface components of the vegetation cover, the rate of change can either be dramatic and or abrupt—as exemplified by fire, subtle and or gradual (Lund, 1983; Milne, 1988). This is a common global scenario, especially, with regard to forest cover and biomass accumulation or mass deforestation (Hansen et al., 2013).

Africa has been observed to be experiencing the fastest rate of vegetation change and most of its segments are already evidently been impacted and plagued with diverse ecological problems (Erika et al., 2015; Ofori, Owusu, & Attuquayefio, 2014; Yeshaneh, Wagner, Exner-Kittridge, Legesse, & Blöschl, 2013). Especially, as land use remains a significant driver of habitat degradation and removal and associated losses in biodiversity and vegetation resources (Lucas et al., 2015). According to Odjugo and Ikhuoria (2003), these problems, though anthropogenic in context, do have a direct link with the ongoing global climatic and environmental change. Nevertheless, natural vegetation equally suffers from these consequences, as they are profoundly been altered by human activities, and so, few natural stands remain (FAO, 2012).

In Nigeria, a wide range of vegetation types exist, and they reflect past and present climatic variations (Federal Republic of Nigeria, 1977; Igbawua et al., 2016; Odjugo, 2010). Generally, the southern part of the country is flanked by maritime stretches where the sandy uniformity is occasionally broken by mangrove ecosystem, bushes, lush vegetation, hardy trees and putrid water, while the northern part displays segments of croplands and rangelands that are heavily populated by grasses of varied species (Igbozurike, 1975). The global land cover map by DIVA-GIS (2016) reveals Nigeria's current vegetation cover belts (see Figure 1(A)). From top to bottom indicate Sahel Savannah, Sudan Savannah, Guinea Savannah and Forest vegetation (Igbawua et al., 2016).

Based on the obvious reasons of regional vegetation dynamics which has been influenced by global environmental change (Igbawua et al., 2016), the natural vegetation of Nigeria has ever since been under considerable threat like those of most other parts of tropical Africa (Adekunle, Olagoke, & Ogundare, 2013; FAO, 2012; Federal Republic of Nigeria, 1977). The World Resources Institute (WRI) did identify Nigeria's humid tropical rainforests as one of the most ecologically vibrant places on the planet because it is home to over 4,850 different plants and tree species, 1,340 species of animals, among which are 274 mammals, 860 birds (Meduna, Ogunjimi, & Onadeko, 2009; World resources, 1987). Despite these rich and abundant vegetation resources, the country is highly rated for unsustainable exploitations, deforestation and forest degradation among others (Momoh, 2014). Yet, the country lack adequate monitoring framework (Erika et al., 2015) especially, in those steadily urbanizing locations of the humid tropical rainforest belts (Ofori et al., 2014) including southeastern Nigeria i.e. Umuahia.

Prior to Umuahia been declared a headquarter city in Nigeria by the national government on the 27 August 1991, Umuahia was previously known to be home to ecological parks, protected forests, unprotected and undeveloped forest areas and trees growing amidst scattered residential settlements. These natural green covers provide different benefits, ecosystem services and support to well-being, ecosystem health, urban livelihoods and other advantages to the society at large. Notwithstanding these recognized benefits to sustainable development and environmental sustainability to the area, spatial quantification of local vegetation changes has been lacking for this part of Nigeria.

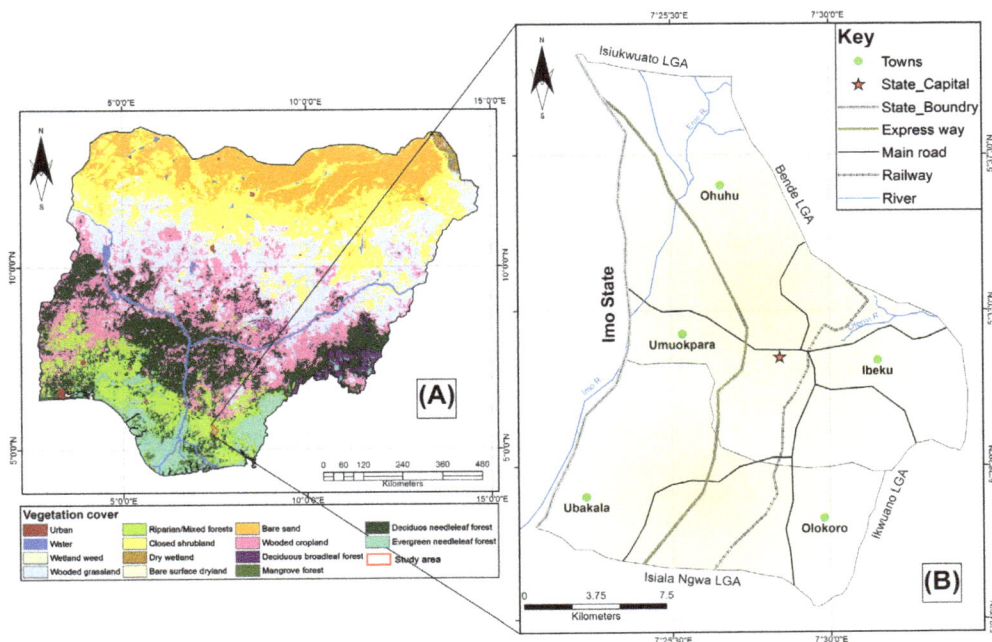

Figure 1. (A) Nigeria's vegetation cover (DIVA-GIS, 2016) and (B) the study area.

Remote sensing technology has proven essentially relevant in establishing land use and land cover monitoring frameworks at different scales (Hansen et al., 2013; Loveland & Dwyer, 2012). It is the science by which information about an object is obtained from electromagnetic radiation reflected from the surface of that object (Jensen, 2014; Lillesand, Kiefer, & Chipman, 2015). Its mechanisms have revolutionized traditional mapping methods (Ariti, van Vliet, & Verburg, 2015; Newbold et al., 2014; Okpala-Okaka & Igbokwe, 2010; Scull et al., 2016; White & Oates, 1999), by advancing and characterizing the spatiotemporal distribution of environmental phenomena using air-borne sensor platforms and image processing and interpretation techniques or packages (Chen, Lu, Luo, & Huang, 2015; Dubula, Tesfamichael, & Rampedi, 2016; Forkel et al., 2013; Hansen et al., 2013; Loveland & Dwyer, 2012; Luo, Zhou, Chen, & Li, 2008). The advent of remote sensing paved way for improved methodological mapping of vegetation cover changes since the establishment of Landsat mission in 1972 (Coppin & Bauer, 1996; Coppin, Jonckheere, Nackaerts, & Muys, 2004; Loveland & Dwyer, 2012).

Several researchers have comprehensively used Landsat imageries, with its extensive data archive at no cost and suitable spectral and spatial resolutions to detect and quantify vegetation cover changes (Chen et al., 2015; Luo & Dai, 2012; Luo et al., 2008; Omo-Irabor et al., 2011; Zhu, Liu, & Chen, 2012). But literature search, with focus on the Umuahia segment of the tropical rainforest belt of Nigeria shows a dearth of information about the dynamics of vegetation cover changes in the area, probably due to limited accessible monitoring data and a lack of appropriate research methods.

This paper, therefore, initiated a remote sensing-based vegetation baseline assessment that is nonexistent in the study area (Erika et al., 2015), as a strategy for informing the non-technical and expert policy makers involved in the sustainable governance and development of the region.

This is necessary to understand the dynamic nature of emerging regional and local vegetation and land cover types that may have been impacted over time. The information will be useful in streamlining policy efforts towards sustainable urban growth in the long-term, land cover recovery, agricultural resilience and carbon storage/sequestration in the face of increasing changing climate. The following specific objectives have been addressed:

(1) to estimate the spatial extent of the vegetation cover changes using remotely sensed satellite data;

(2) to quantify the temporal changes based on spatial extent of the vegetation cover, before and after 1991 (the year the study area was officially created) to a near present date—2014; and,

(3) to identify and document responsible factors for the changes in terms of areal extent of the study area.

The results generated in this study would, thus, show the spatial variations and pattern of vegetation cover distribution of different vegetation classes. While their validation helped to reveal interclass confusion than could be resolved with the use of other biased information (Foody, 2002), especially as it had been established that the region lacked adequate vegetation monitoring data (see, for example, Erika et al., 2015). In this way, appropriate place-specific reinvestigations and or decisions can be achieved, such as rapid identification of alternatives where necessary, and to incorporate same in the formulation of sound policies required in the governance and natural resources management framework of the region (see, for example, Larcom, van Gevelt, & Zabala, 2016; Obydenkova, Nazarov, & Salahodjaev, 2016).

2. Materials and method

2.1. Case study area

Geographically referred to as Umuahia, the study area (latitude 5°26′06.00″N to 05°36′04.00″N and longitude 07°21′50.00″E to 7°34′03.00″E) covered the administrative boundary of the present day Umuahia North and Umuahia South local government areas in Abia State. It lies entirely in the southeastern segment of the humid tropical rainforest in Nigeria (Figure 1(B)). It covers an area of about 363.13 km^2 amidst a constellation of scattered villages and towns within a 15 km radius, while the total area covered by the metropolis is about 71 km^2 (Ejenma, 2013).

The climate is humid tropical rainforest—Koppen Af (Kottek, Grieser, Beck, Rudolf, & Rubel, 2006). Daily average insolation is generally low—4.8 h, nevertheless the area experiences mean annual maximum temperature of 31°C with little daily variations (Iloeje, 2007). Meteorologically, Umuahia experiences an annual mean rainfall of 2,278 mm, with eight months of precipitation, which extends from early March to late October (Figure 2). Meaning, that, the study area witnesses two major seasons—dry season and rainy season. The dry season is dominated by a period of short spell of dry/cool season referred to as *harmattan*. Usually, the heaviest of monthly total of rains → 363 mm (Figure 2), is experienced more in September, while the month with the lowest rainfall fluctuates between January and December, except for recent climatic anomalies with random variance of precipitation in other locations (e.g. Nsukka & Ogbudu) (Campling, Gobin, & Feyen, 2001).

The soils of Umuahia have been greatly influenced by different ancient geologic formations, climate, vegetation, and general topographic configuration of the area (Onu, Opara, & Ehirim, 2012). The major rivers that drain the catchment area of Umuahia are the Imo river on the west axis, Kwa-Ibo river draining southwardly and Enyong creek on the east axis. River Eme with it various tributaries in Ohuhu flows into the Imo river while the Ofenyi river tributaries transverse the Ibeku landscape to deposit into the Enyong creek. These rivers and streams are ephemeral and dry up after the cessation of rainfall. The study area currently serves as one of the nation's commercial hubs for economic development, with high potentials for growth and development.

2.2. Field observations

Validation of remotely sensed data through field surveys, documented information and discussions with the local people has proven useful in several land use/land cover (LU/LC) studies (see, for example, Ariti et al., 2015; Scull et al., 2016; Yeshaneh et al., 2013). In this study, vegetation of the study area was carefully observed by extensive field surveys conducted during rainy season, and the early dry season in 2011 and 2014 respectively. Ground truth of relevant land cover types in the study area were collected alongside plant species and other landscape features. A Garmin 62s Global Positioning System (GPS) with 3 m accuracy was used to identify infrastructural facilities that were relocated to

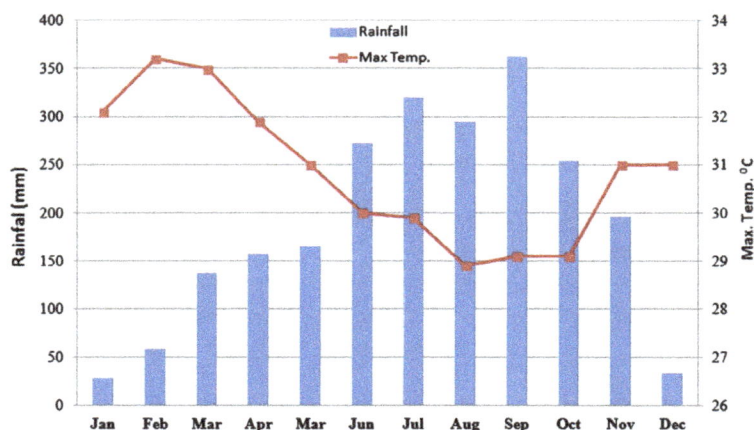

Figure 2. Monthly distribution of rainfall and temperature over Umuahia.

previously vegetated areas. Likewise, certain geographical features, ranging from landscape configurations to vegetation types, were equally identified. This was necessary, to complement and affirm the relationship and variations that may be there, would have occurred and the rationale behind the changes.

As such, the study area lies in the lowland rainforest vegetation belt typified by almost continuous cover of *riparian* forest along stream valleys. Tilling, in most cases, for agriculture and road construction, has resulted in an ecological situation where the normal development of vegetation is markedly retarded (Onu et al., 2012). Although, most of these locations have been drastically affected by human activity so much, large sections of the upper rain forest zone may be called an "oil palm bush". However, the observed vegetation of the area is characterized by an abundance of seed-bearing plants species, such as *Brachustegia eurycoma*, *Carya* spp., *Steroulia*, *Canarium*, *Cassia siame*, *Triplochitan scleroxylong*, *Meliaceae*, *Mitragyna ciliate*, *Khaya* spp., *Entandrophrayma* spp., *Lovia trichiloides*, *Nauclea diderrichii*, *Gmelina arborea*, *Neliaceae*, *Acacia nilotica* and *Gum arabi*. Others include *Cauarea and Tarminalia*. Many times, they occur in about 100 species per hectare. It is this great abundance of species that makes the area rich in terms of biomass productivity of all terrestrial ecosystems, therefore, requiring a periodic assessment.

2.3. Satellite remote sensing data

Remote sensing offers priceless source of geospatial information for ecological and vegetation resource management (Hagenlocher, Lang, & Tiede, 2012; Hansen et al., 2013; Lillesand et al., 2015; Luo & Dai, 2012). With reference to complex terrain, data unavailability, area coverage, data obtained by remote sensing can be timely, cost-effective, and objectively presented and demonstrated for either specific or post assessment—spatiotemporal investigations (Hagenlocher et al., 2012; Omo-Irabor et al., 2011). In this study, vegetation variation in Umuahia was mapped and visualized from Landsat data covering the total extent occupied by the study area.

Given the different software algorithm-based platforms available for ecological studies (Coppin et al., 2004), ERDAS Imagine 2014 and ArcGIS 10.1 were used for image pre-processing so as to maintain data spatial reference consistency to the georectified format—the Universal Transverse Mercator (UTM) zone 32 North and the World Geodetic System 1984 datum, at the various stages. As such, classification challenges (Huang, Lu, Zhang, & Plaza, 2014; Lillesand et al., 2015) were minimized by selecting suitable dates—1984, 1991 and 2014. This was necessary for maintaining seasonal uniformity as an a priori for increased ground vegetation and cloud cover during the wet season (Churches, Wampler, Sun, & Smith, 2014; Schwartz, 2003). More importantly, the 1991 data equally served as a second baseline date for the reason of the study area's official pronouncement as headquarter to a major state in southeastern Nigeria. A scan line corrector (SLC) error experienced by the Landsat 7/ETM+ was corrected with a gap fill function (Zhu et al., 2012) in ERDAS Imagine 2014.

Datasets of 1984, 1991 and 2014 (path 188, row 56 with a spatial resolution of 30 × 30 m) were obtained from the United States Geological Survey (USGS) archives at www.earthexplorer.usgs.gov. After image acquisition, bands 5 (mid-IR): 1.55–1.75 µm, 4 (NIR): 0.76–0.90 µm and 3 (red): 0.63–0.69 µm in each of the image scene were stacked together to form a single multispectral image data-set using the "layer stack" function in ERDAS Imagine.

The United Nations Fund for Population Activities (UNPFA) had recognized that population growth and its resultant human influence constitute serious pressures on global natural resources especially on local and regional forests ecosystems and natural vegetation (United Nations Population Fund, 1991). Using the annual population growth rate of 2.83%, (Abia State of Nigeria, 2005), the population of the area was projected to present to determine the influence of population growth on vegetation resources.

So, the resulting multispectral images based on the layer-stacked data-sets were used as the baseline data—before impact imagery (1984–1991—*marked year of population influx in the study*

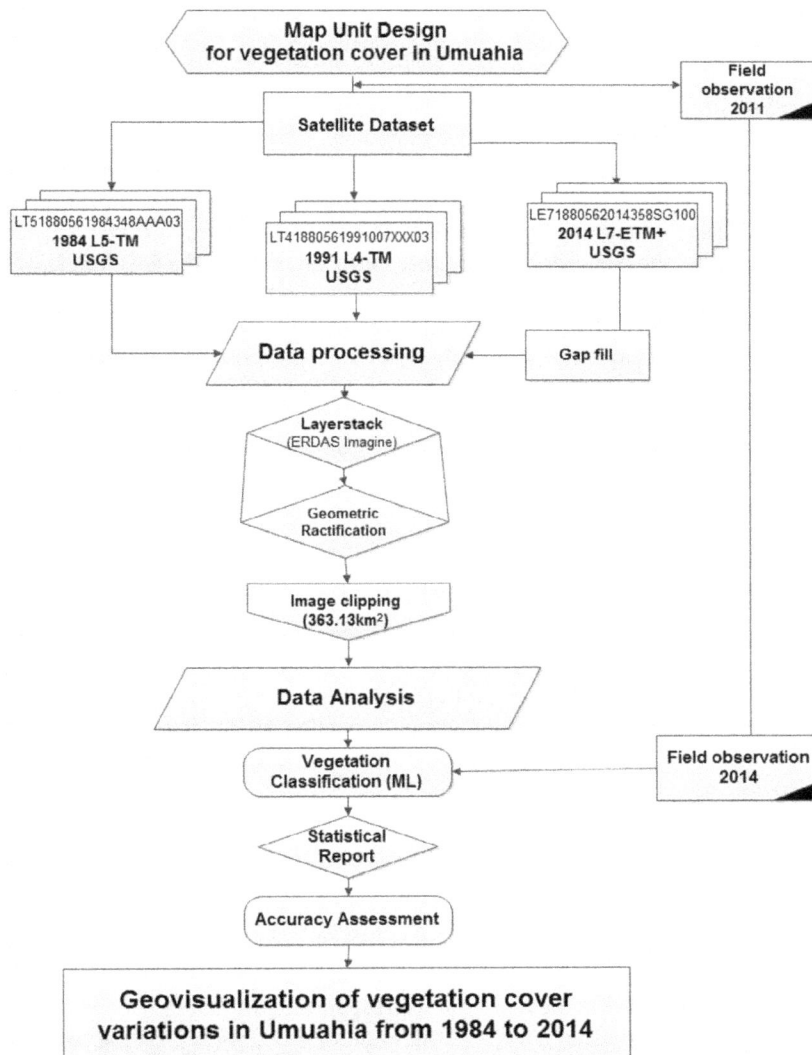

Figure 3. Vegetation cover mapping workflow.

area). While the analysed data-sets served as the post impact assessment imagery (1991–2014). The process allowed for straightforward detection of changes and human-induced impacts on the landscape/vegetation cover over the period of study—30 years. The vegetation classification work-flow for the study consists of several other steps and stages, as illustrated in Figure 3.

3. Data analysis

3.1. Image classification

Since land use cover change is a continues issues (Watson et al., 2000), detecting spatial and temporal patterns of vegetation cover changes with satellite data depends largely on pixel-based spectral signatures or vegetation indices (Chen et al., 2015). Based on their spectral signatures or vegetation intensity values, the Landsat image pixels covering the study area were organized into a finite set of classes that represents surface types. This can be done in two ways; supervised and unsupervised classification (Lillesand et al., 2015). This study used the maximum likelihood classification (MLC) method of the set of the supervised classification algorithms in grouping vegetation cover changes. Several other methods do exist (Minimum distance technique, Mahalanobis distance technique and Parallelepiped classification methods (Soofi, 2005), but the maximum likelihood method is preferred because it required field observation to aid the classification procedure which tends towards accurate data analysis. The analysis was performed with ERDAS Imagine.

Using the (MLC) on ERDAS Imagine (Congalton & Green, 1999), the enhanced false colour composite bands 5, 4 and 3 of the different years depicting the vegetation image pixels were trained and categorized into appropriate classes. A total of five classes were discriminated (i.e. densely vegetated and sparsely vegetated areas, water bodies, congested residential and built-up industrial areas). The same five classes have been adopted in this work. This classification pattern is in accordance with the 2010 updated version of the global ecological zones for forest reporting by FAO (2012).

The supervised classification maps, showing the spatial extent and variations in forest cover across the study area, are presented in Figure 5. For effective visual interpretation, suitable colour patterns have been used to identify and show the various classes. A class name is assigned to a colour. Thick green represents dense vegetation, light green is for sparse vegetation (Ochege, 2014a), ox-blood is used to show congested sections in the study area, light brown represents built-up areas while blue is used to show the presence of any kind of water body in the area. This pattern is acceptable according to the vegetation classification standard as modified by Anderson, Hardy, Roach, and Witmer (1976).

3.2. Accuracy assessment

It is usually very important to ensure the correctness of the classification analyses because it seeks to measure the quality of results shown on the classification maps (Banko, 1998; Congalton & Green, 1999). The accuracy assessment can either be quantitative or qualitative. In this study, we evaluated the accuracy process of vegetation classification by visual inspection of the classified image in ERDAS Imagine using Kappa statistics function on a scale range of −0.1 to 1 (Congalton, 1991). Kappa statistics is calculated as in Equation (1):

...

$$K = \frac{N \sum_{i=1}^{x} x_{ii} - \sum_{i=1}^{r} (x_{i+} \times x_{+i})}{N^2 - \sum_{i=1}^{r} (x_{i+} \times x_{+i})} \tag{1}$$

where N = is the total number of samples in the matrix, r = corresponds to the number of rows in the matrix, x_{ii} = is the number in row i and column i, x_{+i} = is the total for row i, and x_{i+} = is the total for column i.

We had generated 150 (30 for each class) reference sites (Figure 4) which were based on the simple random sampling technique (Lins & Kleckner, 1996).

Each sample point was assigned the ideal class value during field observations, and was used to enhance geovisual inspections. These points were further superimposed on the classified images. Features of the representing land use pixels (class) that correspond with each point were compared with the features that existed on the ground by also using the federal government approved national vegetation atlas base map of Nigeria (Federal Republic of Nigeria, 1977), Google Earth pro web-based GIS, in addition to other reference information obtained from field observations. Then, the following accuracy assessment estimators: the error matrix, overall accuracy, producer's accuracy, user's accuracy and the kappa coefficient, were computed in ERDAS for each of the year understudied—1984, 1991, 2014 (see Section 4.5).

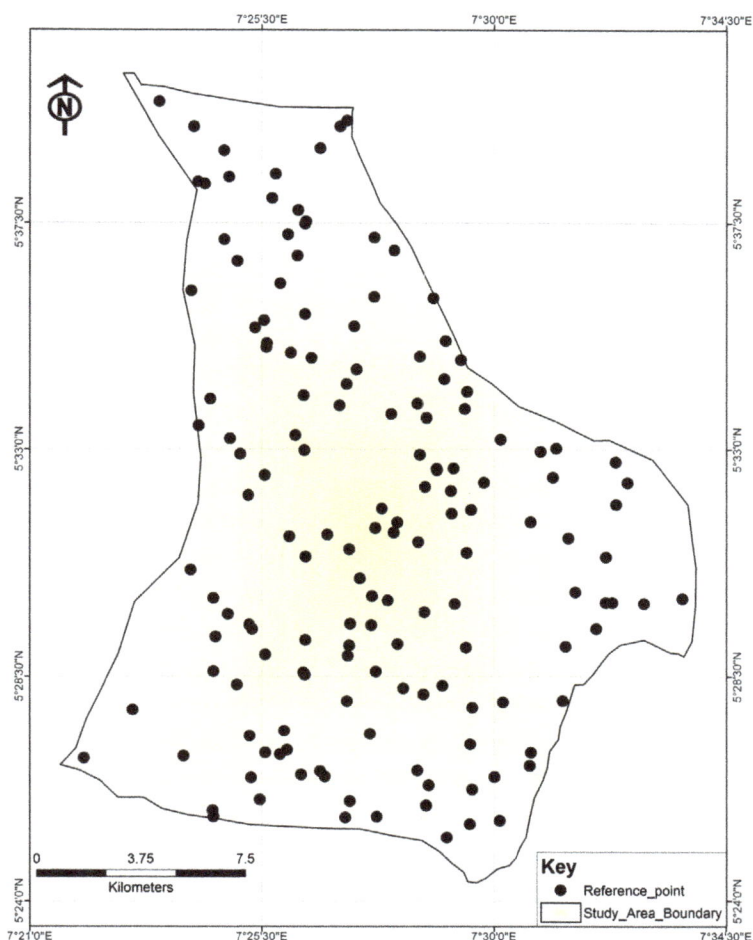

Figure 4. Simple random points generated for vegetation classification validation.

4. Results and discussion

4.1. Spatio-temporal dynamics of vegetation cover changes
Results from this study include a vegetation cover statistics and land-cover classification maps of the years understudied—1984, 1991 and 2014. From 1984 to 1991, dense vegetation (healthy vegetation) had reduced in percentage by 21.2%. Between 1991 and 2014, it further reduced by 22.8%. While sparse vegetation (disturbed or unhealthy vegetation cover) increased by 18.32% (1984–1991) and 6.64 (1991–2014). Likewise, residential and built up areas increased by 0.09% and 2.76% in 1991, and 5.13 and 8.94% in 2014, respectively (Table 1).

4.2. Vegetation and other land cover stands as at 1984
The 1984 classification map clearly shows that dense vegetation covered most of the fragments in the study area followed by patches of other cover types—sparse vegetation, water bodies, congested and built-up areas (Figure 5(A)). As such, vegetation resources in the study area may not be adjudged to be entirely untouched or referred to as pristine because of the scattered pattern of sparse vegetation which indicates disturbances and stress on vegetation canopies and phenology (Ochege, 2014a). In this regard, there are chances that most vegetation fragments in the humid tropical rainforests of southeastern Nigeria have experienced one form of human or natural interruptions.

Often times, fragments of tropical forests are considered primary or virgin, whereas, in the actual sense, they have passed through a number of stages to become secondary forests (Aubreville, 1938;

Land cover	Total area covered			% change		
	1984 (%)	1991 (%)	2014 (%)	1984–1991	1991–2014	1984–2014
Dense vegetation	330.63 (91.05)	253.65 (69.85)	170.87 (47.05)	−21.2	−22.8	−44
Sparse vegetation	6.86 (1.8)	73.05 (20.12)	97.16 (26.76)	18.32	6.64	24.87
Congested (residential)	9.69 (2.67)	10.04 (2.76)	28.66 (7.89)	0.09	5.13	5.22
Built-up (industrial)	15.07 (4.15)	25.09 (6.91)	54.94 (15.13)	2.76	8.94	10.98
Water body	0.87 (0.24)	1.30 (0.36)	10.46 (2.88)	0.12	2.52	2.64

Table 1. Changes in vegetation and other land cover types in Umuahia

Notes: Total area occupied by each cover type was calculated as follows: (CT/SA) × 100%; where CT = area occupied by each cover type of year under consideration, SA = 363.13 km² is the area covered by the study area; while the percentage change (1984–2014) for each cover type was derived by subtracting the percentage change of each cover type in 1984 from those in 2014, likewise for time interval of 1984–1991 and 1991–2014. In furtherance of this exploratory analysis, Table 1 is summarized as follows: dense vegetation and sparse vegetation covers represent vegetation resources, congested (residential) and Built-up(industrial) areas represent urban encroachment, while water body represents Water (see Appendix 2).

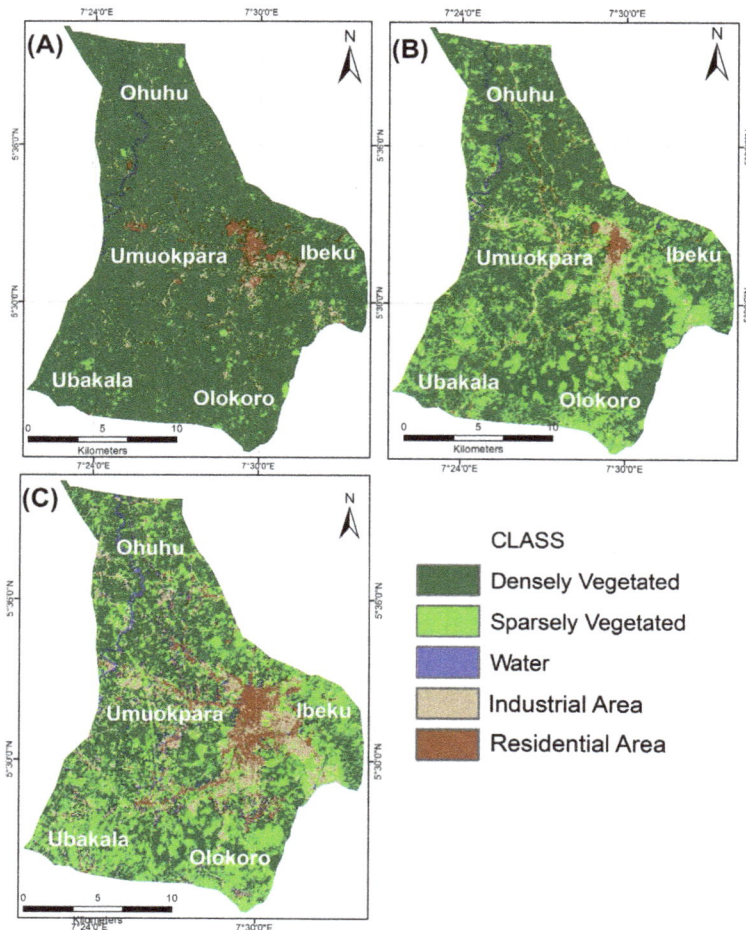

Figure 5. Vegetation classification of the study area (A) in 1984, (B) in 1991, and (C) in 2014.

Figure 6. Harvested Woodstock within the study area.

Budowski, 1970; Bush & Colinvaux, 1994). For instance, the Okomu Forest Reserve in southwest Nigeria was considered to be a primary forest by Richards (1939), but later studies by Jones (1956) revealed extensive charcoal and pottery deposits and a tree population structure reflecting "second growth" vegetation. Recent studies now provide evidence that the forests of Okomu can be traced back to a period soon after 700 years ago, following a period of intensive human use (White & Oates, 1999).

Conducting interviews or discussions with the local people is one way of ascertaining age long anthropogenic impacts on past vegetation status, vegetation loss and their root causes (Ariti et al., 2015; Ofori et al., 2014; Yeshaneh et al., 2013). Critical information obtained through informal discussions with some indigenous people of the area during the field observations, suggests that: *most rural dwellers in the study area depend to a large extent on wood fuel energy for cooking and timber logging for furniture and other kinds of woodwork.* Figure 6 show harvested wood fuel from the Ibeku axis of the study area, intended for domestic energy consumption. This practice is neither new nor exclusive to the Umuahia ecological zone of the humid tropical rainforest, but is obtainable in most parts of the country (Anyiro, Ezeh, Osondu, & Nduka, 2013; Food & Agriculture Organization of the United Nations, 1981; Jones, 1956; Tee, Ancha, & Asue, 2009; White & Oates, 1999).

Likewise, forests and vegetation of the study area may be of similar secondary growth vegetation with those of Okomu forests, as shown by the scattered distribution of stressed vegetation which is an indication of human influence (Figure 5). To say the least, wood fuel harvests are on the increase and constitute a major factor to vegetation cover changes in the humid tropical rainforests of Nigeria and the entire sub-Saharan Africa (Sulaiman, Abdul-Rahim, Mohd-Shahwahid, & Chin, 2017).

Statistical report generated from the 1984 image pixel training show that, dense vegetation dominated a greater portion and occupied a spatial extent of 330.63 km^2 out of the 385 km^2 of the entire study area, while sparse vegetation fragments occupied a spatial extent of about 6.86 km^2. As at 1984, built-up and congested areas occupied 15.07 and 9.69 km^2, respectively. Given the total study area size of 385 km^2, the classification analysis accounted for 94.32% of the total land cover. About 5.68% of the study area's land cover was unclassified, and so is unaccounted for. Out of the new land area (i.e. 363.13 km^2), dense vegetation totalled 91%, while sparse vegetation, congested and built-up areas occupied 2, 3 and 4%, respectively (Figure 7(A)).

4.3. Vegetation and other land cover changes in 1991

Figure 5(B) shows that increased population among other factors have had significant impact and probably, some negative implication on the spatial extent of vegetation cover changes on the Umuahia segment of the humid tropical rainforest. The statistical report generated from the

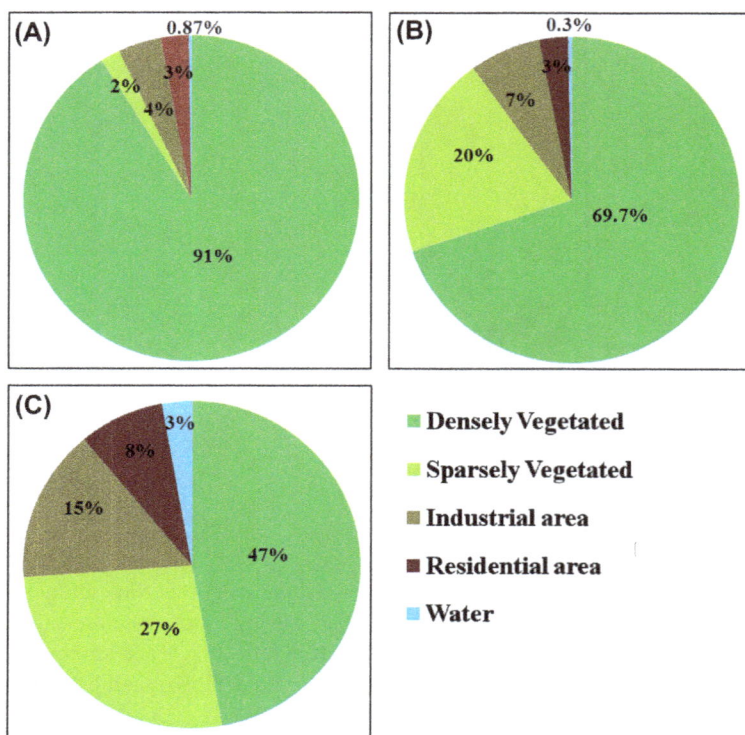

Figure 7. Spatiotemporal variations of vegetation cover of the study area (A) as at 1984, (B) as at 1991, (C) as at 2014.

classification analysis carried out on the 1991 data-set shows that land area covered by dense vegetation decreased from 330.63 km^2 (91%) in 1984 to 253.65 km^2 (69.7%) in 1991, while every other land cover types, i.e. sparse vegetation, congested and built up areas increased in spatial extent (Table 1). As such, 76.98 km^2 which is about 20% of dense vegetation cover was lost between 1984 and 1991. Then fragments occupied by sparse vegetation increased by 18% given the initial 2% as at 1984. Though, congested sections maintained its size, but built-up areas gained by 3% (Figure 7(B)).

It is obvious that population growth is associated with increased demand for land and economic spaces and or exploitation of natural resources—including forest and non-forest products (Obeta, Aujara, Ochege, & Shehu, 2013; United Nations Population Fund, 1991). In this study, data analyses reveal a strong correlation between population growth and vegetation resource depletion. Projected population of Umuahia, which is currently 438,992 rose from 220,104 in 1991 to 359,230 in 2006 (Federal Republic of Nigeria, 2007) show that the area is rapidly urbanizing. Consequently, observations from field surveys revealed serious spread of urban infrastructure and socio-economic activities to previously vegetated sections.

Pieces of information gathered from the indigenous people by preliminary random discussions in 2014 confirm that: (1) urban encroachments (Figure 8) were often not coordinated, (2) have been ongoing since the 1980s, and (3) did significantly accelerate after the study area became state capital in 1991. The people further indicated that: "past methodologies that saw to the sitting and relocation of urban infrastructure to forest-rich zones derided adequate ecological unit accounting and or impact assessment. They maintain that some ecological units of already destroyed forest corridors (e.g. Ibeku and Olokoro) were natural habitats to certain endemic vegetation resources (e.g. Nauclea diderrichii, Myrianthus arboreus P. Beauv., K. Schum., Canarium schweinfurhii, Dialium guineense Wild)". Unfortunately, most of the species are now difficult to come by even within other sections of the study area (See, for example, Meregini, 2005).

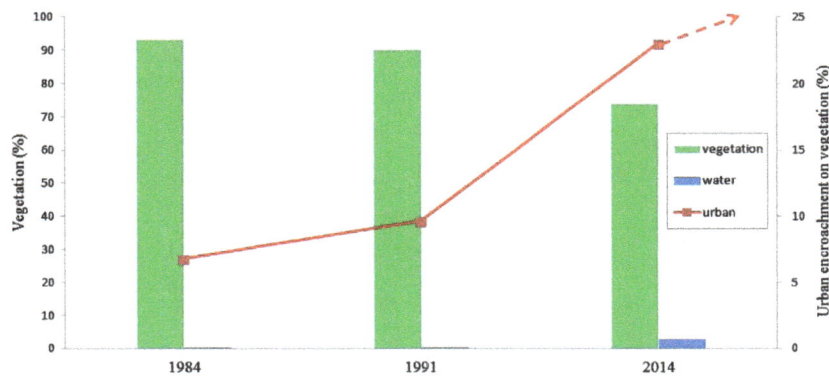

Figure 8. Temporal pattern of urban encroachment on vegetation resources.

These assertions by the local people shows that heterogeneity in the spatial distribution of land resources, in addition to weak concerns towards preserving the accruing benefits of vegetation resources made certain vegetated sites in the study area more attractive for anthropogenic functions of population and urbanization dynamics (Osemeobo, 1988). Thereby, leading to "mixed patterns of land-use and land-cover changes" which are the resultant effect of long-term gaps and or ambiguity in the implementation of urban development plan (Niemelä, 1999; Rojas, Pino, Basnou, & Vivanco, 2013). Suggesting that, before now, it is not unlikely that urban planning actions in the humid tropical rainforest of southeastern Nigeria were perceived as ecologically unsustainable and developmentally biased.

4.4. Vegetation and other land cover changes in 2014
Like the previous result, the 2014 classification map yet shows reduced fragments of dense vegetation on one hand, and on the other hand increased fragments of sparse vegetation, built-up and congested areas and even the water bodies. Indeed, the 2014 classification result is a near-perfect representation of the current vegetation status in the study area as observed during field surveys. It portrays some distinct variations from the previously classified images of 1984, 1991 and 2014, respectively (Figure 5(C)).

The 2014 image classification analysis indicates a serious decline in the phonological quantity and characteristics of general vegetation resources, judging from the spatial dominance of the sparse vegetation. Since 1984, dense vegetation has been decreasing, while other land cover types gain from these losses (Appendices 1 and 2, Table 1, and Figure 8). Initially, dense vegetation occupied about 330.63 km^2, but today it only maintains just about half of that green space i.e. 170.87 km^2. This shows a 44% decrease in its spatial extent. Sparse/unhealthy vegetation on the other hand rose from 2% in 1984 to 27% in 2014; thereby occupying as much as 97.161 km^2 in spatial extent (Figure 7(C)). Built-up and congested areas—*urbanization* have also grown to sustain a steady growth pattern (Figure 8), especially because of the rising population and increasing quest to satisfy human need for urban occupancy in the area.

In 1984, built-up and congested areas occupied 15 and 9.69 km^2, respectively, but the 2014 results show that they currently occupy about 54.94 and 28.66 km^2, in that order. Consequently, the built up area increased by 14% over a period of 30 years while the congested central business district experienced a steady 5% increase over the same period of years. It therefore suggests that Umuahia axis of the humid tropical rainforest is experiencing more of urban diffusion as shown by fragmented patches of other land cover types.

The marked influx of people since 1991 induced urban congestion and increased need for living spaces. Thereby, introducing opportunities for urban planning interventions aimed at sustainable governance of natural resources. The state government did relocate some urban infrastructural facilities e.g. markets, to previously vegetated areas. Yet, the level of increased disturbances as shown

Vegetation cover	Water	Sparsely vegetated	Densely vegetated	Industrial	Residential	Row total	User accuracy (%)
Table 2. Standard summaries of accuracy assessment report							
(a) Confusion matrix for 1984 classification							
Water	32	1	0	2	0	35	91
Sparsely vegetated	1	38	1	1	0	41	93
Densely vegetated	0	2	36	1	0	39	92
Industrial	0	1	2	34	0	37	92
Residential	0	0	0	2	21	23	91
Column total	33	42	39	40	21	175	
Producers accuracy (%)	97	90	92	85	100		
(b) Confusion matrix for 1991 classification							
Water	31	0	2	0	0	33	94
Sparsely vegetated	0	37	1	1	0	39	95
Densely vegetated	1	0	38	1	0	40	95
Industrial	0	0	1	34	1	36	94
Residential	1	0	0	1	25	27	93
Column total	33	37	42	37	26	175	
Producers accuracy	94	100	90	92	96		
(c) Confusion matrix for 2014 classification							
Water	35	0	0	0	0	35	100
Sparsely vegetated	0	42	0	0	0	42	100
Densely vegetated	0	0	38	0	0	38	100
Industrial	0	0	1	32	0	33	96
Residential	0	1	0	2	24	27	89
Column total	35	43	39	34	24	175	
Producers accuracy (%)	100	98	97	94	100		

(a) Overall accuracy = 92%; Kappa coefficient = 0.92.
(b) Overall accuracy = 94%; Kappa coefficient = 0.94.
(c) Overall accuracy = 97%; Kappa coefficient = 0.97.

by the classification analyses of 1991 and 2014 suggests increased human impacts on the region's ecological landscape. Regrettably, deforestation and habitat loss are the greatest threat to terrestrial biodiversity and ecosystem health which equally leads to species extinction at the time of occurrence and in the future (Millennium Ecosystem Assessment, 2005).

4.5. Accuracy assessment
Using the mapped data against 150 reference data obtained during field observations, the image analysis in this study were subjected to automated quantitative accuracy assessment using the Cohen Kappa's statistics function in ERDAS Imagine, on the scale of 0 to 1 (i.e. Kappa is a value less than or equal to 1, where 1 corresponds to a perfect agreement) (Congalton & Green, 1999). The acceptable standard of overall accuracy for land cover map is set between 80% (Anderson et al., 1976) and 100% (Lins & Kleckner, 1996). In this study, the results of the accuracy assessment obtained for the classified images of 1984, 1991 and 2014 are presented in a standard summaries report in Table 2(a)–(c). The error matrices quantitatively compared the relationship between the classified images with the reference data obtained from field observations. All the classified images—1984, 1991 and 2014 returned high percentage of overall accuracy and Kappa values as follows: 92% (0.92), 94%

(0.94) and 97% (0.97), respectively. This show a high level of conformity with the supervised classification analysis carried out in the study (Table 2).

4.6. Factors responsible for vegetation cover changes in the study area

This study identified four major factors responsible for vegetation cover changes in the humid tropical rainforest segment of Umuahia. (1) Unchecked and increased local demand for forest and non-forest products, (2) Rapidly growing population induced by the reason of the study area's centrality and official recognition as state capital in 1991, (3) Urban diffusion influenced by increased demand for land and economic spaces (i.e. urbanization), (4) Long-term gaps in the implementation of vegetation monitoring frameworks (see, for example, Erika et al., 2015).

Field observations conducted in 2011 and 2014 reveal that the humid tropical rainforest of southeastern Nigeria is a biodiversity rich belt, with abundance of several endemic vegetation resources that provides various benefits to the indigenous people and uphold useful potentials for posterity. Yet, this rich ecologically vibrant area is considered one of the most rapidly urbanizing and endangered areas in Nigeria.

In 1991, the study area was officially declared a headquarter city in southeastern Nigeria. The action trickled-down a ripple cause-effect response of vegetation resources to anthropogenic phenomenon of urbanization. The rising population initiated increased demand for land and economic spaces, which in turn, affected vegetated areas by increased exploitation of natural resources—including forest and non-forest products.

Data extracted from the analyses in this study (Table 1 and Appendix 2) showed increased human impact on vegetation cover. This is indicated by fragmented sparsely vegetated areas, growth of industrial and residential land uses (Figure 8). One of the understudied geographies of population impact on natural resources is the role of institutional governance in monitoring deforestation and vegetation loss in Africa (Erika et al., 2015; Larcom et al., 2016). Nevertheless, like most other regions, the study area continues to experience unrestrained exploitation of natural resources by virtue of domestic need and industrial demand for fuel wood and timber, respectively.

Also, the spatiotemporal extent maps generated in this study show that built-up and residential areas had become fragmented given the accelerated rate of vegetation reduction since 1991 to 2014. Generally, this kind of situation is often attributed to the concentration of built-up patches or new infrastructural developments along emerging economic corridors (Müller, Griffiths, & Hostert, 2016; Simmons et al., 2016).

Based on the forgoing, Umuahia section of the tropical rainforest is currently witnessing its middle phase of urbanization process. This is exemplified by the area experiencing more of urban diffusion, urban growth (Figure 8) and less of vegetation resilience and forest recovery (Table 1, Appendix 1). As such, vegetation reduction in the area is highly correlated with anthropogenic functions from population dynamics.

5. Conclusion and recommendation

This study shows that remote sensing of vegetation is a consistent methodology in ascertaining changes and phenological characteristics in the humid tropical rainforests. The change detection covered a 30-year period, starting from 1984 to 2014, and revealed a reduction in size of healthy vegetation by 159.76 km^2 which indicates a 44% vegetation loss. Composite replacement of healthy vegetation cover in the area is impacted by other land use cover types as follows; 90.3 km^2 (unhealthy vegetation), 18.97 km^2 (congested/residential), 39.87 km^2 (built-up/industrial area), 9.59 km^2 (Water-body). Thereby, indicating a gross encroachment of urban land use of about 23.02% into vegetated areas, and reduced it from 93.03% in 1984 to 73.8% in 2014.

The most significant changes in the spatiotemporal dynamics of land use cover in the study area accelerated after the area became capital city in 1991. Similarly, the much higher percentage of vegetation loses recorded before 1991 (i.e. mid-1980s) is attributed to the persistent unsustainable exploitation of vegetation resources resulting from lack of adequate vegetation monitoring frameworks. Implying that, unsustainable exploitation of vegetation resources, increased economic activities in need of industrial and residential occupancies, in addition to uncoordinated urban expansions constitutes the anthropogenic functions of urbanization currently experienced in the study area. These, therefore suggest that, there is need to strengthen institutional monitoring frameworks that should account for all ecological units in the humid tropical rainforests.

Though limited by data availability, this study provides the baseline information about vegetation depletion in the humid tropical rainforests of Nigeria, and recommends periodic integration of high-resolution satellite images with urban afforestation strategies in natural resources governance through public engagements. Stakeholders and the local people, whom the vegetation resources are domiciled in their communities, can be adequately consulted, sensitized and integrated into every activity that may directly or indirectly affect ecological heritage. This will increase rates of vegetation resilience and forest recovery, such that, developments that disregard ecological losses and urban greening can be monitored effectively.

Acknowledgements
Prof R.N.C. Anyadike (Late) significantly initiated the process that led to this study. We thank our local field guides, and all the reviewers for their helpful comments.

Funding
The authors received no direct funding for this research.

Author details
Friday Uchenna Ochege[1,2]
E-mail: uchenna.ochege@uniport.edu.ng
ORCID ID: http://orcid.org/0000-0002-6661-0966
Chukwunonyelum Okpala-Okaka[2]
E-mail: chukwunonyelum.okpala-okaka@unn.edu.ng
[1] Laboratory for Cartography and GIS, Department of Geography & Environmental Management, University of Port Harcourt, Choba, Rivers State, Nigeria.
[2] Faculty of Environmental Studies, Department of Surveying and Geoinformatics, University of Nigeria, Enugu, Nigeria.

Cover image
Source: Authors.

References
Abia State of Nigeria. (2005). Abia state economic empowerment and development strategy. *Abia State Government Publication*. Retrived June 14, 2014, from http://web.ng.undp.org/documents/SEEDS/Abia_State.pdf

Adekunle, V. A. J., Olagoke, A. O., & Ogundare, L. F. (2013). Timber exploitation rate in tropical rainforest ecosystem of southwest nigeria and its implications on sustainable forest management. *Applied Ecology and Environmental Research, 11*, 123–136. http://dx.doi.org/10.15666/aeer

Anderson, J. R., Hardy, E. E., Roach, J. T., & Witmer, R. E. (1976). *A land-use and land cover classification system for use with remote sensor data* (Professional Paper 964). Washington, DC: United State Geological Survey.

Anyiro, C. O., Ezeh, C. I., Osondu, C. K., & Nduka, G. A. (2013). Economic analysis of household energy use: A rural urban case study of Abia State, Nigeria. *Research and Reviews: Journal of Agriculture and Allied Sciences, 2*, 20–27.

Ariti, A. T., van Vliet, J., & Verburg, P. H. (2015). Land-use and land-cover changes in the Central Rift Valley of Ethiopia: Assessment of perception and adaptation of stakeholders. *Applied Geography, 65*, 28–37. http://dx.doi.org/10.1016/j.apgeog.2015.10.002

Aubreville, A. (1938). La forêt coloniale: les forêts de l'Afrique occidentale française [The colonial forest: The forests of French West Africa]. *Annuales Academie des Sciences Coloniales, 9*, 1–245.

Banko, G. (1998). A review of assessing the accuracy of classifications of remotely sensed data and of methods including remote sensing data in forest inventory. In *International Institution for Applied Systems Analysis*, Laxenburg, Austria. IIASSA Interim Report No. IR-98-081. Retrived from http://pure.iiasa.ac.at/5570/1/IR-98-081.pdf

Budowski, G. (1970). The distinction between old secondary and climax species in tropical Central American lowland forests. *Tropical Ecology, 11*, 44–48.

Bush, M. B., & Colinvaux, P. A. (1994). Tropical forest disturbance: Paleoecological records from Darien, Panama. *Ecology, 75*, 1761–1768. http://dx.doi.org/10.2307/1939635

Campling, P., Gobin, A., & Feyen, J. (2001). Temporal and spatial rainfall analysis across a humid tropical catchment. *Hydrological Processes, 15*, 359–375. http://dx.doi.org/10.1002/(ISSN)1099-1085

Chen, Y., Lu, D., Luo, G. P., & Huang, J. (2015). Detection of vegetation abundance change in the alpine tree line using multitemporal Landsat Thematic Mapper imagery. *International Journal of Remote Sensing, 36*, 4683–4701. doi:10.1080/01431161.2015.1088675

Churches, E. C., Wampler, P. J., Sun, W. & Smith, A. J. (2014). Evaluation of forest cover estimates for Haiti using supervised classification of Landsat data. *International Journal of Applied Earth Observation and Geoinformation, 30*, 203–216. doi:10.1016/j.jag.2014.01.020

Congalton, R. (1991). A review of assessing the accuracy of classifications of remotely sensed data. *Remote Sensing of Environment, 37*, 35–46. http://dx.doi.org/10.1016/0034-4257(91)90048-B

Congalton, R. G., & Green, K. (1999). *Assessing the accuracy of remotely sensed data: Principles and practices*. Boca Raton, FL: Lewis Publishers.

Coppin, P., Jonckheere, I., Nackaerts, K., Muys, B., & Lambin, E. (2004). Digital change detection methods in ecosystem monitoring: A review. *International Journal of Remote Sensing, 25*, 1565–1596. http://dx.doi.org/10.1080/0143116031000101675

Coppin, P. R., & Bauer, M. E. (1996). Digital change detection in forest ecosystems with remote sensing imagery. *Remote Sensing Reviews, 13*, 207–234.

DIVA-GIS. (2016). *Land cover map of Nigeria*. Retrieved December, 2016, from http://biogeo.ucdavis.edu/data/diva/cov/NGA_cov.zip

Du, J., Shu, J., Yin, J., Yuan, X., Jiaerheng, A., Xiong, S., ... Liu, W. (2015). Analysis on spatio-temporal trends and drivers in vegetation growth during recent decades in Xinjiang, China. *International Journal of Applied Earth Observation and Geoinformation, 38*, 216–228. doi:10.1016/j.jag.2015.01.006

Dubula, B., Tesfamichael, S. G., & Rampedi, I. T. (2016). Assessing the potential of remote sensing to discriminate invasive *Asparagus laricinus* from adjacent land cover types. *Cogent Geoscience, 2*(1), 1–17.

Ejenma, E. (2013). *Trends and patterns of house rents in Umuahia* (Unpublished PhD Seminar 1). Choba: Department of Geography and Environmental Management, University of Port Harcourt.

Erika, R., Celso, B. L., Martin, H., Erik, L., Robert, O., Arief, W., Daniel, M., & Louis, V. (2015). Assessing change in national forest monitoring capacities of 99 tropical Countries. *Forest Ecology and Management, 352*, 109–123.

FAO. (2012). *State of the world's forests 2012*. Italy: Author.

FAO. (2013). *The latest state of the World's Forest Report of the Food and Agricultural Organisation (FAO) indicate a dangerous cutback in the overall forest reserves in Africa especially between 1990 and 2012*. Italy: Author.

Federal Republic of Nigeria. (1977). *The national atlas of the Federal Republic of Nigeria* (FRN 1st ed.). Lagos: Federal Surveys.

Federal Republic of Nigeria. (2007). *Official gazette: Legal notice on publication of the details of the breakdown of the national and state provisional totals 2006 census*. Lagos: Author.

Fisher, B., & Turner, R. (2008). Ecosystem services: Classification for valuation. *Biological Conservation, 141*, 1167–1169. http://dx.doi.org/10.1016/j.biocon.2008.02.019

Food & Agriculture Organization of the United Nations. (1981). *Map of fuelwood situation on developing countries* (pp. 1–10). Rome: Author.

Foody, G. M. (2002). Status of land cover classification accuracy assessment. *Remote Sensing of Environment, 80*, 185–201. http://dx.doi.org/10.1016/S0034-4257(01)00295-4

Forkel, M., Carvalhais N., Verbesselt J., Mahecha M. D., Neigh S. R. C., & Reichstein M. (2013). Trend change detection in NDVI time series: Effects of inter-annual variability and methodology. *Remote Sensing, 5*, 2113–2144. doi:10.3390/rs5052113

Friberg, N., Bonada, N., Bradley, D. C., Dunbar, M. J., Edwards, F. K., Grey, J., ... Woodward, G. (2011). Biomonitoring of human impacts in freshwater ecosystems: The good, the bad and the ugly. *Advances in Ecological Research*, 1–68. http://dx.doi.org/10.1016/B978-0-12-374794-5.00001-8

Hagenlocher, M., Lang, S., & Tiede, D. (2012). Integrated assessment of the environmental impact of an IDP camp in Sudan based on very high resolution multi-temporal satellite imagery. *Remote Sensing of Environment, 126*, 27–38. http://dx.doi.org/10.1016/j.rse.2012.08.010

Hansen, M. C., Potapov, P. V., Moore, R., Hancher, M., Turubanova, S. A., Tyukavina, A., ... Townshend, J. R. G. (2013). High-resolution global maps of 21st-century forest cover change. *Science, 342*, 850–853.

Huang, X., Lu, Q., Zhang, L., & Plaza, A. (2014). New postprocessing methods for remote sensing image classification: A systematic study. *IEEE Transactions on Geoscience and Remote Sensing, 52*, 7140–7159. http://dx.doi.org/10.1109/TGRS.2014.2308192

Igbawua, T., Zhang, J., Chang, Q., & Yao, F. (2016). Vegetation dynamics in relation with climate over Nigeria from 1982 to 2011. *Environmental Earth Sciences, 75*, 1–16.

Igbozurike, M. U. (1975). Vegetation types. In G. E. K. Ofomata (Ed.), *Nigeria in maps* (pp. 30–32). Benin City: Ethiope Publishing House.

Iloeje, N. P. (2007). *A new geography of Nigeria*. Ikeja: Longman Nigerian PLC.

Jensen, J. R. (2014). *Remote sensing of the environment: An earth resource perspective* (2nd ed.). Harlow: Pearson.

Jones, E. W. (1956). Ecological studies of the rain forest of southern Nigeria IV. The plateau forest of the Okomu Forest Reserve, Part 2. The reproduction and history of the forest. *The Journal of Ecology, 44*, 83–117. http://dx.doi.org/10.2307/2257155

Kottek, M., Grieser, J., Beck, C., Rudolf, B., & Rubel, F. (2006). World map of the Köppen-Geiger climate classification updated. *Meteorologische Zeitschrift, 15*, 259–263.

Larcom, S., van Gevelt, T., & Zabala, A. (2016). Precolonial institutions and deforestation in Africa. *Land Use Policy, 51*, 150–161. http://dx.doi.org/10.1016/j.landusepol.2015.10.030

Lillesand, T., Kiefer, R., & Chipman, J. (Eds.). (2015). *Remote sensing and image interpretation* (7th ed.). Hoboken, NJ: John Wiley & Sons.

Lins, K. S., & Kleckner, R. L. (1996). Land cover mapping: An overview and history of the concepts. In J. M. Scott, T. H. Tear, & F. Davis (Eds.), *Gap analysis: A landscape approach to biodiversity planning* (pp. 57–65). Bethesda, MD: American Society for Photogrammetry and Remote Sensing.

Loveland, T. R., & Dwyer, J. L. (2012). Landsat: Building a strong future. *Remote Sensing of Environment, 122*, 22–29. http://dx.doi.org/10.1016/j.rse.2011.09.022

Lucas, R., Blonda, P., Bunting, P., Jones, G., Inglada, J., Arias, M., ... Charnock, R. (2015). The earth observation data for habitat monitoring (EODHaM) system. *International Journal of Applied Earth Observation and Geoinformation, 37*, 17–28. http://dx.doi.org/10.1016/j.jag.2014.10.011

Lund, H. G. (1983). Change: Now you see it—now you don't! In *Proceedings of the international conference on renewable resource inventories for monitoring changes and trends* (pp. 211–213). Corvallis, OR: Oregon State University.

Luo, G. P., & Dai, L. (2012). Detection of alpine tree line change with high spatial resolution sensed data. *Journal of Applied Remote Sensing, 2*. doi:10.1117/1.JRS.7.073520

Luo, G. P., Zhou, C. H., Chen, X., & Li, Y. (2008). A methodology of characterizing status and trend of land changes in oases: A case study of Sangong River watershed, Xinjiang, China. *Journal of Environmental Management, 88*, 775–783.

Matlack, G. R. (1994). Vegetation dynamics of the forest edge—Trends in space and successional time. *The Journal of Ecology*, 113–123. http://dx.doi.org/10.2307/2261391

Mckey, D., Waterman, P. G., Gartlan, J. S., & Struhsaker, T. T. (1978). Phenolic content of vegetation in two African rain forests: Ecological implications. *Science (Washington, DC); (United States), 202*.

Meduna, A. J., Ogunjimi, A. A., & Onadeko, A. (2009). Biodiversity conservation problems and their implications on Kainji Lake National Park, Nigeria. *Journal of Sustainable Development in Africa, 10*, 59–73.

Meregini, A. O. (2005). Some endangered plants producing edible fruits and seeds in Southeastern Nigeria. *Fruits, 60*, 211–220. http://dx.doi.org/10.1051/fruits:2005028

Millennium Ecosystem Assessment. (2005). *Ecosystems and human well-being: Synthesis*. Washington, DC: Island Press. Retrieved November 10, 2011, from http://www.maweb.org/documents/document.356.aspx.pdf

Milne, A. K. (1988). Change direction analysis using Landsat imagery: A review of methodology. In *Proceedings of the IGARSS'88 Symposium Edinburgh, Scotland, ESASP-284* (pp. 541–544). Noordwijk: ESA.

Momoh, S. (2014). Endangered timber business. *Vanguard, 2*, 4.

Müller, H., Griffiths, P., & Hostert, P. (2016). Long-term deforestation dynamics in the Brazilian Amazon—Uncovering historic frontier development along the Cuiabá-Santarém highway. *International Journal of Applied Earth Observation and Geoinformation, 44*, 61–69. http://dx.doi.org/10.1016/j.jag.2015.07.005

Newbold, T., Hudson, L. N., Phillips, H. R., Hill, S. L., Contu, S., Lysenko, I., … Purvis, A. (2014). A global model of the response of tropical and sub-tropical forest biodiversity to anthropogenic pressures. *Proceedings of the Royal Society B: Biological Sciences, 281*, 20141371. http://dx.doi.org/10.1098/rspb.2014.1371

Niemelä, J. (1999). Ecology and urban planning. *Biodiversity & Conservation, 8*, 119–131. http://dx.doi.org/10.1023/A:1008817325994

Obeta, M. C., Ochege, F. U., Aujara, I. Y., & Shehu, S. M. (2013). Socio-economic significance of woodstocks and non-timber forest products in Jigawa State, Nigeria. In L. Popoola, F. O. Idumah, O. Y. Ogunsanwo, & I. O. Azeez, (Eds.). *Forest industry in a dynamic global environment. Proceeding of the 35th Annual conference of the forestry Association of Nigeria* (pp. 727–735). Sokoto: Forestry Association of Nigeria.

Obydenkova, A., Nazarov, Z., & Salahodjaev, R. (2016). The process of deforestation in weak democracies and the role of Intelligence. *Environmental Research, 148*, 484–490. http://dx.doi.org/10.1016/j.envres.2016.03.039

Ochege, F. U. (2014a). *Spatio-temporal variations of vegetation cover over Umuahia, Abia state Nigeria from 1984 to 2014* (MSc thesis). University of Nigeria, Nsukka.

Ochege, F. U. (2014b). *Geospatial assessment of forest degradation in Sagbama, Niger delta region of Nigeria* (MSc thesis). University of Edinburgh, Scotland.

Odjugo, P. A. O. (2010). General overview of climate change impacts in Nigeria. *Journal of Human Ecology, 29*, 47–55.

Odjugo, P. A. O., & Ikhuoria, A. I. (2003). The impact of climate change and anthropogenic factors on desertification in the semi-arid region of Nigeria. *Global Journal of Environmental Science, 2*, 118–126.

Ofori, B. Y., Owusu, E. H., & Attuquayefio, D. K. (2014). Ecological status of the Mount Afadjato-Agumatsa range in Ghana after a decade of local community management. *African Journal Ecology, 53*, 116–120.

Okpala-Okaka, C., & Igbokwe, J. I. (2010). Revision of Nsukka N.E Topographic Map Sheet 287 1:50,000 (1964) using Nigeria SAT-1 Imagery. *Nigerian Journal of Space Research, 7*, 13–24.

Omo-Irabor, O., Olobaniyi, S. B., Akunna, J., Venus, V., Maina, J. M., & Paradzayi, C. (2011). Mangrove vulnerability modelling in parts of Western Niger Delta, Nigeria using satellite images, GIS techniques and Spatial Multi-Criteria Analysis (SMCA). *Environmental Monitoring and Assessment, 178*, 39–51. http://dx.doi.org/10.1007/s10661-010-1669-z

Onu, N. N., Opara, A. I., & Ehirim, C. N. (2012). Delineation of Active fractures in a gully erosion area using geophysical methods: Case study of the Okigwe—Umuahia Erosion Belt, Southeastern Nigeria. *International Journal of Science and Technology, 2*(4). ISSN 2224-3577.

Osemeobo, G. J. (1988). The Human causes of forest depletion in Nigeria. *Environmental Conservation, 15*, 17–28. http://dx.doi.org/10.1017/S0376892900028411

Richards, P. W. (1939). Ecological studies on the rain forest of southern Nigeria. 1. The structure and floristic composition of primary forest. *The Journal of Ecology, 27*, 1–61. http://dx.doi.org/10.2307/2256298

Rojas, C., Pino, J., Basnou, C., & Vivanco, M. (2013). Assessing land-use and land-cover changes in relation to geographic factors and urban planning in the metropolitan area of Concepción (Chile). *Implications for biodiversity conservation. Applied Geography, 39*, 93–103.

Schwartz, M. D. (2003). *Phenology: An integrative environmental science* (vol. 39). Netherlands: Springer.

Scull, P., Cardelús, C. L., Klepeis, P., Woods, C. L., Frankl, A., & Nyssen, J. (2016). The resilience of Ethiopian church forests: Interpreting aerial photographs, 1938–2015. *Land Degradation & Development*.

Simmons, C., Walker, R., Perz, S., Arima, E., Aldrich, S., & Caldas, M. (2016). Spatial patterns of frontier settlement: Balancing conservation and development. *Journal of Latin American Geography, 15*, 33–58. http://dx.doi.org/10.1353/lag.2016.0011

Smith, B. D. (2007). The ultimate ecosystem engineers. *Science, 315*, 1797–1798. http://dx.doi.org/10.1126/science.1137740

Soofi, K. (2005). *Remote sensing lecture: Image classification. UCSD: Satellite remote sensing - SIO 135/SIO 236.* Retrieved from http://topex.ucsd.edu/rs/classification.pdf

Sulaiman, C., Abdul-Rahim, A. S., Mohd-Shahwahid, H. O., & Chin, L. (2017). Wood fuel consumption, institutional quality, and forest degradation in sub-Saharan Africa: Evidence from a dynamic panel framework. *Ecological Indicators, 74*, 414–419. http://dx.doi.org/10.1016/j.ecolind.2016.11.045

Tee, N. T., Ancha, P. U., & Asue, J. (2009). Evaluation of fuelwood consumption and implications on the environment: Case study of Makurdi area in Benue state, Nigeria. *Journal of Applied Biosciences, 19*, 1041–1048.

UNEP, IISD (Ed.). (2004). *Exploring the links: Human well-being poverty and ecosystem services*. Nairobi and Winnipeg, Manitoba: Author.

United Nations Population Fund. (1991). *Population, resources and the environment: The critical challenges* (154 p.). New York, NY. ISBN 0-89714-101-6.

Watson, R. T., Noble, I. R., Bolin, B., Ravindranath, N. H., Verardo, D. J., & Dokken, D. J. (2000). *Land use, land-use change and forestry IPPC Report*. Retrieved from www.ipcc.ch/pdf/special-reports/spm/srl-en.pdf

White, L. J. T., & Oates, J. F. (1999). New data on the history of the plateau forest of Okomu, southern Nigeria: An insight into how human disturbance has shaped the African rain forest. RESEARCH LETTER. *Global Ecology and Biogeography, 8*, 355–361. http://dx.doi.org/10.1046/j.1365-2699.1999.00149.x

World resources. (1987). *An assessment of the resource base that supports the global economy*. New York, NY: International institute of environment and development and the world resources institute Basic Books, 369 pp. https://www.wri.org/our-work/project/world-resources-report/publications?page=2

Yeshaneh, E., Wagner, W., Exner-Kittridge, M., Legesse, D., & Blöschl, G. (2013). Identifying land use/cover dynamics in the Koga catchment, Ethiopia, from multi-scale data, and implications for environmental change. *ISPRS International Journal of Geo-Information, 2*, 302–323. http://dx.doi.org/10.3390/ijgi2020302

Zhu, X., Liu, D., & Chen, J. (2012). A new geostatistical approach for filling gaps in Landsat ETM+ SLC-off images. *Remote Sensing of Environment, 124*, 49–60. http://dx.doi.org/10.1016/j.rse.2012.04.019

Appendix 1

Trend of land cover changes (%) in the study area

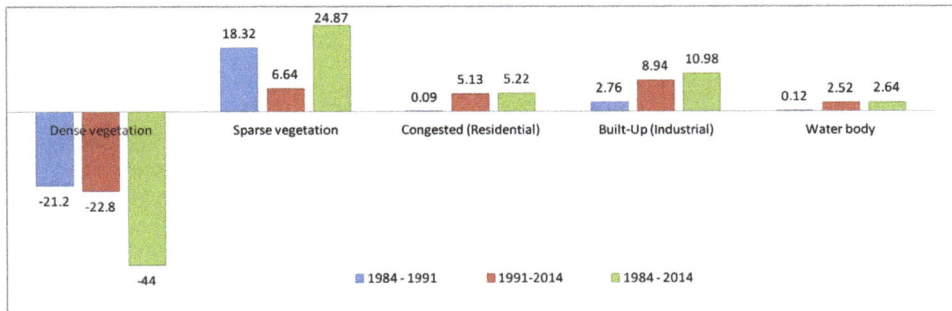

Appendix 2

Temporal pattern

Land use cover	1984	1991	2014
Vegetation resources	93.03	89.97	73.81
Urban encroachment	6.82	9.67	23.02
Water	0.24	0.36	2.88

Sedimentological study of Lake Nasser; Egypt, using integrated improved techniques of core sampling, X-ray diffraction and GIS platform

Hussien ElKobtan[1], Mohamed Salem[2], Karima Attia[3], Sayed Ahmed[2] and Islam Abou El-Magd[4]*

*Corresponding author: Islam Abou El-Magd, Environmental Studies Department, National Authority for Remote Sensing and Space Sciences, 23 Josef Tito St., El-Nozha El-Gedida, P.O. Box 1564 Alf-Maskan, Cairo, Egypt
E-mail: imagd@narss.sci.eg

Reviewing editor:
Xiangming Tang, Nanjing Institute of Geography and Limnology Chinese Academy of Sciences, China

Abstract: Lake Nasser is one of the largest man-made reservoirs, that is located on the Nile River. To understand the sedimentation process of the lake, bottom sediments from the bottom-surface of the lake core samples from the top 1.25 m of the bottom layer were collected. These samples were mechanically analysed in the laboratory. The analysis of statistical parameters of the sediment samples has generally classified the lake into two depositional environments that reflect the sedimentation process; (1) the riverine environment that exist at the entrance of the lake between El-Daka and CC stations, (2) the lacustrine environment that extend along the rest of the lake to the High Aswan Dam. Along the riverine environment, the river processes were the prevailing, which being reflected on the bottom sediments that are nearly free from clay and composed mainly of sand (>87%) mixed with small ratios of silt (<10%). Further downstream to the end of the lake the lacustrine environment is dominating with slow deposition from quite water with bottom sediments free of sand and the bottom sediments composed mainly of clay (>57%). X-ray analysis indicated that montmorillonite, kaolinite and illite are the dominant clay minerals. GIS was used to spatially simulate the bottom sediment distribution at the bottom of the lake.

Subjects: Geology-Earth Sciences; Geomorphology; GIS, Remote Sensing & Cartography; Sedimentology & Stratigraphy

Keywords: sedimentation process; X-ray diffraction; core sampling; grain size analysis; GIS; Lake Nasser

ABOUT THE AUTHOR

Islam Abou El-Magd is an associate professor working for the National Authority for Remote Sensing and Space Sciences and acting as the head of the Environmental Studies Department. Abou El-Magd research area of interest is remote sensing and GIS modelling in environmental related issues. Abou El-Magd has obtained his PhD from the School of Civil Engineering and The Environment, University of Southampton, UK where he also worked there for few years. Currently, he is managing few research projects that are directly function remotely sensed data to study the environmental pollution and coastal zone management and climate changes. He is also teaching both undergraduate and postgraduate courses in remote sensing and its environmental applications, within Egyptian universities and the region.

PUBLIC INTEREST STATEMENT

Lakes are important water mass bodies for local livelihoods communities, food security and environmental balance. Lake Nasser is one of the largest man-made lake that support Egypt community in water storage and other socio-economic activities. The lake has huge area that very vulnerable to flooding season that carry water and sediments. This research was undertaken to enable for more understanding of the environmental status of the Lake and the interaction between the lake and the hydrological parameters. For example, the huge amount of sediment that arrives every year to the Lake during the flooding season creates threat on the storage capacity of Lake. Such sediment load also creates changes in the shape of the bottom of the lake and could be extended to the morphology of the lake. This research highlights these issues.

1. Introduction

Egypt has constructed a huge human controlled structure "High Aswan Dam" to manage the Nile water and rescue the Nile Delta and flood plain from flooding. At the upstream side of this dam a huge reservoir was formed, which is called Lake Nasser. The lake stores water budget of Egypt (55.5 BCM per year), to secure the water usage in the country (Abul-Atta, 1978). It also receives huge amount of sediments that carried with the water flow from the catchments of the Nile River. The tunnels of the High Aswan Dam allow water to drain downstream into the Nile River, however sediments continuously settle down and accumulate in the lake deforming and reducing the storage capacity.

It is estimated that more than 134 million tons of Nile sediments are deposited annually in Lake Nasser (Shalash, 1980). Depending on this estimation, it could be estimated that there were about 6.8 Milliard tons of sediments deposited along the lake since the construction of the dam till 2015. Owing to the strategic importance of the preservation of the lake storage capacity and the priority of sediments as an essential component of the lake system, there is an essential need for understanding the sedimentation processes. Up till now, there is no precise information about the material beneath the surface of the sediments of Lake Nasser.

For accurate representation of the lake sediments, sampling and measuring processes took place. The locations in respect to the cross section's vertex were determined earlier based on optical survey instruments like Levels and Total Station (Dahab & EL-Moattassem, 1994; Makary, 1982; Shalash, 1980). However, El-Kobtan (2007) used global positioning system (GPS) technique to accurately position the sampling-measuring stations. The bottom sediment samples were mechanically analysed in order to determine their graphical grain size statistical parameters and their significances, mineralogical and chemical composition (Abdel-Aziz, 1997; Dahab & EL-Moattassem, 1994; El-Kobtan, 2007; El-Manadely, 1991; Makary, 1982; Philip, Hassan, & Khalil, 1978; Shalash, 1980). Minerology of sand fraction included in the bottom sediments was investigated using the polarized light microscopic technique (Philip et al., 1978). Whereas, the clay mineralogy was investigated using X-ray Florescence Diffraction Technique (XRFD) (El-Kobtan, 2007; Makary, 1982).

In this research, an integrated approach of using improved techniques: (1) core sampler to collect core samples from the bottom sediments of Lake Nasser, (2) X-ray diffraction and laboratory analysis and (3) geographic information systems to provide comprehensive knowledge and understanding of the sedimentation processes in Lake Nasser.

1.1. Description of the study area

Lake Nasser is located on the border between Egypt and Sudan, lies between Latitudes 20°27′N–23°58′N and Longitudes 30°07′–33°15′E. Its northern two-third part is located in Egypt (always called Lake Nasser) whereas the southern one-third part is located in Sudan (sometimes called Lake Nubia).

In the case of full storage capacity, the water level reaches 182 m (ASL), and the lake extends from the Aswan High Dam in Egypt to Dal Cataract in Sudan with length of about 500 km (Figure 1). At this level, the lake occupies an area of about 6,500 km² whereas its storage capacity reaches about 162 billion m³ (Abul-Atta, 1978). This creates a maximum width of about 24 km and a maximum depth of about 110 m.

Lake Nasser is located in the Nubian Desert that surrounded by barren desert and hilly areas. The climatic condition is typical arid environment with high temperature that reaches as high as 45°C in June and as low as 5°C in January. It also has nearly no precipitation that classified this region as one of the driest areas in the world. Such conditions have classified this region as hyper-arid (Springuel, Hassan, Sheded, El-Soghir, & Ali, 1991).

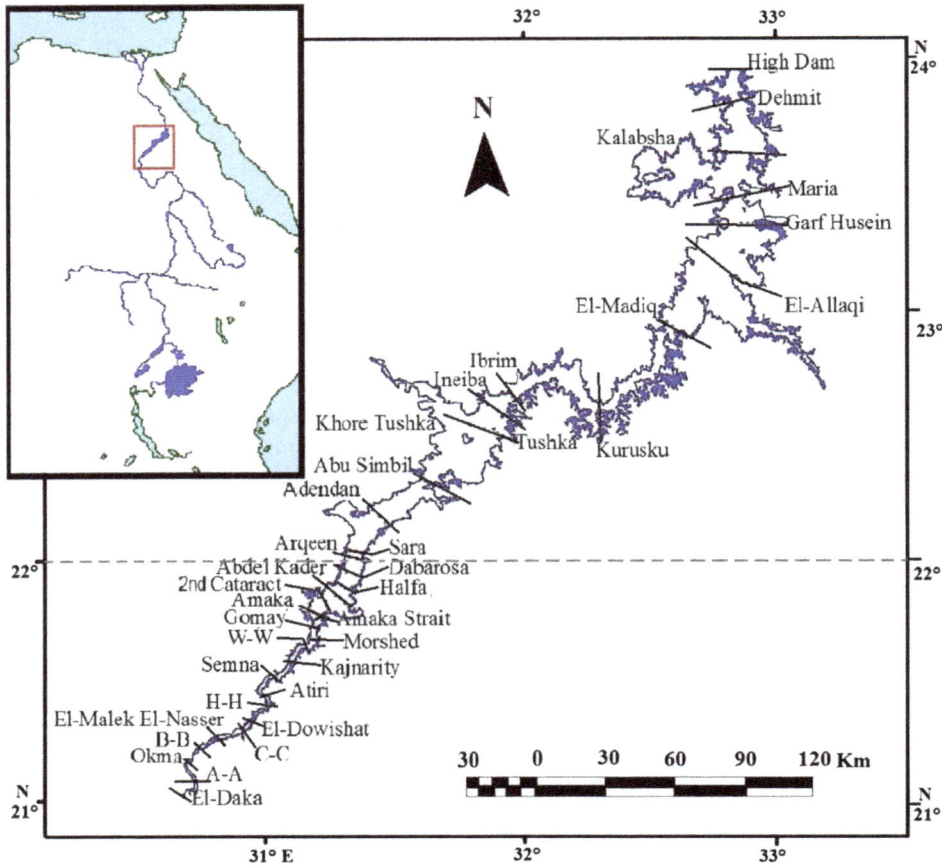

Figure 1. Lake Nasser as appears in its full storage case and the traces of the investigated profiles.

Lake Nasser is located in a complex geological area; therefore, it is underlain and surrounded by a wide variety of lithology like granite, granitoids, gneisses, schists, sandstones, conglomerates and shales. The area is highly affected by structural elements like folds, faults and fractures, which highly controlled the lake path. The lithological variation around the lake body controlled the differentiation of the lake into its southern and northern parts around Gomay. The hard basement rocks surrounding the lake along the southern part may be responsible for the narrowing of the lake course. However, the softer sedimentary rocks surrounding the lake along the northern part may be responsible for the widening of the lake course. Therefore, the changes in the hydro-morphologic features (depth, width and hence, profile area), to a large extent, appear to be litho-structurally controlled.

2. Materials and methods

Some measuring and sampling processes were carried out through two field trips. The first field trip was of two phases covered almost by the lake. The first phase between 1 and 18 December 2006 covered the Egyptian part of the lake, whereas, the second phase between 2 and 14 February 2007 covered the Sudanese part. Through this field trip, the measurements included some bathymetric and hydro-morphologic measurements in addition to some hydrographic (physicochemical) measurements. The collected bottom sediments and surface water samples were analysed in the laboratory.

The second field trip was carried out during November 2011 to collect core samples between Latitudes 30°58′31.64″ E & 31°26′25.10″ E and Longitudes 21°27′42.07″ N & 22°10′15.11″ N. The hydro-morphologic and bathymetric measurements were carried out using a combination of GPS and the Ecosounding system. This technique was used throughout 32 horizontal cross sections. These cross sections were selected to geographically cover most of the lake.

Some hydrographic (physicochemical) measurements were carried out at three main levels (surface, middle and near bottom). The current velocity was measured using a currentmeter model VALEPORT BFM 108 MK11. The TDS and EC were measured using a portable electrical conductivitymeter (WTW Model LF 197). The pH was also measured using a portable pH meter (WTW Model pH 197). The temperature was measured using a portable instrument (WTW Model Oxi 197). The turbidity was measured using a Hach 2100P portable turbidity meter. In addition, 93 water samples were collected from the lake and were filtered to determine the suspended sediments concentration.

Eighty-eight grab sediment samples were collected from the bottom of Lake Nasser that comprises 32 profiles (cross sections) (Figure 1). The samples are geographically representing the eastern, middle and western parts of each profile. The exact location of the sampling stations was carried out using the GPS.

The lack of advanced resources and fund to obtain such equipment to collect core samples has challenged to design and produce a low-cost local core sampler. A core sampler (gravity corer) was designed and manufactured for this research to collect bottom sediments samples from the lake. The core sampler was made of a 1.50 m length and 3″ diameter steel pipe with a plastic (PVC) liner. Using this corer, 13 cores were collected along the distance between Atiry (30°58′31.64″ E–21°27′42.07″ N) and Adendan (31°26′25.10″ E–22°10′15.11″) to represent the mid-line of the lake.

2.1. Laboratory analysis
The collected bottom samples were air-dried, which consequently fractionated into two main types according to their texture; coarser than 0.0625 mm (less than 4Ø) and finer than 0.0625 mm (more than 4Ø). For the first type, wet and dry sieving were applied using standard set of sieves with mesh opening 1, 0.5, 0.25, 0.125 and 0.0625 mm. However, pipette analysis was carried out upon the second type applying the method described by Krumbein and Pettijohn (1938). Na hexameta phosphate and Na carbonate with concentrations of 1.65 and 0.35 gm/L, respectively, were used as peptizer. In this method, 20 gm of washed dried sediment sample is flocculated in a 1,000 ml measuring cylinder charged with peptizer. Suctions at intervals of time according to stokes law were taken and the weight per cent for each suction was calculated in weight per cent and cumulative per cents.

Histograms and cumulative curves were constructed. Depending on these histograms, sand, silt and clay components were calculated in percentage using Wentworth (1922) that modified by Friedman and Sanders (1978). Depending on the cumulative curves, key statistical parameters included median diameter (MdØ), mean size (MzØ), inclusive sorting (σ_1), skewness (Sk_1) and kurtosis (K_G) were calculated according to the equations proposed by Folk and Ward (1957).

In addition to the above analysis, the collected cores were divided into sub-core samples based on depths ranged 0–25, 25–50, 50–75, 75–100 and 100–125 cm. The produced 47 core samples were air-dried and analysed to determine the grain size distribution and mineralogical composition. The grain size distribution indicated that silt and clay sized particles are the main component.

2.2. X-ray diffraction analysis
X-ray diffraction is an efficient tool for the mineralogical analysis of fine-grained sediments (Worden & Morad, 2003). Therefore, the mineralogical composition of the bottom sediments was determined for 19 selected core samples of the bottom sediment using the X-ray diffraction technique. The selected core samples were grinded to less than 63 μm grain sized powder and investigated using Philips X- ray Vertical diffractometer (type PW 1373, Holland). For studying the clay mineralogical composition, the clay fraction of each sample (particle size < 2 μm) was separated by sedimentation using glass slide method (Bish & Reynolds, 1989; Hughes, Moore, & Glass, 1994), after the removal of carbonate, iron oxide, organic matter and soluble salts. The clay fraction on the glass slides was

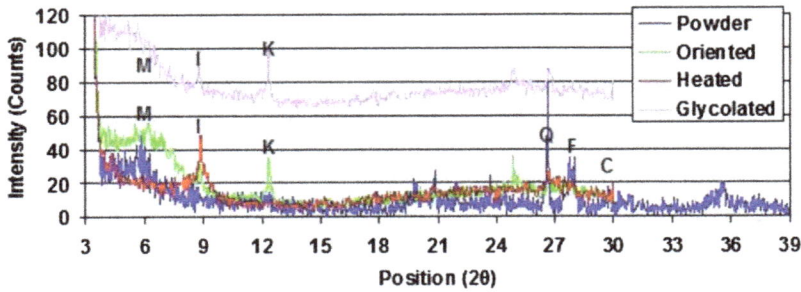

Figure 2. The X-ray diffraction patterns of a powdered, clay oriented, heated and glycolated sample showing the peaks of montmorillonite (M), illite (I), kaolinite (K), quartz (Q), feldspar (F) and calcite (C).

analysed in each of the air-dried, heated and ethylene glycol-solvated conditions. Figure 2 shows the typical X-ray diffraction patterns of a powdered, clay oriented, heated and glycolated core sample.

The relative abundance of each mineral was estimated from the intensities of the diffraction peaks, measured by the peak heights or peak areas (Ruhe & Olson, 1979; Tucker, 1988). Semi-quantitative comparisons were made between samples by means of various ratios of peak heights or/and peak areas (Biscaye, 1965).

A simple mathematical procedure was applied for roughly calculating the relative proportion of each of the composing non-clay minerals as a part of the unit using the following equation:

$$P_{M1} = \left(\frac{I_{M1}}{I_{M1} + I_{M2} + I_{M3}} \right), \; P_{M2} = \left(\frac{I_{M2}}{I_{M1} + I_{M2} + I_{M3}} \right), \; P_{M3} = \left(\frac{I_{M3}}{I_{M1} + I_{M2} + I_{M3}} \right) \tag{1}$$

whereas

$$P_{M1} + P_{M2} + P_{M3} = 1 \tag{2}$$

where P_{M1}, P_{M2} and P_{M3} are the proportions of the non-clay minerals 1, 2 and 3, respectively. I_{M1}, I_{M2} and I_{M3} are the peak intensities of the non-clay minerals 1, 2 and 3, respectively.

2.3. GIS

GIS was used for spatial representation and analysis of the data. The sampling locations were converted into a point layer with all the results of the laboratory analysis embedded attributes. It enabled for simulating the data spatially to understand the spatial distribution of the sediments and then understand the sedimentological processes influenced the deposition of these sediments. It also enabled for understanding the hydro-morphology of the lake.

3. Results

The bottom sediment samples were mechanical analysed to be statistically represented. The statistical parameters determined found to be widely varied along the lake.

MdØ ranges between 1.66Ø (medium sand) and 11.36Ø (clay) with an average of 8.14Ø (very fine silt). MzØ ranged between 1.65Ø (medium sand) and 10.84Ø (clay) with an average of 7.97Ø (fine silt). σ_I between 0.29Ø (very well sorted) and 3.64Ø (very poorly sorted) with an average of about 1.94Ø (poorly sorted). Sk$_I$ ranged between −0.76 (strongly coarse skewed) and 0.75 (strongly fine skewed) with an average of −0.10 (coarse skewed). K_G ranged between 0.58 (very platykurtic) and 4.41 (extremely leptokurtic) with an average of about 1.28 (leptokurtic).

To understand the mechanisms of deposition along the lake, bivariant plots between the average values of median diameter MdØ and each of the inclusive sorting σ_I were illustrated in Figure 3 and skewness Sk_I in Figure 4. The study of the bivariant plot diagrams allows us to distinguish between two main depositional mechanisms, river processes and slow deposition from quiet water (Stewart, 1958).

Distribution of statistical parameters of the bottom sediments along the lake was represented in Figure 5. It proved the occurrence of two depositional environments of riverine (the most upstream 50 km at the entrance of the lake) and lacustrine (the rest of the lake). In turn, the lacustrine environment was differentiated at the 2nd Cataract into two distinctive sedimentological regions, southern and northern.

Along the riverine environment, the average values of MdØ increased northward from 1.79Ø (medium sand) at El-Daka to 2.85Ø (fine sand) at CC with an average of 2.53Ø (fine sand). MzØ increased northward (R = −0.767) from 1.08Ø (medium sand) to 2.88Ø (fine sand). However, σ_I abruptly increased from 0.36Ø (well sorted) at AA to 1.13Ø (poorly sorted) at Okma, then decreased northward to 0.31Ø (very well sorted) at the site CC. Meanwhile, Sk_I ranged between 0.14 (fine skewed) and 0.65 (strongly fine skewed) with no direction to change (i.e.) the sediments tend to skew towards the fine fraction. K_G decreased northward from 0.86 (platykurtic) at AA to 2.29 (very leptokurtic) at CC.

Along the southern part of the lacustrine environment (which is the transitional from riverine to lacustrine), the average value of MdØ increased northward from 5.83Ø (coarse silt) at El Dowishat to 9.26Ø (clay) at the Second Cataract. MzØ increased northward from 6.16Ø (medium silt) to 9.07Ø (clay). σ_I ranged between 2.3 and 2.75Ø (very poorly sorted) with no direction for change. Sk_I decreased northward from 0.28 (fine skewed) to −0.17 (coarse skewed). K_G ranged between 0.75 and 0.85 (platykurtic) with no direction for change.

Along the northern part of the lacustrine environment, the average value of MdØ slightly decreased northward from 10.76Ø (clay) at Abdel Kader to 9.45Ø (clay) at Kalabsha. MzØ slightly decreased from 10.29Ø (clay) to 9Ø (clay). σ_I increased from 1.88Ø (poorly sorted) to 2.57Ø (very poorly sorted). Sk_I sharply decreased to range between −0.48 and −0.66 (strongly coarse skewed) between Abdel Kader and Dabarosa, then decreased to range between −0.16 (coarse skewed) and −0.32 (strongly coarse skewed) further north, with nearly no direction for change. K_G ranged between 1.11 and 1.50 (leptokurtic) with no direction for change.

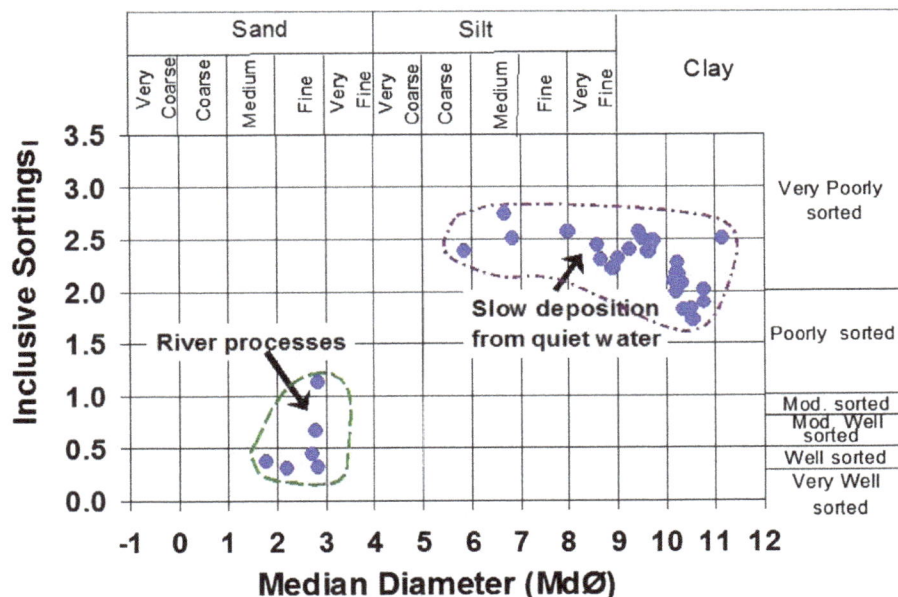

Figure 3. Bivariant plot between MdØ and σ_I.

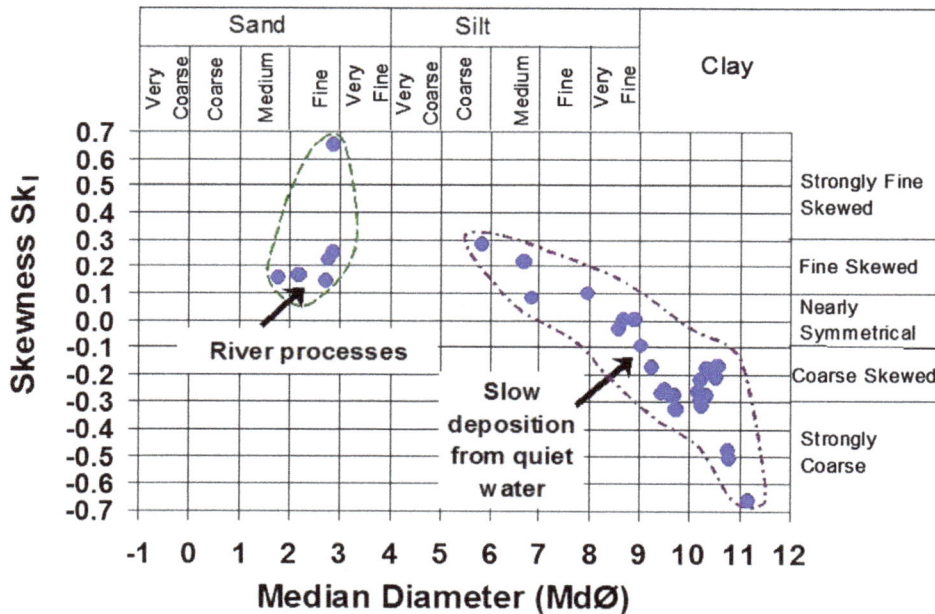

Figure 4. Interrelation between MdØ and Sk$_I$.

To understand the factors affecting the sediments distribution along the lake, a correlation between the mean diameter (MzØ) of the bottom sediments and each of the current velocities, total
dissolved salts TDS, pH and each of the hydro-morphologic features was illustrated (Figure 6).

The interrelation between MzØ and the suspended sediments concentration (Figure 6(A)) showed
a reverse relation along the part of the lake south of the site WW ($R = -0.94$), i.e. the riverine environment and the entrance of the lacustrine environment. This indicates the northward decrease in the
grain size of the bottom sediments with the continuity of settlement of the suspended sediments.
Along the distance between the site WW and Abu Simbil, MzØ continued to show inverse relation
with the suspended sediments concentration ($R = -0.90$). This indicates the continuity of the northward decrease in the grain size of the bottom sediments with the continuity of settlement of the
suspended sediments. North of Abu Simbil, MzØ showed a direct relation with the suspended sediments concentration ($R = 0.93$). This indicates that the grain size of the bottom sediments increased
northward with the continuity of the settlement of the suspended sediments from the water column. This may indicate, in turn, the northward increase in the grain size of the sediments in suspension to be coagulated to settle down.

The interrelation between MzØ and TDS (Figure 6(C)) showed no distinct correlation along the riverine and southern part of lacustrine environment. However, it showed a reverse correlation along
the northern part of the lacustrine environment indicating the increase in the grain size with the increase in the TDS concentration. This may be due to the increase in the clay particle size in suspension which coagulate to settle down with the increase in the TDS concentration along the northern
part of the lacustrine environment.

The interrelation between MzØ and pH (Figure 6(D)) showed no distinct correlation along the riverine environment and along the upstream entrance of the southern part of the lacustrine environment (till reaching Atiry). North of Atiry, as the MzØ value exceeded 8 (i.e. clay content was more
than 45% of the bottom sediment), there was a direct relationship between the MzØ and pH. This
means that the grain size of the clay bottom sediments decreases with increasing pH. This may be
due to the decrease in the clay particles size in suspension to coagulate and settle down with the
increase in pH values.

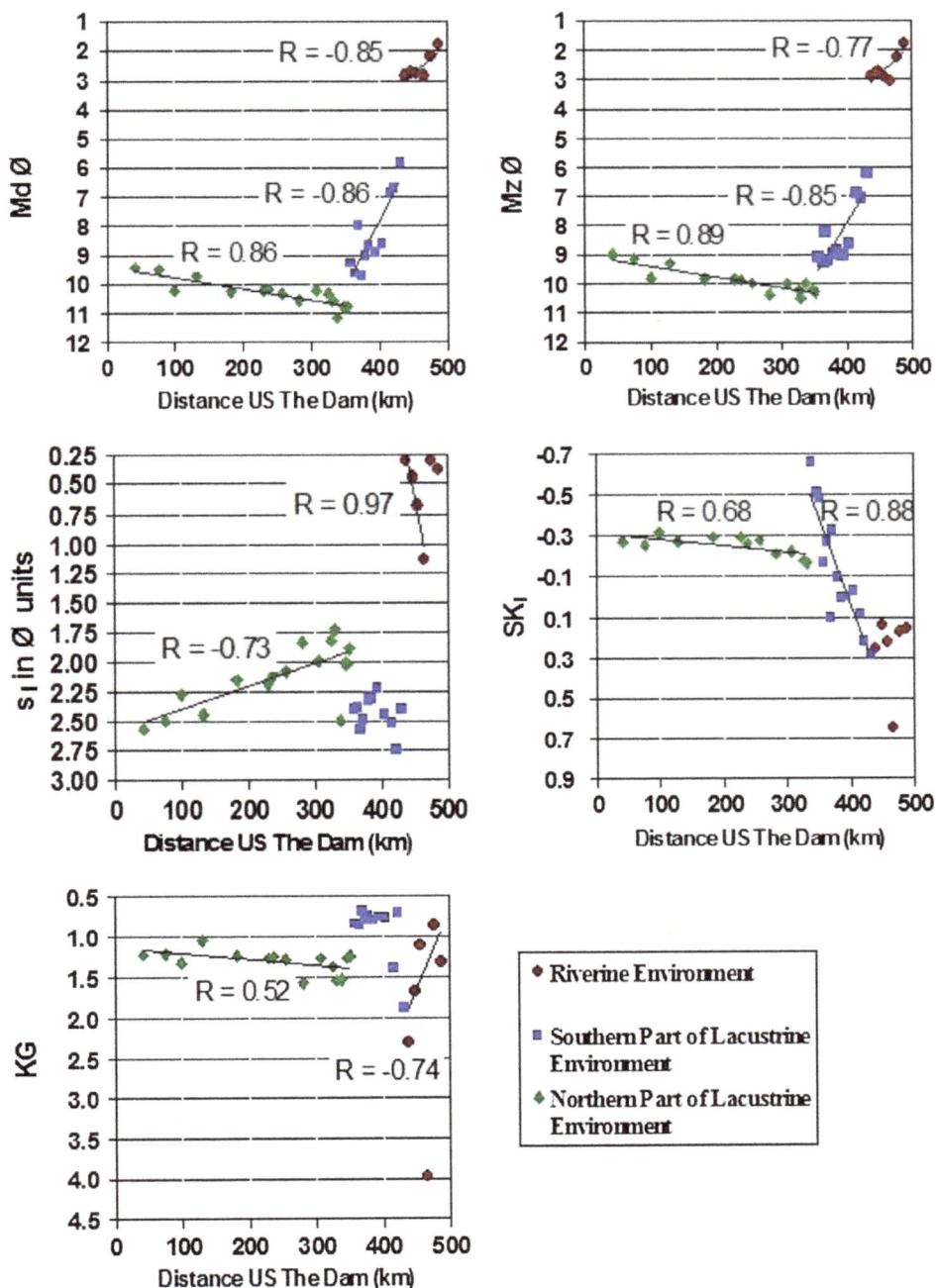

Figure 5. The variations of statistical parameters of the surface layer of bottom sediments along the lake.

The interrelation between MzØ and hydro-morphologic features including the width, depth and profile area (Figure 6(E)–(G), respectively) showed no distinct relation. There was an exception for the depth that showed an inverse relation with MzØ along the northern part of the lacustrine environment ($R = -0.736$) meaning that the grain size of the bottom sediments increased with the increase in depth.

Sediment types strongly control the mineralogical distribution along the lakes. Such distribution of sediment types in reservoirs (lakes) is required to be evaluated before studying the mineralogical or elemental distribution (Horowitz, 1991). Therefore, the core samples were mechanically analysed to determine the grain size distribution. It indicated that the core sampled sediments composed of

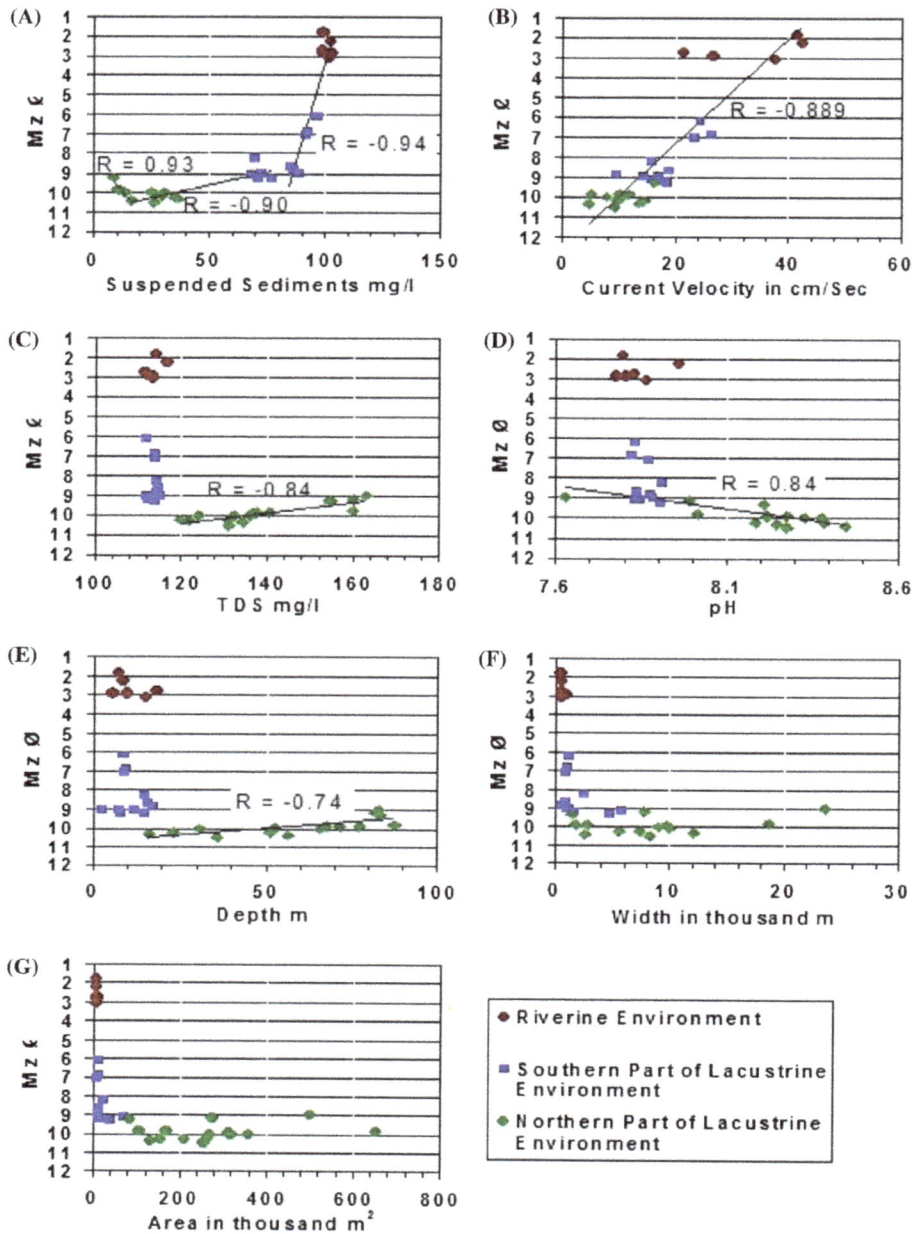

Figure 6. Interrelations between MzØ and each of (A) current velocities, (B) total dissolved salts TDS, (C) pH and each of the hydro-morphologic features including (D) depth, (E) width and (F) profile area.

mixed sand, silt and clay grain mixture. The part of the lake that was sampled during the coring process located in the lacustrine environment. The riverine environment has not included; therefore, there was no sample classified as sand.

For more understanding of the vertical distribution of the bottom sediment of the lake, the vertical distribution of 125 cm depth of the bottom sediments below the water-bottom sediments interface is shown in Figure 7. The figure shows the realistic sediment distribution that respond to the depositional environment. The sandy silt and silty sand classes were limited to the most southern part of the studied area (at Atiry). Northward, reaching Halfa, the investigated thickness of the bottom sediments composed of inter-bedded silt and clayey silt classes. North of Halfa, the silt variety

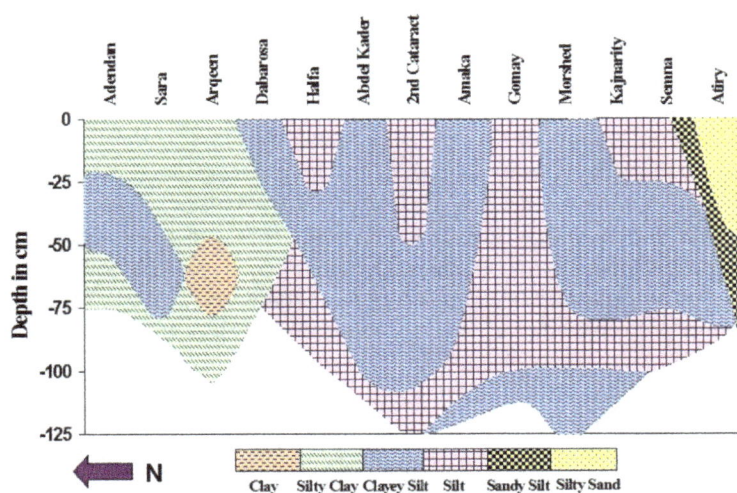

Figure 7. The vertical distribution of sediment classes.

withdrawn to lower depths giving rise to the clayey silt, silty clay and clay varieties to spread out in this area. This completely reflects the two depositional environments of riverine to the south with coarse sediments and lacustrine environment downstream where the sediment getting finer.

X-ray diffraction patterns of the samples' powder showed that the bottom sediments along the studied locality are composed mainly of clay minerals including montmorillonite, kaolinite and illite in addition to some non-clay minerals including quartz, feldspar and calcite.

Table 1. Peak height of the non-clay minerals and its relative peak height proportions in the analysed core samples along the studied area

Site name	Sample no.	Peak height in counts			Relative peak height proportions		
		Quartz	Feldspar	Calcite	Quartz	Feldspar	Calcite
Atiry	2	333	85	15	0.77	0.20	0.03
	3	167	57	30	0.66	0.22	0.12
Semna	5	36	16	14	0.55	0.24	0.21
Kajnarity	8	22	11	5	0.58	0.29	0.13
	10	51	30	16	0.53	0.31	0.16
Morshed	12	24	10	14	0.50	0.21	0.29
	15	18	11	6	0.51	0.31	0.17
Gomay	17	55	83	18	0.35	0.53	0.12
Amaka	20	22	12	9	0.51	0.28	0.21
2nd Cataract	25	43	18	9	0.61	0.26	0.13
	26	25	14	10	0.51	0.29	0.20
Abdel Kader	31	21	11	13	0.47	0.24	0.29
Halfa	34	88	35	17	0.63	0.25	0.12
Dabarosa	35	12	6	5	0.52	0.26	0.22
	37	52	13	9	0.70	0.18	0.12
Arqeen	38	15	11	7	0.45	0.33	0.21
	40	26	12	10	0.54	0.25	0.21
Sara	44	22	12	13	0.47	0.26	0.28
Adendan	47	27	12	8	0.57	0.26	0.17

The peak height of the non-clay minerals and its relative peak height proportions is listed in Table 1. The peak height was used as the measure of peak intensity. Applying this simplified procedure, used not to quantify the mineral content in each sample, but significantly to look for the trends of the changes in the minerals relative abundance.

Velde and Meunier (2008) specified the X-ray diffraction as the most important analytical technique used to identify and quantify the clay minerals present in a sample. Peaks created by clay minerals in the X-ray diffraction patterns, especially mixed-layer varieties, are often very broad. Therefore, a clay mineral present is more closely related to peak area than to peak height because of the natural variation in clay mineral peak shapes and widths (Klug & Alexander, 1954). The X-ray diffraction patterns of the glycolated oriented clay fraction were used in this aspect.

Montmorillonite, kaolinite and illite are the main constituent clay minerals. The quantification of the abundance of each clay mineral constituent was carried out through two steps. The first is the semi-quantitative estimation of the relative abundance of the clay minerals using Equations (1) and (2). In these equations, the variants P_{M1}, P_{M2} and P_{M3} are the proportions of the peak intensities of clay minerals 1, 2 and 3, respectively; whereas, the variants I_{M1}, I_{M2} and I_{M3} are the peak intensities of the clay minerals 1, 2 and 3, respectively. The peak intensity is calculated by multiplying the mineral's peak area by its specific power factor (1 for montmorillonite, 2 for kaolinite and 4 for illite). The second step is the quantitative estimation based on calculating the percentage of each of the clay minerals in a sediment sample. The second step was carried out by considering that the clay minerals integrated together to form the percentage of the clay sized portion in a sediment sample. The relative proportions (semi-quantitative) and the percentage (quantitative) of the composing clay minerals along the studied locality were listed in Table 2.

Table 2. Relative peak height proportions of the clay minerals montmorillonite (M), kaolinite (K) and illite (I) in the analysed core samples along the studied area

Site name	Sample no.	Relative proportions			Percentage from total clay		
		M	K	I	M	K	I
Atiry	2	0.858	0.041	0.101	2.81	0.14	0.33
	3	0.819	0.090	0.090	8.89	0.98	0.98
Semna	5	0.887	0.075	0.038	21.95	1.86	0.93
Kajnarity	8	0.841	0.101	0.058	29.15	2.59	1.82
	10	0.869	0.077	0.054	16.39	1.98	1.13
Morshed	12	0.881	0.062	0.057	36.81	2.04	4.49
	15	0.849	0.047	0.104	28.93	2.04	1.86
Gomay	17	0.875	0.054	0.072	15.84	0.97	1.30
Amaka	20	0.881	0.068	0.051	32.62	2.51	1.88
2nd Cataract	25	0.873	0.101	0.025	29.61	2.88	1.20
	26	0.879	0.085	0.036	27.67	3.21	0.80
Abdel Kader	31	0.857	0.070	0.073	34.39	2.79	2.92
Halfa	34	0.839	0.078	0.083	16.27	1.52	1.60
Dabarosa	35	0.917	0.064	0.020	23.79	1.66	0.51
	37	0.816	0.076	0.109	41.65	3.87	5.55
Arqeen	38	0.685	0.107	0.207	46.36	7.27	14.02
	40	0.872	0.040	0.087	64.97	3.00	6.51
Sara	44	0.874	0.069	0.057	41.87	3.30	2.75
Adendan	47	0.724	0.142	0.134	39.35	7.72	7.27

Depending on the peak area of the X-ray diffraction patterns of the glycolated oriented clay samples, montmorillonite, was estimated as the main constituent clay mineral (0.685–0.917 parts of the unit) followed by kaolinite (0.040–0.142) and illite (0.025–0.207).

4. Discussion

The variation of sediment characteristics and composition along the lake provided information about the efficiency of the transporting agent and the depositional environment along Lake Nasser. Figure 8 shows GIS simulation of the spatial distribution of the sediment types along the bottom layer of the lake.

The river processes prevailed along the riverine environment at the entrance of the lake (Figures 3and 4). Therefore, the bottom sediments are nearly free from clay and composed mainly of coarse sediments including sand (87–100%) (Figure 8(A)) mixed with small ratios of silt (0–10%) (Figure 8(B)). MzØ of the bottom sediments mainly exceeded the limit of MdØ (Figure 5(A) and (B)) that influenced with its skewness towards the fine fraction (Figure 5(D)). The distribution of each of the median and mean diameter values (MdØ and MzØ) described the capability of the sedimentation processes to deposit sediments grains like coarse, medium and fine sand. In addition, the well sorting, the fine skewness and the leptokurtic grain size distribution represent a high degree of texture maturity along this part.

The transitional part between riverine and lacustrine environment (the southern part of the lacustrine environment) was partially influenced by the riverine processes with gradual increase in the slow deposition from quite water to operate (Figures 3 and 4). Accordingly, sand composed 83 and 61% at the middle part of El-Dowishat and HH and 78% at the western part of Atiry, north of which it formed no more than 5% of the bottom sediments (Figure 8(A)). Along this segment of the lacustrine environment, silt composed 36–58% as average values (Figure 8(B)), whereas clay composed 18–59% of the bottom sediments (Figure 8(C)). The continuity of the fine skewness (Figure 5(D)) caused the MzØ to exceed MdØ till reaching the site W-W (Figure 5(A) and (B)). North of W-W, the MdØ exceeded MzØ influencing with the change towards the coarse skewing.

Along the southern part of the lacustrine environment (Figures 3 and 4), the distribution of the MdØ and MzØ described the decrease in the sedimentation processes capability which deposited sediment particles as fine as silt and clay (Figure 5(A) and (B)). In addition, the very poorly sorting (Figure 5(C)), coarse to strongly coarse skewness (Figure 5(D)), platykurtic kurtosis (Figure 5(E)) represented a textural immaturity. North of Atiry, as the MzØ value exceeded 8 (i.e. clay composed more than 45% of the bottom sediment), the grain size of the bottom sediments decreased with the increase in pH (Figure 6(D)). Along this portion of the lacustrine environment, the very poorly sorting and the platykurtic particle size distribution indicate a textural immaturity.

Along the northern part of the lacustrine environment, the slow deposition from quite water prevailed (Figures 3 and 4); therefore, the bottom sediments were sand free (Figure 8(A)). Along this segment of the lacustrine environment, the bottom sediments composed mainly of clay, which shared with (57–84%) (Figure 8(C)), in addition to some silt which shared with (14–37%) (Figure 8(B)). However, MdØ exceeded MzØ, the distribution of the MdØ and MzØ described that silt and clay particle size were the main components of sediments (Figure 5(A) and (B)). This indicates the continuity of the deterioration of the sedimentation processes capability. The abundance of clay fraction in addition to the very poorly sorting and coarse skewness indicates a textural immaturity. The leptokurtic distribution along this segment is due to the narrowing of the grain size range to the clay class, indicating the severely decrease in the capability of the transporting agent. Along this segment of the lacustrine environment, the grain size of the clay particles increased with the increase in each of bottom depth and the total dissolved salts concentration and decreases with the increase in pH.

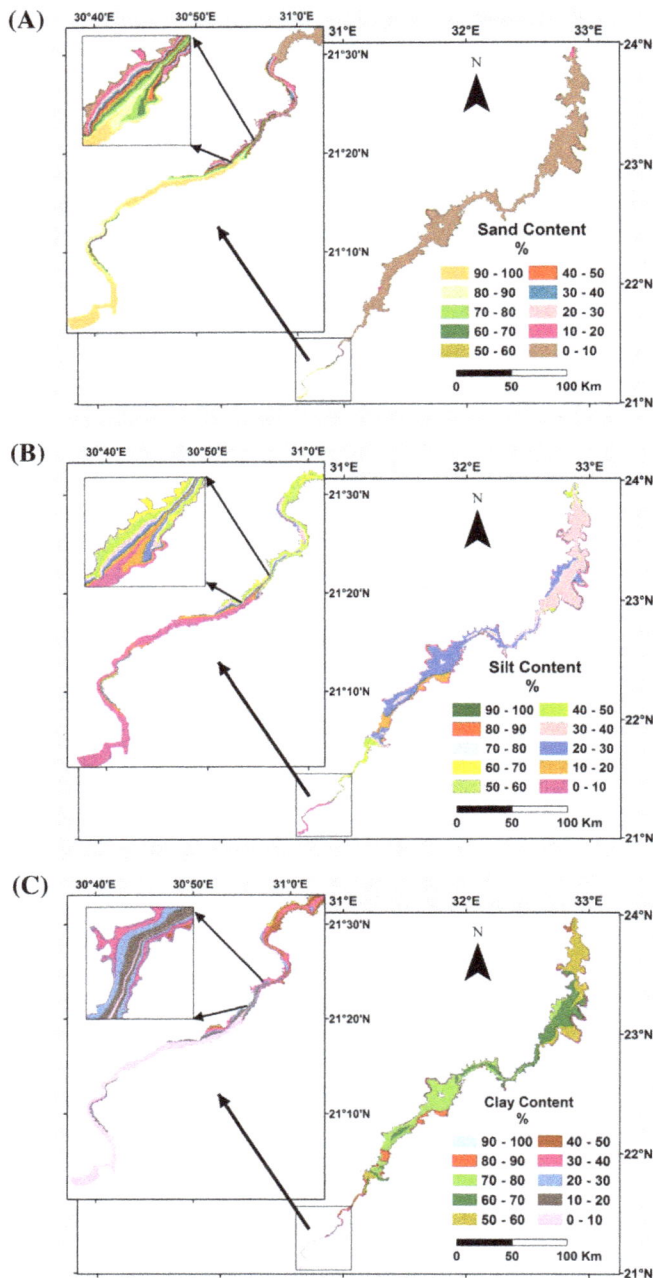

Figure 8. Distribution of the sediment type components.

The mineralogical composition of the bottom sediments may provide stratigraphic and sedimentological indicators. Interrelation was plotted between the mean grain size MzØ and the calculated relative peak height of each of quartz, feldspar and calcite along the studied locality.

The relative peak height of quartz decreased with the increase in MzØ (Figure 9(A)) indicating a reverse relation ($R = -0.883$). In other words, the abundance of quartz decreased with the decrease in grain size of the bottom sediments along the studied area.

The interrelation between relative peak height of calcite and MzØ (Figure 9(B)) showed that the studied locality may be distinguished into two regions; Atiry-Dabarosa and Arqeen-Adendan

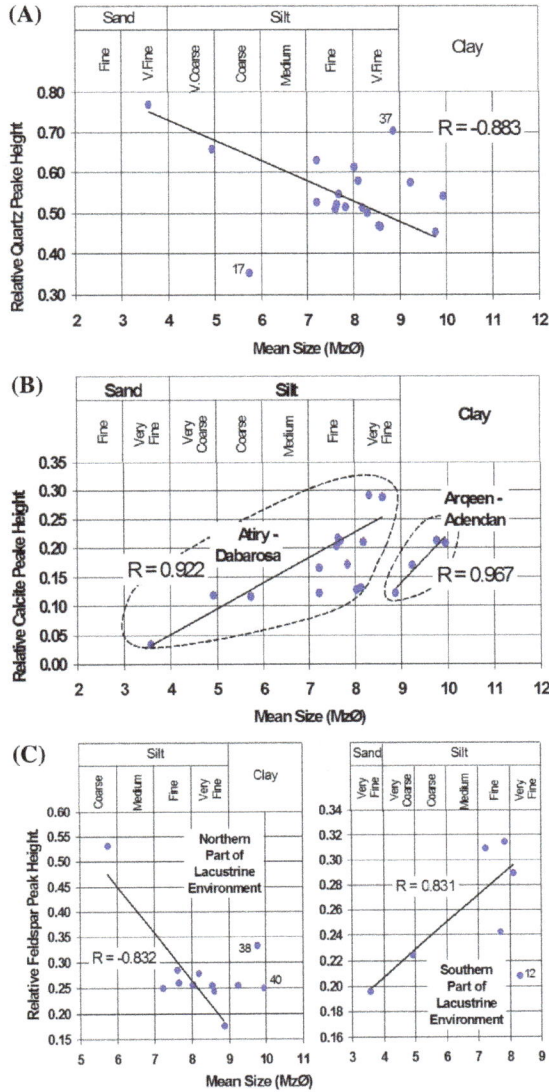

Figure 9. Interrelation between MzØ and the relative peak height of (A) quartz, (B) calcite and (C) feldspar along the studied part of the lake.

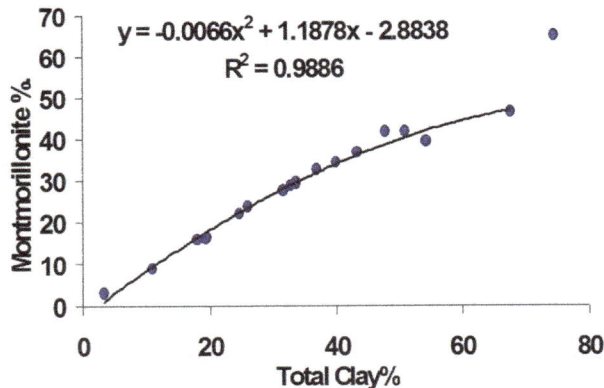

Figure 10. Interrelation between the estimated percentage of the montmorillonite concentration and total clay size content along the studied locality.

(A)

$y = 0.0021x^{3.3582}$

$R^2 = 0.97$

Kaolinite vs **MzØ**

(B)

$y = 0.00005x^{5.149}$

$R^2 = 0.9375$

Kaolinite vs **MzØ**

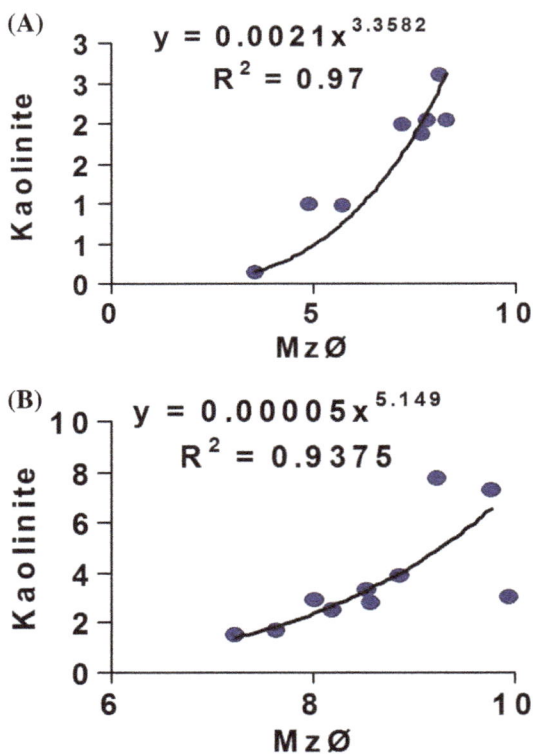

Figure 11. Interrelation between the estimated percentage of kaolinite concentration and MzØ; (A) south of Gomay and (B) north of Gomay.

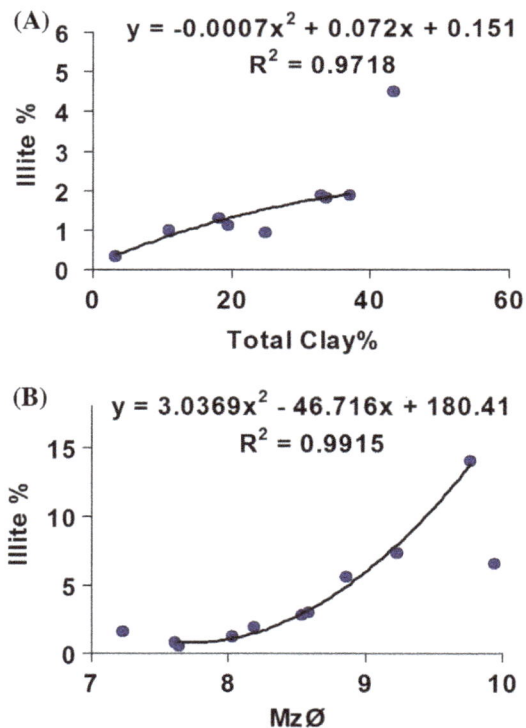

(A)

$y = -0.0007x^2 + 0.072x + 0.151$

$R^2 = 0.9718$

Illite % vs **Total Clay%**

(B)

$y = 3.0369x^2 - 46.716x + 180.41$

$R^2 = 0.9915$

Illite % vs **MzØ**

Figure 12. Interrelation between the estimated percentage of the illite content and total clay size content south of Gomay (A) and with MzØ north of Gomay (B).

regions. Along the two regions, the relative peak height of calcite increased with the increase in MzØ indicating a direct relation (R = 0.922 and 0.967, respectively). In other words, the abundance of

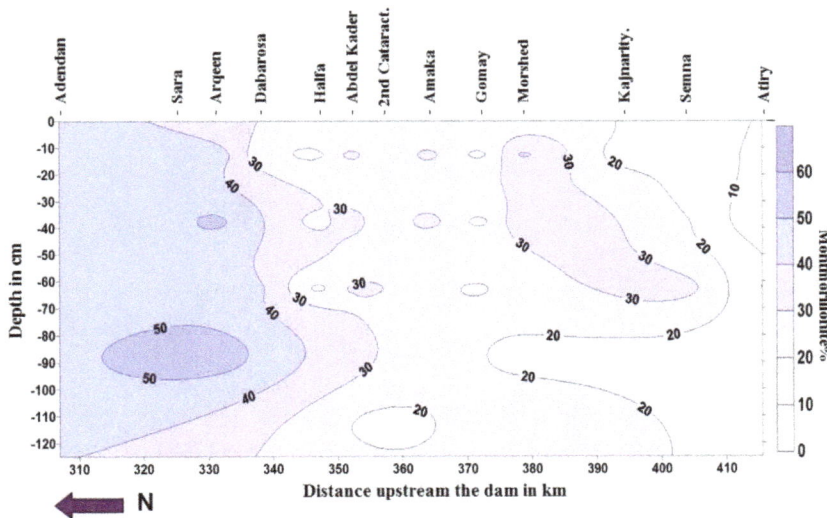

Figure 13. The vertical distribution of the montmorillonite abundance.

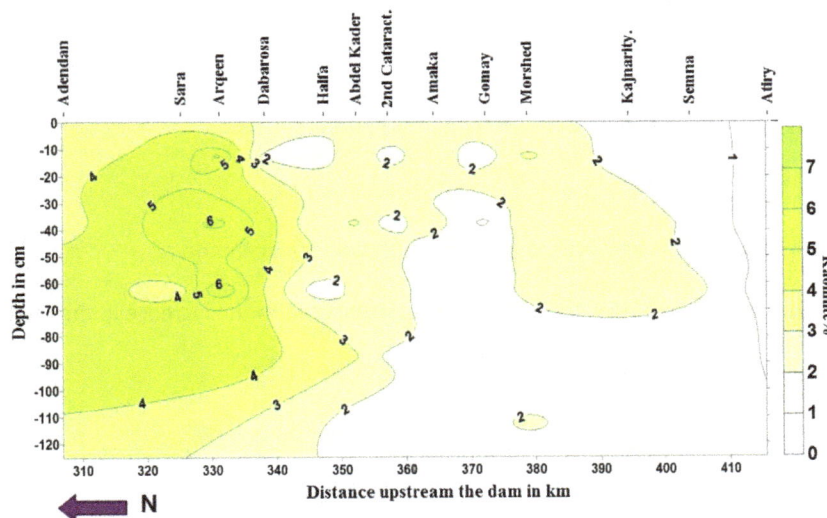

Figure 14. The vertical distribution of the kaolinite abundance.

calcite increased with the decrease in grain size of the bottom sediments along the studied area of the lake.

The interrelation between relative peak height of feldspar and MzØ (Figure 9(C)) showed that the studied locality may be classified into its southern and northern parts of lacustrine environment around Gomay. Along the southern part, the relative peak height of feldspar increased with the increase in MzØ indicating a direct relation (R = 0.831). Conversely, along the southern part, the relative peak height decreased with the increase in MzØ indicating a reverse relation (R = −0.832). In other words, the abundance of feldspar increased along the southern part with the decrease in grain size of the bottom sediments. However, the feldspar abundance decreased with the decrease in grain size along the northern part.

The montmorillonite content found to increase with the increase in the total clay content (R² = 0.99) along the studied area (Figures 10 and 11) through the correlation equation:

$$y = -0.0066\,x^2 + 1.1878\,x - 2.8838 \qquad (3)$$

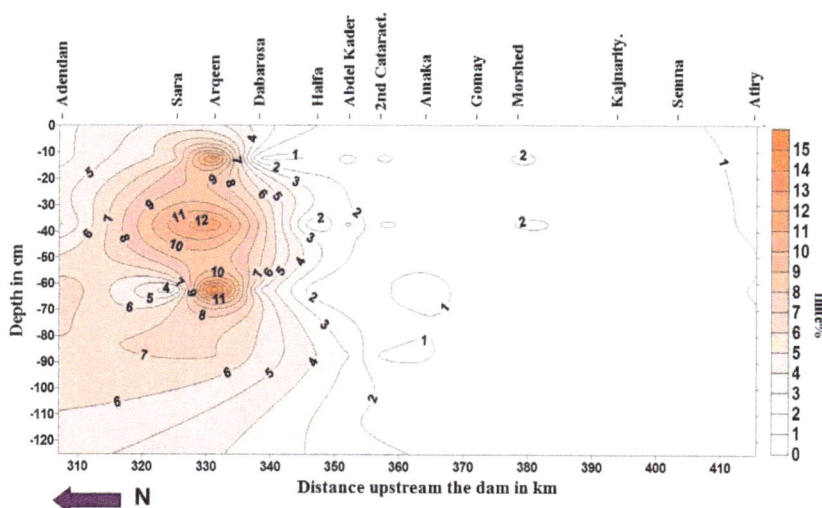

Figure 15. The vertical distribution of the illite abundance.

where x is the total clay content in percentage and y is the montmorillonite concentration in percentage.

The illite content found to increase with the increase in the total clay content ($R^2 = 0.97$) along the distance between Atiry and Gomay (Figure 12(A)) through the correlation equation:

$$y = -0.0007\,x^2 + 0.072\,x + 0.151 \tag{4}$$

where x is the total clay size content and y is the illite concentration in percentage.

However, north of Gomay (Figure 12(B)), the illite content increased with the increase in the MzØ ($R^2 = 0.99$) through the equation:

$$y = 3.0369\,x^2 - 46.716\,x + 180.41 \tag{5}$$

where x is the MzØ value and y is the illite concentration in percentage.

Applying Equations (3)–(5) on each of the collected core samples, the percentages of montmorillonite, kaolinite and illite content were estimated. The margins of error were repaired depending on the proportion of each mineral in a sample. The vertical distribution of the abundance of each of the clay minerals along 1.25 m depth of the bottom sediments was illustrated in the Figures (13–15).

5. Conclusion

In-situ measurements and sampling processes were carried out along Lake Nasser between latitudes 21°02′33″ N and 23°38′55″ N and longitudes 30°38′42″ E and 32°54′23″ E. The integration between some improved techniques, Core Sampling, X-ray diffraction and GIS, was used to comprehensively study the sedimentation processes and its relation to the hydrographic physicochemical conditions.

The grain size analysis and mineralogical composition of the core sediments reflect the linkage between the sedimentation process and the depositional environment with the river morphology and both water and sediment flow. This has enabled to understand that Lake Nasser could be classified into two main sedimentation processes and two depositional environments (riverine and lacustrine) each of them has its specific sedimentological properties and sedimentation process:

(1) The riverine environment, located between latitudes 21°01″ 54′and 21°18″57.92′and longitudes 30°36″50′and 30°53″ 12.99′. It occupies the first part of the lake where the river processes are still the dominant mechanism of deposition. The lake is narrow and its morphology is controlled by the geology of geomorphology of the surrounding area enabling for such sedimentation process which reflected in the relatively coarse bottom sediment grain size (fine to medium sand) compared with the rest of the lake. With the distance northward as the lake get wider and bigger the grain size of the bottom sediments decreased as the current velocity and the suspended sediments concentration decreased. This created a transitional sedimentation process between the riverine and the lacustrine. The cross-sectional area, pH, TDS and EC increased in the same direction.

(2) The lacustrine environment located between latitudes 21°18″57.92′and 23°38″55′and longitudes 30°53″12.99′and 32°54″23′. It occupies the larger area of the lake that is more wide and calm. The sedimentation process of this area was sub-categorized into two main parts (Southern and Northern). Along this lacustrine environment, the grain size of the bottom sediments decreased northward as each of the current velocity and suspended sediment concentration decreased; whereas, each of the profile area, TDS and EC increased in the same direction. As MdØ gradually elevated compared with MzØ to exceed it by Gomay, the slow deposition from quite water became the prevailing mechanism. Along this part of the lacustrine environment, as the particle size decreased, the abundance of quartz decreased, feldspar increased, whereas, calcite and clay minerals (montmorillonite, kaolinite and illite) increased.

Along the northern part of the lacustrine environment, the bottom sediments composed mainly of clay size particles, MdØ continued to exceed MzØ and the slow deposition from quite water continued to be the prevailing mechanism. Along this segment of the lacustrine environment, the grain size of the clay particles increased with the increase in each of bottom depth and the total dissolved salts concentration and decreases with the increase in pH. Along this part of the lake, with the increase in grain size of the bottom sediments, the abundance of quartz and feldspar increased, whereas, each of the feldspar and clay minerals decreased.

Funding
The authors received no direct funding for this research.

Author details
Hussien ElKobtan[1]
E-mail: H_Elkobtan@Gmail.com
Mohamed Salem[2]
E-mail: mohamedsalem199373@gmail.com
Karima Attia[3]
E-mail: Karima_Attia@yahoo.com
Sayed Ahmed[2]
E-mail: sayed.ahmed@fsc.bu.edu.eg
Islam Abou El-Magd[4]
E-mail: imagd@narss.sci.eg
[1] Nile Research Institute, National Water Research Center, Cairo, Egypt.
[2] Geology Department, Benha University, Benha, Egypt.
[3] Water Resources Research Institute, National Water Research Center, Cairo, Egypt.
[4] Environmental Studies Department, National Authority for Remote Sensing and Space Sciences, 23 Josef Tito St., El-Nozha El-Gedida, P.O. Box: 1564 Alf-Maskan, Cairo, Egypt.

Cover image
Source: Authors.

References
Abdel-Aziz, T. M. (1997). *Prediction of bed profile in the longitudinal and transverse directions in Aswan High Dam Reservoir* (PhD thesis). Cairo University, Giza.
Abul-Atta, A. A. (1978). *Egypt and Nile after the high dam.* Cairo: Ministry of Irrigation and Land Reclamation.
Biscaye, P. E. (1965). Mineralogy and sedimentation of Recent Deep Sea clay in Atlantic Ocean and adjacent seas and oceans. *Geological Society of America Bulletin, 76,* 803–832. http://dx.doi.org/10.1130/0016-7606(1965)76[803:MASORD]2.0.CO;2
Bish, D. L., & Reynolds, R. C. (1989). Sample preparation for X-ray diffraction. In D. J. Bish & J. E. Post (Eds.) *Modern powder diffraction* (Reviews in mineralogy, 20, pp. 73–100). Washington, DC: Mineralogical Society of America.
Dahab, A. H., & EL-Moattassem, M. (1994). *Land forms of High Dam Lake area.* International Symposium on River Waterfront Development, Nile Research Institute, C-2-4, Cairo, pp. 129–141.
El-Kobtan, H. M. (2007). *Geological studies on the recent sediments of Lake Nasser (Southern Part) as a sign reflecting its evolution* (MSc thesis). Benha University, Egypt.
El-Manadely, M. S. (1991). *Simulation of sediment transport in the high aswan dam lake* (PhD thesis). Cairo University: Giza.
Folk, R. L., & Ward, W. C. (1957). Brazos River Bar: A study of the significance of grain size parameters. *Journal of Sedimentary Research, 27,* 3–26. http://dx.doi.org/10.1306/74D70646-2B21-11D7-8648000102C1865D

Friedman, G. M., & Sanders, J. E. (1978). *Principles of sedimentology*. New York, NY: Wiley.

Horowitz, A. J. (1991). *A primer on sediment-trace element chemistry*. Ann Arbor, MI: Lewis.

Hughes, R. E., Moore, D. M., & Glass, H. D. (1994). Qualitative and quantitative analysis of clay minerals in soils. In J. E. By Amonette & L. W. Zelazny (Eds.), *Quantitative methods in soil mineralogy* (pp. 330–359). Madison, WI: Soil Science Society of America.

Klug, H. P., & Alexander, L. E. (1954). *X-ray diffraction procedures for polycrystalline and amorphous materials*. New York, NY: Wiley.

Krumbein, W. C., & Pettijohn, F. J. (1938). *Manual of sedimentary petrography*. New York, NY: Appleton-Century-Crofts.

Makary A. Z. (1982). *Sedimentation in the high aswan dam reservoir* (PhD thesis). Ain Shams University, Cairo.

Philip, G., Hassan, F., & Khalil, I. (1978). *Mechanical analysis and mineral composition of Lake Nasser* (Lake Nasser and River Nile project, Report prepared to Academy Science Research and Technology). Egypt.

Ruhe, R. V., & Olson, C. G. (1979). Estimate of clay-mineral content: Additions of proportions of soil clay to constant standard. *Clays and Clay Minerals, 27*, 322–326. http://dx.doi.org/10.1346/CCMN

Shalash, S. (1980). *Effect of sedimentation on storage capacity of high aswan dam reservoir* (Nile Research Institute Report, National water Research Center). Cairo.

Springuel, I. V., Hassan, L. M., Sheded, M., El-Soghir, M., & Ali, M. M., (1991) *Plant ecology of Wadi Allaqi and Lake Nasser. 3. Flora of the Wadi Allaqi basin* (Allaqi Project Working Paper No. 10). Glasgow: University of Glasgow.

Stewart, Jr., H. B. (1958). Sedimentary reflections on depositional environments in San Migue Lagoons. Baja, California, Mexico. *AAPG Bulletin, 42*, 2567–2618.

Tucker, M. (1988). *Techniques in sedimentology*. London: Blackwell Scientific Publication Osney Mead Oxford OX2 0EL.

Velde, B., & Meunier, A. (2008). *The origin of clay minerals in soils and weathered rocks*. Berlin Heidelberg: Springer-Verlag. http://dx.doi.org/10.1007/978-3-540-75634-7

Wentworth, C. K. (1922). A scale of grade and class terms for clastic sediments. *The Journal of Geology, 30*, 377–392. http://dx.doi.org/10.1086/622910

Worden, R. H., & Morad, S. (2003). *Clay cements in sandstones*. Oxford: International Association of Sedimentologists, Blackwell Science.

11

Effect of organics, biofertilizers and crop residue application on soil microbial activity in rice – wheat and rice-wheat mungbean cropping systems in the Indo-Gangetic plains

Geeta Singh[1]*, D. Kumar[2] and Pankaj Sharma[3]
*Corresponding author: Geeta Singh, Division of Microbiology, Indian Agricultural Research Institute, New Delhi 110012, India
E-mail: geetasinghkartik@gmail.com
Reviewing editor:
Nir Krakauer, City College of New York, USA

Abstract: The aim of this study was to investigate the response of soil microbial parameters to nutrient management practices involving organic amendments, farmyard manure (FYM), vermicompost (VC), crop residues (CR) and biofertilizers (BF) in rice–wheat and rice–wheat–mung bean cropping system of the Indo-Gangetic Plains, India. Soil microbial biomass C (C_{mic}), basal respiration, ergosterol, glomalin, soil enzymes (glucosidases, phosphatases and dehydrogenases), FDA activity, organic carbon (C_{org}), C_{mic}-to-C_{org} ratio and metabolic quotient (qCO_2) were estimated in soil samples collected at 0–15 cm depth. The highest C_{org} (0.64%) and C_{mic} (103.8 µg g^{-1}) soil levels occurred in the treatment receiving a combination of VC, CR and BF. Soil respiration, C_{org} and C_{mic}-to-C_{org} ratio were significantly enhanced by the input of CR to plots receiving FYM and VC. The qCO_2 was the highest in plots receiving a combination of FYM, CR and BF followed by control (no nutrient input) and least in plots receiving a combination of VC, crop residue and biofertilizer. These results indicate that the organic practices involving VC, CR and BF improved soil microbial characteristics and C_{org} in rice–wheat systems.

ABOUT THE AUTHORS

The activity of our research group at the Division of Microbiology, Indian Agricultural research institute (New Delhi), is mainly related to the improvement of rice–wheat-based cropping systems in the Indo-Gangetic plains by integrated resource management (IRM) practice to minimize cost of production, reduce environmental damage in developing countries and improving food security in the developing countries. The research reported in the paper pertains to observe the effect of use of mixed organic fertilizers on soil enzyme activity, and hence soil fertility during rice–wheat cultivation. High input of resources like chemical fertilizers pollutes the environment and harms soil fauna. The present research demonstrates that integrated use of various organic fertilizers improves soil enzymatic activity and overall microbial activity of soil and thus fertility of soil.

PUBLIC INTEREST STATEMENT

Public concerns against resource-intensive agriculture include excessive use of chemical fertilizers, which reportedly leach into soil and water and pollute them. The present study demonstrates that mixed use of various organic fertilizers can reduce the dependence on chemical fertilizers besides improving soil enzyme fertility. It was observed in our study that important soil enzyme activities like glucosidase, alkaline and acid phosphatase, which circulate carbon (C) and phosphorous (P), were increased with mixed use of organic fertilizers. An increase in soil microbial activity was also observed. These observations confirmed increase in soil microbial activity and fertility. To check whether it affects the overall metabolic activity of soil fauna, soil respiration and soil microbial biomass carbon (SMBC), ergosterol content, soil glomalin content and FDA hydrolysis activity were calculated. It was observed that the overall enzyme activities of soil increased when organic fertilizers were used as compared to control suggesting improvement of soil fertility.

Subjects: Agriculture; Bioscience; Conservation - Environment Studies; Environmental Studies & Management

Keywords: crop residue management; Indo-Gangetic Plains; metabolic quotient; rice–wheat cropping system

1. Introduction

Rice–wheat cropping systems (RWCS) are the main source of food and income for millions of people in India, but crop productivity is either stagnating (wheat) or declining (rice) despite the use of higher yielding cultivars (Padre-Tirol & Ladha, 2006). This raises major concerns over the long-term sustainability of current farming practices and poses a threat to future food security against a background of climate change. Key factors responsible for deterioration in soil fertility and crop productivity include decline in soil organic matter (SOM) due to reduced inputs of bioresources and lack of an adequate rotation (Shibu, Van Keulen Leffelaar, & Aggarwal, 2010), negative macro and micro-nutrient balances, leading to depletion of soil fertility and nutrient deficiencies (Timsina & Connor, 2001), declining water availability and poorer quality water (Farooq, Kobayashi, Wahid, Ito, & Basra, 2009), and deterioration in soil structure of continuously puddled soils in rice paddies (Saharawat et al., 2010). The decline in the soil fertility, mainly due to the inadequate organic carbon (C_{org}) levels in soil, seems to be the most significant factor for decreased sustainability of the system.

The practice of adopting a cereal–cereal cropping system on the same piece of land over years has led to soil fertility deterioration, and questions are being raised about its sustainability (Prasad, 2005). However, introduction of summer legumes, such as mung bean, in the RWCS after the harvest of wheat and before the transplanting of rice can increase the productivity of these crops, besides improving the carbon and nitrogen status of soil (Prasad, 2011). After picking matured pods of mung bean, the plant biomass (3–4 t ha^{-1} dry matter) can be used for in situ green manuring.

Organic nutrient sources and crop residues (CR) are the primary source of C inputs (Lal, 2004), and the ways in which these are managed have a significant effect on soil's physical, chemical and biological properties (Kumar & Goh, 2000). The incorporation of CR alters the soil's physicochemical environment (Prasad & Power, 1991) which in turn influences the microbial population/activity in the soil and subsequent nutrient transformations. In general, soil enzymes are good markers of soil fertility since they are involved in the cycling of the most important nutrients. Keeping the above facts in view, the objective of the present study was to investigate the effects of different organic manure, crop residue and biofertilizer applications on soil biological functionality as described by enzyme activities, microbial biomass carbon (C_{mic}) and microbial signature molecules such as ergosterol and glomalin content in RWCS and rice–wheat–mung bean cropping (RWMCS) systems.

2. Materials and methods

2.1. Study site and experimental design

The field experiment was conducted in the main block 14-C of the research farm of the Indian Agricultural Research Institute, New Delhi, India, during 2003–2012. This is situated at 28.4° N and 77.1° E at an elevation of 228.6 m above mean sea level (Arabian Sea). New Delhi has a semi-arid and sub-tropical climate with hot and dry summers and cold winters. It falls under the Trans-Gangetic Plains' agro-climate zone. Summer months (May and June) are the hottest with the maximum temperature ranging between 41 and 48°C, while January is coldest with the minimum temperature ranging between 3 and 7°C. The temperature rises gradually through the months of February and March and reaches a maximum during June, then falls slightly with the advent of south-west monsoon rain. The mean precipitation of Delhi is 650 mm which is mostly received during July–September with occasional rain during winter. The soil of the experimental field is a sandy clay loam (typical Ustochrept) in texture, having 52.06% sand, 22.54% silt and 25.40% clay (pH 8.18, organic matter 1.25%). The physiochemical properties of the experimental field are given in Table 1.

Table 1. Physicochemical characteristics of soil of the experimental field

Property value	
Mechanical composition	
Sand (%)	52.06
Silt (%)	22.54
Clay (%)	25.40
Textural class Sandy clay loam	
Chemical composition and physical properties	
pH (1:2.5 soil:water ratio)	8.16
Electrical conductivity (dS m21 258C)	0.79
Cation exchange capacity (C.mol kg21 soil)	14.73
Organic C (g kg21 soil)	5.20
Total Kjeldahl N (mg kg21 soil)	580
0.5 M NaHCO3 extractable P (mg kg21 soil)	8.42
Neutral 1 N NH4OAC extractable K (mg kg21 soil)	187
Bulk density (Mg m23)	1.50
Field capacity at 1/3 atmospheric tension (%)	24.57

The experiment was laid out in a strip-plot design with three replications. No chemical pesticide/disease/weed control agent was supplied in the field, and hence study was carried out totally in organic farming conditions. Treatments consisted of 14 combinations of 2 cropping systems, namely rice–wheat and rice–wheat–mung bean, and 7 combinations of organic manures, CR, referring to incorporation of crop residue from the previous crop and biofertilizers (BF):- (T$_1$) farmyard manure (FYM) equivalent to 60 kg N ha^{-1}, (T$_2$) vermicompost (VC) equivalent to 60 kg N ha^{-1}, (T$_3$) FYM + CR, (T$_4$) VC + CR, (T$_5$) FYM + CR + BF and (T$_6$) VC + CR + BF, as well as (T$_0$), a non-amended control.

These treatments were applied to all the crops, i.e. rice, wheat and mung bean, during the period 2003–2012. The cropping history of the experimental field and treatment details are summarized in Table 2. The BF applied to the wheat, rice and mung bean crops consisted of azotobacter + cellulolytic culture + phosphate-solubilizing bacteria (PSB), blue green algae + cellulolytic culture + phosphate-solubilizing bacteria and rhizobium + phosphate-solubilizing bacteria, respectively. For the present study, soil samples were collected at the wheat harvest of 2011–2012, i.e. after completion of six cycles of the rice–wheat or RWMCS system.

Table 2. Cropping history of the experimental field from 2001–2012

Year	Kharif	Rabi	Summer	Remarks
2001–2002	Rice	Wheat	—	Conventional farming
2002–2003	Rice	Wheat		
2003–2004	Rice (Organic)	Wheat (Organic)	—	Transitional period
2004–2005	Rice (Organic)	Wheat (Organic)	—	
2005–2006	Rice (Organic)	Wheat (Organic)	Mung bean (organic)	
2006–2007	Rice (Organic)	Wheat (Organic)	Mung bean (organic)	Organic farming
2007–2008	Rice (Organic)	Wheat (Organic)	Mung bean (organic)	
2008–2009	Rice (Organic)	Wheat (Organic)	Mung bean (organic)	
2009–2010	Rice (Organic)	Wheat (Organic)	Mung bean (organic)	
2010–2011	Rice (Organic)	Wheat (Organic)	Mung bean (organic)	
2011–2012	Rice (Organic)	Wheat (Organic)	Mung bean (organic)	

2.2. Microbiological analysis

Soil was sampled manually from all the plots at 0–15 cm using a tube auger. Fifteen sub-samples per plot were taken and carefully mixed. Soil biological analyses were carried out on moist samples in triplicate and the results were expressed on a dry weight basis. The microbial biomass content of the soil was determined using the fumigation–extraction method of Vance, Brookes, and Jenkinson (1987). The levels of four enzymatic activities in soil were measured: dehydrogenase (EC 1.1.1.) (Casida, Klein, & Santoro, 1964), alkaline phosphomonoesterase (EC 3.1.3.1), acid phosphomonoesterase (EC 3.1. 3.2) (Tabatabai, 1994; Tabatabai & Bremner, 1969) and β-glucosidase (EC 3.2.1.21) (Eivazi & Tabatabai, 1988). The estimation of total glomalin (T-GRSP) was done by the procedure of Wright and Upadhyaya (1998) and the protein content was expressed as µg per g dry weight of soil. Soil microbial activity expressed as fluorescein diacetate (FDA) hydrolysis was determined following the method of Green, Stott, and Diack (2006). The soil respiration (SR) was measured by the alkali entrapment method (Stotzky, 1965) and the metabolic quotient was computed as respiratory activity in relation to micro-bial biomass (Anderson & Domsch, 1993). The C_{mic}-to-C_{org} ratio and the metabolic quotient (qCO_2) were calculated by dividing the C of CO_2 released from sample in 1 h by the C_{mic} content (Šantrucková & Straškraba, 1991). Soils were also analysed for the fungal biomarker, ergosterol. Ergosterol is a mem-brane-bound molecule commonly used as a fungal biomarker (Bååth & Anderson, 2003). Ergosterol was extracted from the samples by the microwave-assisted extraction method and determined by HPLC analysis (Young, 1995). The qCO_2, i.e. the respiration to biomass ratio, was calculated from qCO_2 = Basal respiration × 1000/C_{mic} (Insam & Haselwandter, 1989).

2.3. Statistical analysis

A two-factor analysis of variance (ANOVA) was performed to determine the effects of nutrient man-agement/organic amendments, cropping systems and their interactions on soil biological and bio-chemical properties. Data analysis for all soil parameters was performed using the SAS software. For statistical analysis of data, least significant difference (LSD at $p = 0.05$) was used to determine whether means differed significantly.

3. Results and discussion

3.1. Phosphatases

Alkaline phosphomonoesterase (ALP) activity was higher in control than in the treatments (except in VC + CR + BF treatment in RWCS) with organic amendments, apparently leading to enhanced miner-alization of native organic P fraction in soil (Table 3). Plots receiving a combination of VC + CR + BF in RWCS showed maximum ALP activity (though not significantly greater than control). But in case of all other organic amendment treatments, ALP activity is less than control in both RWCS and RWMCS. The addition of CR affected the ALP significantly. The ALP activity was significantly higher in VC + CR than in VC in RWCS, but significantly lower in RWMCS.

The acid phosphatase (ACP) activity was significantly high in plots receiving FYM + CR + BF and VC + CR + BF in RWCS and VC, CR and VC + CR + BF in RWMCS (Table 3) than in the control plots. The plots receiving VC showed a significantly lower acid phosphatase activity than without CR in RWMCS, but the addition of BF (VC + CR + BF) gave higher ACP activity than VC + CR alone in case of RWMCS. FYM alone or a combination of FYM and CR was at par with the control. Interestingly, ACP activity was stimulated by the application of BF compared with the control in both RWCS and RWMCS except FYM and FYM + CR treatments of RWCS.

3.2. β-glucosidases

In RWMCS, application of all combinations of organic nutrient sources significantly improved the enzyme activity compared with the control, whereas in RWCS, plots treated with FYM or VC were comparable to control plots (Table 3). Plots receiving VC alone or in combination with CR showed the highest stimulation of β-glucosidase activity in the RWMCS. The magnitude of increase in β-glucosidase activity over the control ranged from 43.8 to 55.5% in RWCS. While in RWMCS, the values ranged from 21.5 to 77.4%.

Table 3. Treatment details

Cropping systems

Treat-ment	Rice		Wheat		Treat-ment	Rice		Wheat		Mung bean	
	Organic nutrients		Organic nutrients			Organic nutrients		Organic nutrients		Organic nutrients	
No.	Manures & compost	Biofertilizer	Manures & compost	Biofertilizer	No.	Manures & compost	Biofertilizer	Manures & compost	Biofertilizer	Manures & compost	Biofertilizer
1	FYM[1]	—	FYM	—	2	FYM	—	FYM	—	—	—
3	VC[2]	—	VC	—	4	VC	—	VC	—	—	—
5	FYM + CR[3]	—	FYM + CR	—	6	FYM + CR	—	FYM + CR	—	CR	—
7	VC + CR	—	VC + CR	—	8	VC + CR	—	VC + CR	—	CR	—
9	FYM + CR	BGA + Cellulolytic culture + PSB	FYM + CR	Azotobacter + Cellulolytic culture + PSB	10	FYM + CR	BGA + Cellulolytic culture + PSB	FYM + CR	Azotobacter + Cellulolytic culture + PSB	CR	Rhizobium + PSB
11	VC + CR	BGA + Cellulolytic culture + PSB	VC + CR	Azotobacter + Cellulolytic culture + PSB	12	VC + CR	BGA + Cellulolytic culture + PSB	VC + CR	Azotobacter + Cellulolytic culture + PSB	CR	Rhizobium + PSB
13	— (control)	— (control)	— (control)	— (control)	14	— (control)	— (control)	— (control)	— (control)	— (control)	— (control)

[1]FYM: Farmyard manure (equivalent to 60 kg N ha^{-1}).

[2]VC: Vermicompost (equivalent to 60 kg N ha^{-1}).

[3]CR: Crop residue (incorporation of crop residue of previous crop in succeeding crop.

The observed low activity of the β-glucosidase in FYM-treated plots corresponded with low-soil acid phosphatase in RWCS, dehydrogenase activity in both RWCS and RWMCS and glomalin content in RWMCS (Tables 3–5).

Table 4. Interactive effect of cropping system and nutrient management practices on glucosidase, alkaline phosphatase and acid phosphatase activities in soil

Cropping system/ Nutrient management	Glucosidase (µg pNPG per g^{-1} soil h^{-1})		Alkaline phosphatase (µg pNPP g^{-1} soil h^{-1})		Acid phosphatase (µg pNPP g^{-1} soil h^{-1})	
	Rice-Wheat	Rice-Wheat-Mung bean	Rice-Wheat	Rice-Wheat-Mung bean	Rice-Wheat	Rice-Wheat-Mung bean
Control	13.7	16.7	530	523	120	97
FYM	14.6	20.3	359	296	111	108
Vermicompost (VC)	14.9	27.3	297	411	147	136
FYM + Crop residue (CR)	19.7	24.0	200	386	112	127
VC + CR	20.6	27.8	374	322	139	104
FYM + CR + Biofertilizer (B)	21.0	22.1	253	383	158	123
VC + CR + B	21.3	23.8	536	289	156	165
Mean	18.0	23.2	364	373	135	123
LSD (p = 0.05)	Cropping system (CS): NS Nutrient management (NM): 5.0 CS × NM: 3.2		Cropping system (CS): NS Nutrient management (NM): 69 CS × NM: 50		Cropping system (CS): NS Nutrient management (NM):14.8 CS × NM: 29.3	

Note: Values are mean of the data (n = 3) and are statistically significant at $p < 0.05$. Data analysed by Two-way ANOVA at LSD < 0.05.

Table 5. Interactive effect of cropping system and nutrient management practices on FDA hydrolysis, dehydrogenase activity and microbial biomass in soil

Cropping system/ Nutrient management	FDA hydrolysis (μg fluorescein g^{-1} dry soil h^{-1})		Dehydrogenase activity (μg TPFg^{-1} soil 24 h^{-1})		Microbial biomass (μg MBC g^{-1}soil)	
	Rice–Wheat	Rice–Wheat–Mung bean	Rice–Wheat	Rice–Wheat–Mung bean	Rice–Wheat	Rice–Wheat–Mung bean
Control	319	280	1,285	954	51.6	62.8
FYM	295	291	1,065	881	66.9	87.9
Vermicompost (VC)	291	257	968	1,370	80.5	63.0
FYM + Crop residue (CR)	290	293	1,169	1,157	103.3	67.9
VC + CR	333	331	1,253	905	54.9	102.5
FYM + CR + Biofertilizer (B)	302	347	1,152	1,022	55.4	69.4
VC + CR + B	335	251	1,234	810	105.5	102.0
Mean	309	393	1,161	1,014	74.0	79.3
LSD (p = 0.05)	Cropping system (CS): NS Nutrient management (NM): 13.6 CS × NM: 47		Cropping system (CS): 89 Nutrient management (NM): 133 CS × NM: 109		Cropping system (CS): NS Nutrient management (NM): 7.2 CS × NM: 9.3	

Note: Values are mean of the data ($n = 3$) and are statistically significant at $p < 0.05$. Data analysed by Two-way ANOVA at LSD < 0.05.

3.3. FDA hydrolysis

A significant increase in FDA activity was observed in the VC + CR + BF (5%) in RWCS and FYM + CR + BF (23.9%) in RWMCS over their respective controls, while FDA activity was reduced compared with the control in all organic treatments except with VC + CR and VC + CR + BF in RWCS. On the other hand, FDA activity was increased in RWMCS except with VC and VC + CR + BF (Table 4).

3.4. Dehydrogenase activity

Only soils receiving VC (43.6%) and FYM + CR (21.3%) in RWMCS were found to have significantly higher dehydrogenase activity among the fertilizer treatments over the control, while with RWCS, all dehydrogenase activity levels were lower than the control except VC + CR and VC + CR + BF. Nevertheless, this activity was not consistently correlated with other parameters such as CO_2 production or microbial biomass.

3.5. Microbial biomass C

Overall, the MBC values ranged from 51.6 to 105.5 μg g^{-1} soil in RWCS and 62.8 to 102.5 μg g^{-1} soil in RWMCS (Table 4). The results indicated statistically significant ($p < 0.05$) differences in the level of soil MBC between various combinations of organic fertilizers, their interaction with the cropping systems but not between the two cropping systems. The MBC values were significantly higher in the plots receiving organics (FYM, CR, C, BF and their combinations) than in the control except with VC + CR or FYM + CR + BF in RWCS and VC, FYM + VC or FYM + CR + BF in RWMCS, reflecting possibly qualitative and quantitative differences in the microbial communities, i.e. 6.4 to 104.5% increase as compared to control in RWCS and up to 63.2% increase in RWMCS, with different organic combinations. In RWCS, MBC with FYM was significantly lower than with VC, while in RWMCS, MBC with FYM was significantly higher than with VC. Application of CR significantly enhanced the soil MBC in conjunction with FYM in RWCS and with VC in RWMCS. The magnitude of increase recorded over control by the application of FYM alone and FYM + CR was 6.4 and 99%, respectively, in RWCS and 39.9 and 8.1%, respectively, in RWMCS. However, the increase of MBC by VC alone and VC + CR + BF was 56 and 104.5%, respectively, over control in RWCS and in RWMCS, increase was 63.2 and 62.4% over control with VC + CR and VC + CR + BF, respectively. A combination of VC + CR + BF was the best treatment as it enhanced microbial biomass significantly over the control in both RWCS and RWMCS, though FYM + CR in RWCS and VC + CR in RWMCS showed almost similar increase as observed in case of VC + CR + BF.

3.6. Basal respiration

Microbial biomass alone does not provide information on microbial activity. Therefore, measurements of microbial biomass turnover, such as SR, which is considered to reflect the availability of carbon for microbial maintenance, are required for that assessment. SR, a measure of the total activity of the soil microbial community, was significantly affected by nutrient management and its interaction with the cropping system. Input of organic nutrient sources significantly improved the SR activity over the control (Table 5). A comparison of the two cropping systems revealed a significant difference in soil CO_2 emission following the input of VC, as in RWCS, significant increase (31.6%) was observed, but in RWMCS, it was slightly lower than the respective control. These differences can be explained on the basis of differences in the C:N ratio of the rhizospheric soil. Leguminous crop fixes atmospheric nitrogen and improves the soil N status, thereby lowering the C:N ratio. In our experiment, respiratory activity was significantly increased with all treatments in RWCS. In RWMCS also, all organic treatments showed significant increase in SR except VC. The addition of CR stimulated the soil CO_2 emission in RWMCS. The SR increased significantly by the residue incorporation and the effect was more apparent where the FYM/VC either singly or in combination with CR was applied, though VC alone did not make any improvement in SR in RWMCS. A corresponding increase in the soil MBC content was also recorded. Carbon mineralization is known to be affected by the complexity of chemical constituents (lignocelluloses content) of organic amendments.

3.7. Metabolic quotient (qCO_2)

The elevated qCO_2 values detected with various organic treatments in RWCS except VC + CR and FYM + CR + BF and FYM + CR and FYM + CR + BF in RWMCS suggest less efficient microbial utilization of C compared to control. The treatment VC + CR in RWCS recorded the highest (5.9 µg CO_2_C µg^{-1} biomass C h^{-1}) (Table 6) followed by FYM + CR + BF (5.3 µg CO_2_C µg^{-1} biomass C h^{-1}) (Table 6), while in RWMCS, the highest value of qCO_2 is recorded in treatment FYM + CR followed by FYM + CR + BF.

Table 6. Interactive effect of cropping system and nutrient management practices on SR, ergosterol and glomalin content activities in soil

Cropping system/ Nutrient management	SR (mg CO_2 (100 g)$^{-1}$soil/week)		Ergosterol (µg g^{-1}soil)		Glomalin content (µg/kg)	
	Rice–Wheat	Rice–Wheat–Mung bean	Rice–Wheat	Rice–Wheat–Mung bean	Rice–Wheat	Rice–Wheat–Mung bean
Control	43.3	44.3	15.96	1.10	43.3	94.3
FYM	50.0	50.3	13.48	7.12	72.0	62.3
Vermicompost (VC)	57.0	43.2	2.97	2.35	61.0	103.3
FYM + Crop residue (CR)	57.1	59.0	10.29	1.04	92.3	102.0
VC + CR	54.3	56.3	7.50	3.14	85.3	103.0
FYM + CR + Biofertilizer (B)	49.6	50.8	1.30	8.46	91.3	64.0
VC + CR + B	53.0	54.2	3.53	6.63	76.7	102.7
Mean	52.0	51.2	7.83	4.26	74.6	90.2
LSD (p = 0.05)	Cropping system (CS): NS Nutrient management (NM): 4.3 CS × NM: 6.0		Cropping system (CS): 0.05 Nutrient management (NM): 0.38 CS × NM: 0.72		Cropping system (CS): NS Nutrient management (NM): 15.8 CS × NM: 30.7	

Note: Values are mean of the data (n = 3) and are statistically significant at p < 0.05. Data analysed by Two-way ANOVA at LSD < 0.05.

3.8. Total glomalin content

Glomalin content in the soil samples showed a significant effect of nutrient management and its interaction with the cropping systems. In RWCS, the highest value of glomalin content was observed in treatment FYM + CR (92.3 µg kg^{-1}) followed by FYM + CR + BF (91.3 µg kg^{-1}). FYM + CR and FYM + CR + BF applications had the maximum and significant ($p < 0.05$) impact in enhancing glomalin content (110.8–113.2%) over the control treatment in RWCS (Table 5). In contrast, in RWMCS, plots receiving FYM alone and FYM + CR + BF caused a significant reduction in this soil protein. Rest all other nutrient management practices in RWMCS recorded statistically identical values to the control. A comparison between the two cropping systems revealed that the quantity of glomalin under RWMCS was significantly higher in control, VC-, FYM + CR- and VC + CR-treated plots over the corresponding treatments in RWCS. These differences can be attributed to the differences in the soil organic carbon status in the two cropping systems. Overall, the nature of organic amendment was found to influence glomalin levels; for instance, application of FYM alone failed to improve soil glomalin content in RWMCS over the control, whereas VC application exerted a positive effect on soil glomalin content in both RWCS and RWMCS (Table 5).

3.9. Ergosterol

The RWCS and RWMCS, nutrient management and their interactions significantly influenced soil ergosterol content. Ergosterol is the main endogenous sterol of fungi, actinomycetes and some microalgae. Its concentration is an important indicator of fungal growth on organic compounds and mineralization activity. In the present study, the application of manure in combination with the CR in RWMCS favours fungal growth as the fungi are dominant decomposers in the soil. However, when bacterial biofertilizer is added along with the FYM + CR and VC + CR in RWCS, a lowered fungal/bacterial ratio may result in the observed decline in the soil ergosterol content.

3.10. Soil organic carbon

The soil organic carbon content, an indicator of soil fertility, was positively and significantly influenced by the cropping system and organic nutrient sources. In the present study, the treatment VC + CR + BF emerges as the best option in improving the soil organic carbon status for our experimental crops: rice, wheat and mung bean. Results indicated that at the end of nine years of crop rotation, application of FYM or VC either alone or in combination with CR increased the SOC (0.56–0.68%) compared to the control plot, where no organics were applied (Table 7). The application of FYM + CR + BF caused 34.04, 35.41 and 32.69% increase over their respective controls in rice, wheat

Table 7. Effect of treatments on organic carbon of soil (2011–2012)			
Treatment	Soil organic carbon (%)	Soil organic carbon (%) After harvest of wheat (RWCS)	Soil organic carbon (%) After harvest of mung bean (RWMCS)
Cropping system			
Rice–wheat	0.56	0.58	0.60
Rice–wheat–mung bean	0.60	0.63	0.68
LSD ($p = 0.05$)	0.02	0.03	0.04
Nutrient sources			
Control	0.47	0.48	0.52
FYM	0.56	0.58	0.61
VC	0.57	0.62	0.65
FYM + CR	0.58	0.61	0.65
VC + CR	0.59	0.65	0.68
FYM + CR + B	0.63	0.65	0.69
VC + CR + B	0.64	0.67	0.71
LSD ($p = 0.05$)	0.03	0.04	0.06

and mung bean crops and VC + CR + BF caused 36.17, 39.58 and 36.54% increase over their controls in rice, wheat and mung bean crops, significantly higher over all other organic sources. A combination of CR with FYM or VC was the next best alternative source of organics. It is very difficult to increase the organic matter content in irrigated soils under sub-tropical climatic conditions due to their very high rates of C mineralization. The present studies suggest that FYM or VC, in combination with CR and BF, could be used as an effective mechanism to sequester SOC and improve soil nutrient status. Further, nature of crop also influences the soil organic carbon content. Although the increase in the amount of soil organic C is important, the increase in the amount of C associated with microbial biomass is more important.

4. Discussion

Among the soil enzyme activities studied, alkaline phosphatase (ALP) activity was the only enzyme activity not stimulated by addition of organic nutrient sources as the values of this activity in soil treated with organics were significantly lower than those found for the control in both RWCS and RWMCS. Phosphatases are a group of enzymes that catalyse the hydrolysis of organic compounds to phosphate. The demand for P by plants and soil micro-organisms can be responsible for the stimulation of the synthesis of this enzyme (Garcia, Hernandez, Roldan, & Albaladejo, 1997). According to Rao and Tarafdar (1992), increase in phosphatase activity indicates changes in the quantity and quality of soil phosphoryl substrates. The observed significant reduction in ALP activity in most organically amended plots may be attributed to the inhibition of phosphatase by an excess of inorganic P (Nannipieri, Grego, & Ceccanti, 1990). The acid phosphatases (ACPs) are reported to be contributed solely by the plant roots (Tarafdar, 1989) and conditions that favour plant root growth may also enhance the secretion of the enzyme. **However, phosphatase activity was found strongly correlated with extractable P (Nottingham et al., 2015), suggesting that increased microbial synthesis of phosphatases was a direct response to low available phosphate (Turner & Wright, 2014).** Additionally, the microbial degradation of CR and metabolic activity of the added BF possibly contribute to the organic acids which perhaps provide optimum pH for the observed high ACP activity.

The hydrolysis products of β-glucosidases are believed to be important energy sources for soil micro-organisms (Tabatabai, 1994). β-glucosidases are key enzymes in the carbon cycle and play a crucial role in hydrolytic processes that take place during organic matter breakdown. Overall, it appears that glucosidase enzyme activity increases with the use of organic nutrients which subsequently results in high available C in the soil and improves the microbial population in soil. Similar results have been reported by Zhang et al. (2010).

The FDA activity is widely accepted as an accurate and simple measurement of total microbial activity in soils, and includes the ubiquitous free and membrane-bound digestion enzymes, such as lipase, protease and esterase enzymes (Green et al., 2006). These differences may be due to the higher levels of organic matter, coupled with the presence of metabolically active micro-organisms (Taylor, Wilson, Mills, & Burns, 2002). A comparison of the two cropping systems revealed a lack of significant differences at all the tested nutrient management levels except VC + CR + BF treatment. However, it is important to interpret the FDA data cautiously because the measured enzyme activities depend on the contribution of both extracellular and intracellular enzyme activities. The enzymes that adhere to the colloids of the organic compost can be another factor to increase the rate of FDA hydrolysis in the organic cultivation (Nannipieri et al., 2003).

Dehydrogenase is involved in the oxidation of SOM and occurs in viable cells and not in stabilized soil complexes. Therefore, the present results are in disagreement with observations where soil amended with organics also exhibits the greatest dehydrogenase activity (Liang, Si, Nikolic, Peng, & Chen, 2005). The observed dissimilar enzymatic activity response to fertilizer treatments (Table 4) may be the result of the resiliency of the respective enzymes to external inputs.

The carbon of the microbial biomass (MBC) is one of the most important variables that reflects differences between organic and conventional areas (Monokrousos, Papatheodorou, & Stamou, 2008).

Microbial biomass is one of the most labile of the pools comprising organic matter. An increase in MBC is likely to better represent changes in the nutrient-supplying capacity of organic matter than an increase in total organic matter (Gunapala & Scow, 1998). The present results are supported by observations of previous workers (Albiach, Canet, Pomares, & Ingelmo, 2000), where they found that organic residues enhanced microbial population, soil microbial biomass and their activity. It has been reported that organic sources like FYM, green manure, CR and BF decompose slowly, resulting in organic carbon accumulation in soil (Sharma, Bali, & Gupta, 2001). Experiments conducted in Punjab, India, in the RWCS showed that the incorporation of CR increased SOC compared to their removal from field (Singh, Singh, Meelu, & Khind, 2000). In contrast, MBC did not result in significant increase over control with FYM + CR + BF in both RWCS and RWMCS. This is contrary to the previous reports where FYM in combination with CR and BF significantly improved the microbial biomass (Banerjee, Aggarwal, Pathak, Singh, & Chaudhary, 2006). The possible reason could be the antagonism among the microflora present in the FYM and the added BF. The present results highlight the importance of the input of CR along with VC in order to increase the microbial biomass carbon in soil. An increase in MBC is linked to changes in the nutrient-supplying capacity of organic matter (Gunapala & Scow, 1998).

The increase in SR activity following the addition of CR to FYM/VC over the FYM/VC alone may be attributed to the enhanced availability of C as an energy source for micro-organisms native to the soil as well as those present in FYM/VC, leading to enhanced mineralization and consequent release of CO_2, though in RWCS, SR decreased slightly in VC + CR than VC alone. CR supplies C as an energy source for micro-organisms and increases the microbial activity (Rousk & Baath 2007; Smith, Papendick, Bezdicek, & Lynch, 1993). Addition of the biofertilizer to the plots receiving FYM + CR caused a significant decline in the SR activity both in RWCS and RWMCS. The observed differential influence of the added bioinoculant may be due to the differences in the C:N of the FYM/VC and also by the interactions among inoculated microbes with native microflora of the FYM/VC. The C/N ratio of VC is much lower (16:1) than that of FYM (30:1). There may be efficient incorporation of C in the microbial biomass and less loss of the CO_2, causing C immobilization in microbial cells. Karmegam and Rajasekar (2012) have reported that microbial population in VC differs qualitatively and quantitatively from that of the compost, and VC is an efficient medium to support the growth of bioinoculants. Interestingly, the highest value of soil microbial biomass carbon was recorded following VC + CR + BF in both RWCS and RWMCS, indicating efficient incorporation of C in the microbial cell mass.

The metabolic quotient (qCO_2) evaluates the efficiency of soil microbial biomass in using the organic C compounds (Anderson & Domsch, 1989). The greater qCO_2 values in these treatments could reflect an increase in the ratio of active:dormant components of the microbial biomass. A low metabolic quotient (qCO_2) in plots receiving FYM may indicate either the presence of microbial populations, which are more efficient in incorporating C compounds, or availability of relatively less labile organic residues.

Application of CR in combination with FYM, VC and/or biofertilizer resulted in high qCO_2 values. This shows that those soils which receive inputs of easily degradable C account for the high qCO_2 values mainly due to more available C present in crop residue. A high microbial quotient generally implies a ready supply of fresh organic residues (Anderson & Domsch, 1989). Additionally, several factors such as low pH, qualitative changes within microbial population (e.g. increase in the proportion of fungi) and prevalence of zymogenous over autochthonous microbiota may explain the differences in metabolic quotient. In the present study, FYM-receiving plots showed the highest level of fungal population as measured by the ergosterol content. Microbiota of the r-strategy ecotype would thrive under such conditions (Insam, 1990). They respire more C per unit of degradable C than K-strategists, which are adapted to more complex C utilization patterns. The low qCO_2 may be due to the occurrence of K-strategists micro-organisms.

This could be possibly due to qualitative and quantitative changes in microbial community structure and function in response to the above ground plant (Patra et al., 2006). The increase of glomalin levels is usually related to greater AMF (arbuscular mycorrhizal fungi) activity in systems with organic substances (Oehl et al., 2004). Overall, the nature of organic amendment was found to influence glomalin levels. Greater availability of mineral nutrients in VC and their rich microbial populations account for the beneficial effects on the mycorrhizal fungi (Arancon, Edwards, Bierman, Welch, & Metzger, 2004). The greater pore volume in VC-amended soils possibly increased the availability of both water and nutrients to micro-organisms including mycorrhizal fungi in soils (Scott, Cole, Elliott, & Huffman, 1996). Addition of organic nutrient sources is known to significantly stimulate mycorrhizal development (Castillo, Rubio, Contreras, & Borie, 2004). However, input of BF in conjunction with FYM + CR led to a significant reduction in soil glomalin content in RWMCS, which was not observed in case of VC + CR. This observation may be attributed to the differences in the native microflora of the VC and FYM and their interaction with the added microbial biofertilizer.

The correlation coefficients between different soil biological properties under RWCS and RWMCS are furnished in Table 8 and 9. It was observed that FDA activity has a significant positive correlation with alkaline phosphatase in RWCS but not in RWMCS. In RWMCS, alkaline phosphatase and DHA showed significant/strong negative correlation with microbial biomass carbon. The correlations between microbial biomass and enzyme activity are influenced by many factors (Stark, Condron, Stewart, Di, & O'Callaghan, 2007).

Table 8. Correlation coefficients between different soil biological properties under rice–wheat cropping system

	GLC	AP	AcP	FDA	DHA	MBC	SR	ERG	GLO
GLC	1.000								
AP	−0.192	1.000							
AcP	0.530	0.108	1.000						
FDA	0.391	0.760*	0.375	1.000					
DHA	0.375	0.539	−0.027	0.766*	1.000				
MBC	0.318	−0.091	0.024	−0.115	−0.156	1.000			
SR	0.397	−0.565	0.163	−0.259	−0.458	0.631	1.000		
ERG	−0.608	0.281	−0.901**	−0.047	0.301	−0.245	−0.507	1.000	
GLO	0.833*	−0.621	0.201	−0.084	0.010	0.288	0.553	−0.471	1.000

*$p < 0.05$.

**$p < 0.01$.

Note: GLC, glucosidase; AP, alkaline phosphatase; AcP, acid phosphatase; FDA, fluorescein diacetate; DHA, dehydrogenase activity; MBC, microbial biomass; SR, soil respiration; ERG, ergosterol; and GLO, glomalin content.

Table 9. Correlation coefficients between different soil biological properties under RWMCS system

	GLC	AP	AcP	FDA	DHA	MBC	SR	ERG	GLO
GLC	1.000								
AP	−0.464	1.000							
AcP	0.383	−0.421	1.000						
FDA	0.040	−0.059	−0.524	1.000					
DHA	0.382	0.419	0.085	−0.168	1.000				
MBC	0.332	−0.817*	0.225	0.047	−0.723*	1.000			
SR	0.339	−0.585	0.182	0.338	−0.348	0.538	1.000		
ERG	−0.091	−0.605	0.287	0.288	−0.468	0.399	0.084	1.000	
GLO	0.485	0.158	0.311	−0.480	0.261	0.092	0.147	−0.724*	1.000

*$p < 0.05$.

Note: GLC, glucosidase; AP, alkaline phosphatase; AcP, acid phosphatase; FDA, fluorescein diacetate; DHA, dehydrogenase activity; MBC, microbial biomass; SR, soil respiration; ERG, ergosterol; and GLO, glomalin content.

A strong negative correlation between soil ergosterol and glomalin content in RWMCS may be explained by the input of organic nutrient sources in the soil which perhaps stimulate the fungal populations, thereby improving the available plant nutrients. These conditions are known to exert a negative effect on the growth and multiplication of the arbuscular mycorrhizal fungi, the source of glomalin protein in soils.

These results indicate that under identical nutrient management conditions, cropping system determines the soil microbial indices. This is supported by the observations that the above ground plant influences the composition and biomass of microbial communities (Jones, Hodge, & Kuzyakov, 2004) because rhizodeposits or organic compounds released by plant roots can be highly specific for a given plant species or even a particular cultivar (Prieto, Bertiller, Carrera, & Olivera, 2011). It indirectly supports the idea that plants also liberate enzymes to the soil through root exudates or after the death and rupture of the cells (Buée, Martin, van Overbeek, & Jurkevitch, 2009). The observed strong positive correlation between FDA activity and alkaline phosphatase as well as with dehydrogenase and in rice–wheat cropping system could be attributed to the fact that these enzymes reflect the hydrolytic and oxidoreductive abilities of the soil microflora. A strong positive correlation between soil glomalin and glucosidase is expected because soil glomalin contains 37% carbon and 3–5% nitrogen, and contributes to the storage of soil carbon (3%), and the glucosidase enzyme catalyses the conversion of the complex carbonaceous polymers into simpler carbon compounds, thereby improving C availability.

The above discussion establishes that organic amendments improve soil microbial activities. Soil microbial activities are directly related to soil biological properties and hence soil fertility. Thus, application of organics, biofertilizer and CR improves soil microbial activity in rice–wheat and rice–wheat–mung bean cropping systems in the Indo-Gangetic plains.

5. Conclusions

The overall microbial activity had been significantly enhanced in soils treated with VC or compost in combination with CR. In conclusion, compost or VC application in combination with CR was found to be beneficial in terms of improving the soil biological parameters in RWCS and RWMCS. The finding from this study possesses specific implications in agricultural, ecological and soil ecosystem restoration perspectives pertaining to maintenance of soil fertility. It is suggested that inclusion of leguminous crop (wheat–mung bean–rice cropping system) is better than wheat–rice cropping system for maintaining soil productivity under semi-arid Indo-Gangetic plains.

Funding
This work was supported by Indian Council of Agricultural Research.

Author details
Geeta Singh[1]
E-mail: geetasinghkartik@gmail.com
D. Kumar[2]
E-mail: dinesh_agro@iari.res.in
Pankaj Sharma[3]
E-mail: pankaj1280@gmail.com
[1] Division of Microbiology, Indian Agricultural Research Institute, New Delhi 110012, India.
[2] Division of Agronomy, Indian Agricultural Research Institute, New Delhi 110012, India.
[3] NRC-DNA (Fingerprinting), NBPGR, Pusa campus, New Delhi, India.

References
Albiach, R., Canet, R., Pomares, F., & Ingelmo, F. (2000). Microbial biomass content and enzymatic activities after the application of organic amendments to a horticultural soil. *Bioresource Technology, 75*, 43–48. http://dx.doi.org/10.1016/S0960-8524(00)00030-4
Anderson, T. H., & Domsch, K. H. (1989). Ratios of microbial biomass carbon to total organic carbon in arable soils. *Soil Biology and Biochemistry, 21*, 471–479. http://dx.doi.org/10.1016/0038-0717(89)90117-X
Anderson, T. H., & Domsch, K. H. (1993). The metabolic quotient for CO_2 (qCO_2) as a specific activity parameter to assess the effects of environmental conditions, such as pH, on the microbial biomass of forest soils. *Soil Biology and Biochemistry, 25*, 393–395. http://dx.doi.org/10.1016/0038-0717(93)90140-7
Arancon, N., Edwards, C., Bierman, P., Welch, C. & Metzger, J. D. (2004). Influences of vermicomposts on field strawberries: 1. Effects on growth and yields. *Bioresource Technology, 97*, 831–840.

Bååth, E., & Anderson, T. H. (2003). Comparison of soil fungal/bacterial ratios in a pH gradient using physiological and PLFA-based techniques. *Soil Biology and Biochemistry, 35*, 955–963. http://dx.doi.org/10.1016/S0038-0717(03)00154-8

Banerjee, B., Aggarwal, P. K., Pathak, H., Singh, A. K., & Chaudhary, A. (2006). Dynamics of organic carbon and microbial biomass in alluvial soil with tillage and amendments in rice-wheat systems. *Environmental Monitoring and Assessment, 119*, 173–189. http://dx.doi.org/10.1007/s10661-005-9021-8

Buée, M., Boer, W. D., Martin, F., van Overbeek, L., & Jurkevitch, E. (2009). The rhizosphere zoo: An overview of plant-associated communities of microorganisms, including phages, bacteria, archaea, and fungi, and of some of their structuring factors. *Plant Soil, 321*, 189–212. http://dx.doi.org/10.1007/s11104-009-9991-3

Casida, Jr. L. E., Klein, D. A., & Santoro, T. (1964). Soil dehydrogenase activity. *Soil Science, 98*, 371–376. http://dx.doi.org/10.1097/00010694-196412000-00004

Castillo, C., Rubio, R., Contreras, A., & Borie, Y. F. (2004). Hongos micorrizógenosarbusculares en un Ultisol de la IX Región fertilizado orgánicamente [Arbuscular mycorrihizal fungi in an organically fertilized ultisol of the Region IX]. *Revista de la Ciencia del Suelo y Nutrición Vegetal, 4*, 39–47.

Eivazi, F., & Tabatabai, M. A. (1988). Glucosidases and galactosidases in soils. *Soil Biology and Biochemistry, 20*, 601–606. http://dx.doi.org/10.1016/0038-0717(88)90141-1

Farooq, M., Kobayashi, N., Wahid, A., Ito, O., & Basra, S. M. A. (2009). Strategies for producing more rice with less water. *Advances in Agronomy, 101*, 351–388. http://dx.doi.org/10.1016/S0065-2113(08)00806-7

Garcia, C., Hernandez, T., Roldan, A., & Albaladejo, J. (1997). Biological and biochemical quality of a semiarid soil after induced devegetation. *Journal of Environment Quality, 26*, 1116–1122. http://dx.doi.org/10.2134/jeq1997.00472425002600040024x

Green, V. S., Stott, D. E., & Diack, M. (2006). Assay for fluorescein diacetate hydrolytic activity: Optimization for soil samples. *Soil Biology and Biochemistry, 38*, 693–701. http://dx.doi.org/10.1016/j.soilbio.2005.06.020

Gunapala, N., & Scow, K. M. (1998). Dynamics of soil microbial biomass and activity in conventional and organic farming systems. *Soil Biology and Biochemistry, 30*, 805–816. http://dx.doi.org/10.1016/S0038-0717(97)00162-4

Insam, H. (1990). Are the soil microbial biomass and basal respiration governed by the climatic regime? *Soil Biology and Biochemistry, 22*, 525–532. http://dx.doi.org/10.1016/0038-0717(90)90189-7

Insam, H., & Haselwandter, K. (1989). Metabolic quotient of soil microflora in relation to plant succession. *Oecologia, 2*, 171–178.

Jones, D. L., Hodge, A., & Kuzyakov, Y. (2004). Plant and mycorrhizal regulation of rhizodeposition. *New Phytologist, 163*, 459–480. http://dx.doi.org/10.1111/nph.2004.163.issue-3

Karmegam, N., & Rajasekar, K. (2012). Enrichment of biogas slurry vermicompost with *Azotobacter chroococcum* and *Bacillus megaterium*. *Journal of Environmental Science and Technology, 5*, 91–108. http://dx.doi.org/10.3923/jest.2012.91.108

Kumar, K., & Goh, K. M. (2000). Crop residue management: Effects on soil quality, soil nitrogen dynamics, crop yield, and nitrogen recovery. *Advances in Agronomy, 68*, 197–319.

Lal, R. (2004). Is crop residue a waste? *Journal of Soil and Water Conservation, 59*, 136–139.

Liang, Y., Si, J., Nikolic, M., Peng, Y., & Chen, W. (2005). Organic manure stimulates biological activity and barley growth in soil subject to secondary salinization. *Soil Biology and Biochemistry, 37*, 1185–1195. http://dx.doi.org/10.1016/j.soilbio.2004.11.017

Monokrousos, N., Papatheodorou, E. M., & Stamou, G. P. (2008). The response of soil biochemical variables to organic and conventional cultivation of *Asparagus* sp. *Soil Biology and Biochemistry, 40*, 198–206. http://dx.doi.org/10.1016/j.soilbio.2007.08.001

Nannipieri, P., Ascher, J., Ceccherini, M. T., Landi, L., Pietramellara, G., & Renella, G. (2003). Microbial diversity and soil functions. *European Journal of Soil Science, 54*, 655–670. http://dx.doi.org/10.1046/j.1351-0754.2003.0556.x

Nannipieri, P., Grego, S., & Ceccanti, B. (1990). Ecological significance of the biological activity in soils. In J. M. Bollag & G. Stotzky (Eds.), *Soil biochemistry* (Vol. 6, pp. 293–355). New York, NY: Marcel Dekker.

Nottingham, A. T., Turner, B. L., Whitaker, J., Ostle, N., McNamara, N. P., Bardgett, R. D., ... Meir, P. (2015). Soil microbial nutrient constraints along a tropical forest elevation gradient: A belowground test of a biogeochemical paradigm. *Biogeosciences Discussions, 12*, 6489–6523. http://dx.doi.org/10.5194/bgd-12-6489-2015

Oehl, F., Sieverding, E., Mader, P., Dubois, D., Ineichen, K., Boller, T., & Wiemken, A. (2004). Impact of long-term conventional and organic farming on the diversity of arbuscular mycorrhizal fungi. *Oecologia, 138*, 574–583. http://dx.doi.org/10.1007/s00442-003-1458-2

Patra, A. K., Abbadie, L., Clays-Josserand, A., Degrange, V., Grayston, S. J., Guillaumaud, N., ... Roux, X. (2006). Effects of management regime and plant species on the enzyme activity and genetic structure of N-fixing, denitrifying and nitrifying bacterial communities in grassland soils. *Environmental Microbiology, 8*, 1005–1016. http://dx.doi.org/10.1111/emi.2006.8.issue-6

Prasad, R. (2005). Rice-wheat cropping system. *Advances in Agronomy, 86*, 285–339.

Prasad, R. (2011). A pragmatic approach to increase pulse production in north India. *Proceedings of the National Academy of Sciences India Section B-Biological Sciences, 81*, 243–249.

Prasad, R., & Power, J. F. (1991). Crop residue management. *Advances in Soil Science, 15*, 205–251. http://dx.doi.org/10.1007/978-1-4612-3030-4

Prieto, L. H., Bertiller, M. B., Carrera, A. L., & Olivera, N. L. (2011). Soil enzyme and microbial activities in a grazing ecosystem of Patagonian Monte, Argentina. *Geoderma, 162*, 281–287. http://dx.doi.org/10.1016/j.geoderma.2011.02.011

Rao, A. V., & Tarafdar, J. C. (1992). Seasonal changes in available phosphorus and different enzyme activities in arid soils. *Annals of Arid Zone, 31*, 185–189.

Rousk, J., & Bååth, E. (2007). Fungal and bacterial growth in soil with plant materials of different C/N ratios. *FEMS Microbiology Ecology, 62*, 258–267. http://dx.doi.org/10.1111/fem.2007.62.issue-3

Saharawat, Y. S., Singh, B., Malik, R. K., Ladha, J. K., Gathala, M., Jat, M. L., & Kumar, V. (2010). Evaluation of alternative tillage and crop establishment methods in a rice–wheat rotation in North Western IGP. *Field Crops Research, 116*, 260–267. http://dx.doi.org/10.1016/j.fcr.2010.01.003

Šantrucková, H., & Straškraba, M. (1991). On the relationship between specific respiration activity and microbial biomass in soils. *Soil Biology and Biochemistry, 23*, 525–532.

Scott, N. A., Cole, C. V., Elliott, E. T., & Huffman, S. A. (1996). Soil textural control on decomposition and soil organic matter dynamics. *Soil Science Society of America Journal, 60*, 1102–1109. http://dx.doi.org/10.2136/sssaj1996.03615995006000040020x

Sharma, M. P., Bali, S. V., & Gupta, D. K. (2001). Soil fertility and productivity of rice (*Oryza sativa*)-wheat (*Triticum* aestivum) cropping system in an Inceptisol as influenced by integrated nutrient management. *Indian Journal of Agricultural Sciences, 71*, 82–86.

Shibu, M. E., Van Keulen, H., Leffelaar, P. A., & Aggarwal, P. K. (2010). Soil carbon balance of rice-based cropping systems of the Indo-Gangetic Plains. *Geoderma, 160*, 143–154. http://dx.doi.org/10.1016/j.geoderma.2010.09.004

Singh, Y., Singh, B., Meelu, O. P., & Khind, C. S. (2000). Long term effects of organic manuring and crop residues on the productivity and sustainability of rice–wheat cropping system of North-West India. In I. P. Abrol, K. F. Bronson, J. M. Duxbury, & R. K. Gupta (Eds.), *Long-term soil fertility experiments in rice-wheat cropping systems* (RWC Paper series 6, pp. 149–162). New Delhi: RWCIGP.

Smith, J. L., Papendick, R. I., Bezdicek, D. F., & Lynch, J. M. (1993). Soil organic matter dynamics and crop residue management. In B. F. Metting (Ed.), *Soil microbial ecology* (pp. 65–95). New York, NY: Marcel Dekker.

Stark, C., Condron, L. M., Stewart, A., Di, H. J., & O'Callaghan, M. (2007). Effects of past and current crop management on soil microbial biomass and activity. *Biology and Fertility of Soils, 43*, 531–540. http://dx.doi.org/10.1007/s00374-006-0132-3

Stotzky, G. (1965). Microbial respiration. In C. A. Black (Ed.), *Methods of soil analysis Part 2* (pp. 1551–1572). Madison, WI: American Society of Agronomy.

Tabatabai, M. A. (1994). Soil enzymes. In R. W. Weaver, J. S. Angle, & P. S. Bottomley (Eds.), *Methods of soil analysis, part 2: Microbiological and biochemical properties* (pp. 775–833). Madison, WI: Soil Science Society of America.

Tabatabai, M. A., & Bremner, J. M. (1969). Use of p-nitrophenyl phosphate for assay of soil phosphatase activity. *Soil Biology and Biochemistry, 1*, 301–307. http://dx.doi.org/10.1016/0038-0717(69)90012-1

Tarafdar, J. C. (1989). Use of electro-focussing technique for characterizing the phosphatases in the soil and root exudates. *Journal of the Indian Society of Soil Science,* 37, 393–395.

Taylor, J. P., Wilson, B., Mills, M. S., & Burns, R. G. (2002). Comparison of microbial numbers and enzymatic activities in surface soils and subsoils using various techniques. *Soil Biology and Biochemistry, 34*, 387–401. http://dx.doi.org/10.1016/S0038-0717(01)00199-7

Timsina, J., & Connor, D. J. (2001). Productivity and management of rice–wheat cropping systems: issues and challenges. *Field Crops Research, 69*, 93–132. http://dx.doi.org/10.1016/S0378-4290(00)00143-X

Tirol-Padre, A., & Ladha, J. K. (2006). Integrating rice and wheat productivity trends using the SAS mixed-procedure and meta-analysis. *Field Crops Research, 95*, 75–88. http://dx.doi.org/10.1016/j.fcr.2005.02.003

Turner, B. L., & Wright, S. J. (2014). The response of microbial biomass and hydrolytic enzymes to a decade of nitrogen, phosphorus, and potassium addition in a lowland tropical rain forest. *Biogeochemistry, 117*, 115–130. http://dx.doi.org/10.1007/s10533-013-9848-y

Vance, E. D., Brookes, P. C., & Jenkinson, D. (1987). An extraction method for measuring soil microbial biomass C. *Soil Biology and Biochemistry, 19*, 703–707. http://dx.doi.org/10.1016/0038-0717(87)90052-6

Wright, S. F., & Upadhyaya, A. (1998). A survey of soils for aggregate stability and glomalin, a glycoprotein produced by hyphae of arbuscular mycorrhizal fungi. *Plant & Soil, 198*, 97–107.

Young, J. C. (1995). Microwave-assisted extraction of the fungal metabolite ergosterol and total fatty acids. *Journal of Agricultural and Food Chemistry, 43*, 2904–2910. http://dx.doi.org/10.1021/jf00059a025

Zhang, Y. L., Chen, L. J., Chen, Z. H., Sun, L. X., Wu, Z. J., & Tang, X. H. (2010). Soil nutrient contents and enzymatic characteristics as affected by 7-year no-tillage under maize cropping in a meadow brown soil. *Revista de la ciencia del suelo y nutrición vegetal, 10*, 150–157.

12

Geological application of ASTER remote sensing within sparsely outcropping terrain, Central New South Wales, Australia

R. Hewson[1]*, D. Robson[2], A. Carlton[2] and P. Gilmore[2]

*Corresponding author: R. Hewson, Faculty of Geo-Information Science & Earth Observation (ITC), University of Twente, P.O. Box 217, 7500 AE, Enschede, The Netherlands
E-mail: r.d.hewson@utwente.nl
Reviewing editor:
Louis-Noel Moresi, University of Melbourne, Australia

Abstract: One of the major problems faced by the application of geological remote sensing is its potential limitation in areas of a temperate climate with agricultural cultivation, limited outcrops and vegetation cover. This was the issue experienced when it was attempted to use the multi-spectral satellite Advanced Spaceborne Thermal Emission Reflectance Radiometer (ASTER) imagery to assist the updating of 1:100,000 geological mapping with the Ardlethan/Barmedman map sheets of central New South Wales (NSW), Australia. Most successful applications of geological remote sensing have been achieved in arid to semi-arid environments where vegetation and cultivation is minimal. Typically, day-time acquired ASTER visible to shortwave surface reflectance derived map products has extracted useful mineral related compositional information in such areas however in the studied areas of central NSW these techniques proved limited, particularly when using large mosaicked products such as the National Australia ASTER Geoscience Maps. Some improvement in geological discrimination was achieved using individual ASTER scenes, masked by high slope angle and processed into spectrally unmixed products. An alternative approach to extracting geoscience related products, utilised, night-time acquired ASTER thermal products. Their surface kinetic temperature products showed some potential for identifying the limited and sparse outcrops useful for field mapping geologists. Overall this study also showed the importance of the image spatial resolution in vegetated and cultivated areas with limited outcrop. Ideally

ABOUT THE AUTHOR

R. Hewson graduated with a MSc study in geophysics at Macquarie University followed by a PhD at the University of NSW within geological remote sensing. He worked from 1998 to 2010 within Australia's CSIRO Exploration and Mining Division, Perth, focusing on the geological case study development of airborne and satellite remote sensing, and its data integration with airborne geophysics. Since CSIRO, Hewson has undertaken remote sensing studies as a honorary research fellow for Professor Simon Jones at RMIT University, Melbourne and actively consulted with Australian mapping agencies. He currently works as an assistant professor in Geological Remote Sensing within ITC at University of Twente.

PUBLIC INTEREST STATEMENT

The present study is of interest for geoscientists attempting to apply remote sensing techniques for mapping in areas of limited outcrops within cultivated temperate terrain. Most published geological remote sensing applications have been typically undertaken within arid to semi-arid environments, however the described study investigates an application in a cultivated flat lying farm land environment with alluvial/soil cover. Attempts were undertaken to overcome these environmental factors by using DEM derived high slope defined masking to target spectrally defined outcrop anomalies from day-time acquisitions, and the application of night-time thermal infrared imagery for mapping limited surface and possible sub surface geological features associated with palaeochannels.

a finer spatial image product than available with ASTER's VNIR-SWIR combined products at 30 m is required.

Subjects: Geology - Earth Sciences; Applied & Economic Geology; Geophysics; Geomagnetics

Keywords: ASTER; geological mapping; night-time satellite; surface temperature; thermal inertia

1. Introduction

1.1. Previous examples of geological remote sensing using Advanced Spaceborne Thermal Emission Reflectance Radiometer (ASTER)

The results of fundamental research into mineral spectroscopy (Adams & Filice, 1967; Hunt & Ashley, 1979; Lyon & Burns, 1963; Vincent, Rowan, Gillespie, & Knapp, 1975; Vincent & Thomson, 1972) laid the basis for later geological remote sensing and prompted the development of such multi-spectral satellite sensors as ASTER, launched in December 1999 by Japan's METI and NASA. Since its operation in 2000, ASTER imagery has proved useful for discriminating and mapping several mineral groups (Rowan & Mars, 2003). However the majority of its applications for geological mapping studies have been in semi-arid to arid environments with minimal vegetation and cultivation. The issue of extending the application of such satellite multi-spectral imagery into less arid environments is important for the exploration of mineral resources in non-traditional areas. This particular study describes different approaches used to process and apply ASTER in such a non-arid environment, within central NSW, Australia. This study also follows on from previous studies by the author comparing ASTER applications in an arid environment with those from a temperate environment, albeit with standard processing methodologies (Hewson & Robson, 2014; Hewson, Robson et al., 2015).

This study also examines the application of seventeen compositional Australia wide map products, released by the Commonwealth Scientific Industrial Research Organisation (CSIRO) and Geoscience Australia (GA) from ASTER imagery (Caccetta, Collings, & Cudahy, 2013; Cudahy, 2012). These map products encompassed a wide variety of climatic environments and cultivation over the entire continental landmass of Australia and were utilised in the previous studies for the comparison between arid and temperate applications of ASTER (Hewson & Robson, 2014; Hewson, Robson et al., 2015). The processed digital values representing the seventeen map products assumed histogram stretched thresholds that at best qualitatively to semi-quantitatively represented mineral (group) composition across the entire Australian mosaicked ASTER data-set (Cudahy, 2012). The best results obtained from such ASTER derived geological mapping is undoubtedly within the arid to semi-arid exposed terrain (e.g. Mt Fitton, South Australia; Cobar, New South Wales–Hewson, Robson et al., 2015; Hewson & Robson, 2014). Hewson, Robson et al. (2015) also highlighted some of the environmental and geomorphological issues when studying temperate, cultivated and floodplain dominated terrain (Hewson, Robson et al., 2015). In particular, the studied temperate Wagga Wagga area of NSW exhibited moderate to limited rock outcrop exposure. The study showed that the generation of ASTER derived map products was improved by filtering or masking them for areas greater than 10% slope, availed by ENVI™ (http://www.harrisgeospatial.com/) processing of Shuttle Radar Topography Mission (SRTM) Digital Elevation Model (DEM) data (Hewson, Robson et al., 2015). This approach was based on a landform classification defined by topographic slope where there was an increased likelihood of erosional scarps or rocky outcrops for hilly slopes greater than 10% (Chen, 1997). Masking such ASTER products for vegetation fractional cover estimates, derived from Landsat (AusCover, http://data.auscover.org.au/), was also found to assist interpretation (Hewson, Robson et al., 2015). In particular, the Ferric oxide content, Fe oxide composition, Silica content and AlOH content products showed geologically associated anomalies either as trends within the alluvium floodplains or highlighted following the application of masks in outcropping areas (Hewson, Robson et al., 2015).

Several other intrinsic data issues can affect the interpretation of day-time ASTER imagery. As described in Hewson, Robson et al. (2015), ASTER's SWIR band spatial resolution of 30 m can be a problem in forested and woodland areas where few pixels would be free of canopy cover and its shadow components. Also, the detection of iron oxide mineralogy can be limited to where the ASTER VNIR spectral signatures are not dominated by those of vegetation (e.g. chlorophyll and pigment). ASTER's SWIR image products can also be affected by the crosstalk issue where stray light from band 4 "leaks" as an additive noise signal into bands 5 and 9 (Iwasaki, Fujisada, Akao, Shindou, & Akagi, 2005). An algorithm and software were developed to correct for this miscalibration issue (Iwasaki & Tonooka, 2005). However, in areas of low ground reflectance (e.g. shadow, thick vegetation, dark surfaces) residual crosstalk effect is apparent and can lead to "false anomalies" using spectral band indices (Hewson & Cudahy, 2011; Hewson, Cudahy, Mizuhiko, Ueda, & Mauger, 2005).

One important additional issue regarding the geological application of remote sensing is the spatial resolution of the imaging sensors. Studies by Kruse (2000), using various airborne hyperspectral VNIR-SWIR sensors, established the importance of image spatial resolution on the mapping of geology with scale dependent variations. In particular resampling airborne imagery of 2.4 m spatial resolution to 20 m showed there is a noticeable loss of discrete mapped occurrences of specific materials (Kruse, 2000). The spatial resolution of the ASTER SWIR sensor is 30 m and used as the overall resolution of many of the Australia ASTER Geoscience Maps. This is a potential limiting factor in its application in vegetated areas with limited geological outcrop.

1.2. Current ASTER mapping study

In this particular study, other additional approaches to aid geological interpretation of ASTER data in temperate areas, have been using on individual scenes, rather than large-scale processed map products, as attempted previously. Such scene specific processing avoids the issue of generalised and inappropriate product thresholds as well as offering greater enhancement of more subtle geological features in a non-arid environment. These approaches utilised a range of processing algorithms and tailored the histogram methods and thresholds for groundcover and geological terrain (Hewson, Cudahy, & Huntington, 2001; Hewson, Koch, Buchanan, & Sanders, 2002). Specifically, this study processed individual ASTER scenes within a cultivated and temperate region into map products, using:

- Colour composites images and band parameter ratios, similar to approaches applied by Cudahy (2012) and Hewson et al. (2005) but assuming different histogram stretches;
- Maximum Noise Fraction (MNF) analysis (Green, Berman, Switzer, & Craig, 1988).
- Surface reflectance and emissivity ASTER imagery with Mixture Tuned Matched Filter (MTMF) techniques (Kruse et al., 1993) to separate the spectral contributions within each image pixel signature into estimated proportions of geologically or vegetation related land cover (Hewson et al., 2002).

In addition to processing individual day-time ASTER imagery, night-time thermal ASTER imagery was also examined in the study area. Night-time thermal imagery, provided as surface kinetic temperature, is capable of (1) identifying areas of limited outcrop, useful for field geologists; and (2) potentially discriminate changes in the physical properties of soil, regolith and exposed outcrops such as density, porosity and thermal capacity when further processed into Apparent Thermal Inertia (ATI) (Kahle, 1987; Price, 1977). This can be useful in areas of extensive weathering and transported alluvial cover where there is limited surface geological outcrop.

Another issue in temperate and cultivated areas is often the limited and sparse rock outcrops for detailed identification and sampling. Access to farmland requires stricter protocols to be followed and targeting known or likely outcrops efficiently within the shortest field trip is a priority. The use of high spatial aerial digital imagery (e.g. 0.5 m resolution ADS40; Sandau et al., 2000) is typically used for identification of likely outcrops to field sample. However, the limited spectral resolution of aerial photography and the restriction of its bands to the visible wavelengths preclude a mineralogical based discrimination of such outcrops. Night-time ASTER TIR products were investigated for its ability to delineate such outcrops as an assistance for geological field survey mapping.

1.3. Study objectives & study outline

The principal aim of this study is to evaluate the potential of using multi-spectral ASTER imagery for geological mapping in a temperate and cultivated environment, with associated sparse outcrops. This required the applying individual scene specific histogram stretches, different processing methodologies and use of night-time ASTER thermal imagery.

A concurrent aim of this study was to provide additional geological information as part of the 1:1,00,000 (1:100 K) mapping update for the Ardlethan and Barmedman area of NSW, undertaken by the Geological Survey of New South Wales (GSNSW). It was hoped that such compositional information could be useful for field mapping activities. This study area was an extension to a previous study at Wagga Wagga (Hewson, Robson et al., 2015) albeit with less geological exposure. A detailed description of this study and its data is also available via the Geological of NSW Report, fulfilled as part of this study (Hewson, 2015), and also outlined in summary in Hewson, Carlton, Gilmore, Jones, and Robson (2015).

In this paper Section 2.0 describes the physiography, environment and geology of the study area of the Ardlethan and Barmedman areas of central NSW. Section 3.0 outlines the data-sets sourced in this project including the ASTER Australian Geoscience map products, and the ASTER data as used as for individual scene processing. The processing methodologies applied to the day- and night-time ASTER imagery are also described in Section 3.0. The results and conclusion to this study are explained in Sections 4.0 and 5.0, respectively.

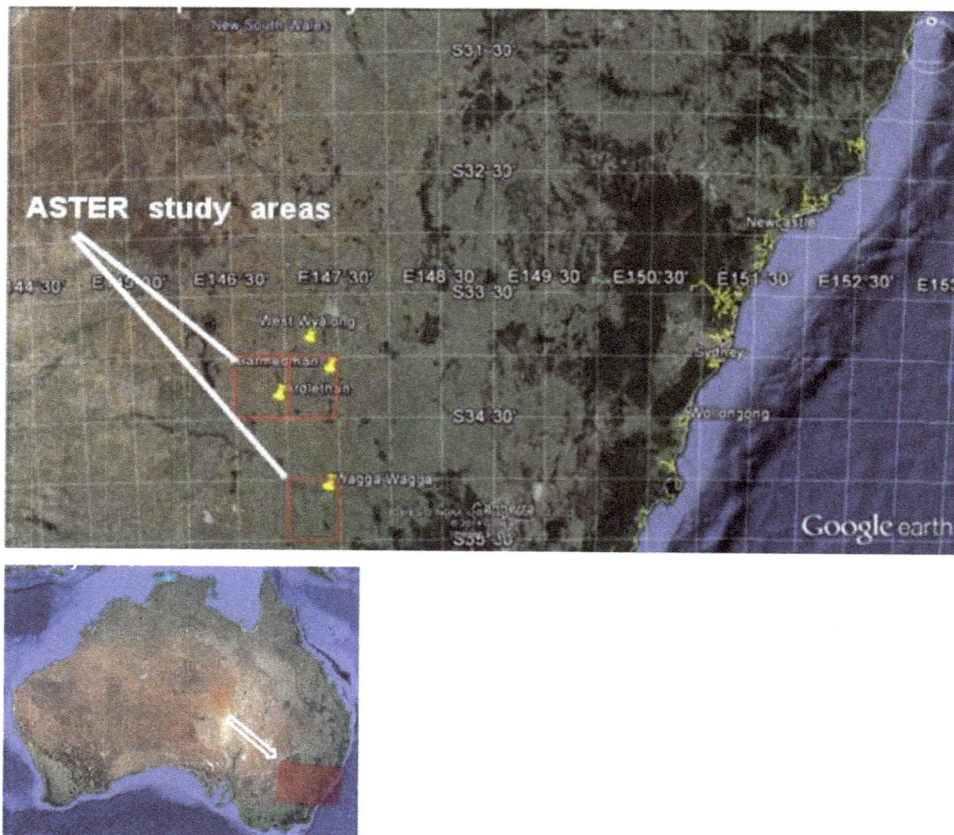

Figure 1. Location of the Ardlethan, Barmedman and Wagga Wagga 1:100,000 map sheet areas.

2. Study area

The study area used in this evaluation of remote sensing within a non-arid environment encompasses the Ardlethan and Barmedman 1:1,00,000 (1:100 K) map sheet areas of south central New South Wales (NSW), Australia (Figure 1). This area is of particular interest for intrusive sourced hydrothermal and alluvial hosted tin and gold deposits (Colquhoun, Meakin, & Cameron, 2005). However limited outcrop within a flat and heavily cultivated terrain has handicapped geological investigations and exploration.

2.1. Location and physiography

The Ardlethan (146.5°–147°E, 34°–34.5°S) and Barmedman (147°–147.5°E, 34°–34.5°S) 1:100 K map sheet areas of central southern NSW lie to the north and north-west of the Wagga Wagga 1:100 K map sheet area studied previously by (Hewson, Robson et al., 2015) (Figure 1). The area comprises extensive cultivated pasture and cropland within the floodplains and catchment of the Lachlan River. Limited hilltops and ridges occur within these areas and typically associated with open woodland or scrubby vegetation. The climate is typically Mediterranean with an average rainfall of 480 mm per year observed at Ardlethan (http://www.bom.gov.au/climate/data/). The False Colour imagery available by ASTER highlights these features (Figure 2). Areas of red hue highlight green photosynthetic vegetation (e.g. crops or trees). Dark red-brown areas within Figure 2(a) typically indicate native vegetated cover. Artificially illuminated SRTM DEM imagery highlights the low relief nature of the environment, varying from 145 to 430 m elevation within the two map sheets (Figure 2(b)). In particular slope analysis of the SRTM DEM reveals only a small proportion exhibited slopes greater than 10% (Figure 2(c)). Past published geological mapping for the Narrandera (Wynn, 1977) and Cootamundra (Warren, Gilligan, & Raphael, 1996) 1:2,50,000 map sheet areas, encompassing the Ardlethan and Barmedman 1:100 K sheets, indicates that outcrops and evidence for geological boundaries lay within mostly moderately topographic relief (Figure 2(c)). Significantly less topographic relief is present in this study area than apparent for the Wagga Wagga 1:100 K map sheet study area (Hewson, Robson et al., 2015).

The environment and native vegetation consists predominantly of eucalyptus/sheoak woodland and scrubland, mostly occupying the limited ridges and hilltops. The ASTER Green Vegetation product, generated as part of the National Australia ASTER Mapping, highlights these native vegetated areas, and to a lesser extent, the cultivated cropland (Figure 3). These woodland areas typically coincide with the areas of higher slope (Figure 2(c)), although they can also be associated with watercourses, tree plantations and crops (Figure 3).

2.2. Geology

The Ardlethan and Barmedman study areas contains a geological sequence of units within the central and western Lachlan Orogen, and consist of Ordovician to Devonian volcanic, sedimentary, meta-sedimentary and plutonic rocks. Tertiary to Quaternary deposits of the eastern Murray Basin are prevalent as cover over much of the Palaeozoic sequences (Clare, Fleming, & Glen, 1997; Colquhoun et al., 2005; Downes, McEvilly, & Raphael, 2004). Widespread unconsolidated Quaternary alluvial, colluvial and aeolian sediments blanket much of the terrain limiting the access to rock outcrops for direct geological mapping. The main units for the Ardlethan and Barmedman 1:100 K study areas are shown within the displayed geological/metallogenetic map (Figure 4) combining the Cootamundra (Fitzpatrick, 1979; Warren et al., 1996) and Narrandera (Heugh, 1979) 1:2,50,000 (250 K) published mapping.

The area has had an extensive history of tin/tungsten mining associated with various Silurian–Devonian intrusives (Colquhoun et al., 2005). In particular, the Ardlethan Tin Mine has been a significant producer (Figure 4). Its porphyry-style deposit has been described as having a magmatic and hydrothermal history associated with the Ardlethan Granite (Figure 4) (Ren, Walshe, Paterson, Both, & Andrew, 1995). Ren et al. (1995) described an alteration zonation of the Ardlethan tin deposit characterised by biotite, chlorite, sericite and tourmaline mineralogy. Scott and Rampe (1984) found mineralisation was a useful pathfinder indicator for the tin deposit than the geochemistry signature from

Figure 2. (a) Ardlethan and Barmedman 1:100 K map sheet areas–ASTER False Colour image; (b) Shaded SRTM DEM relief; (c) and areas of 10% slope (SRTM DEM) shown in red and previously published 1:250 K geological mapped boundaries (white boundaries).

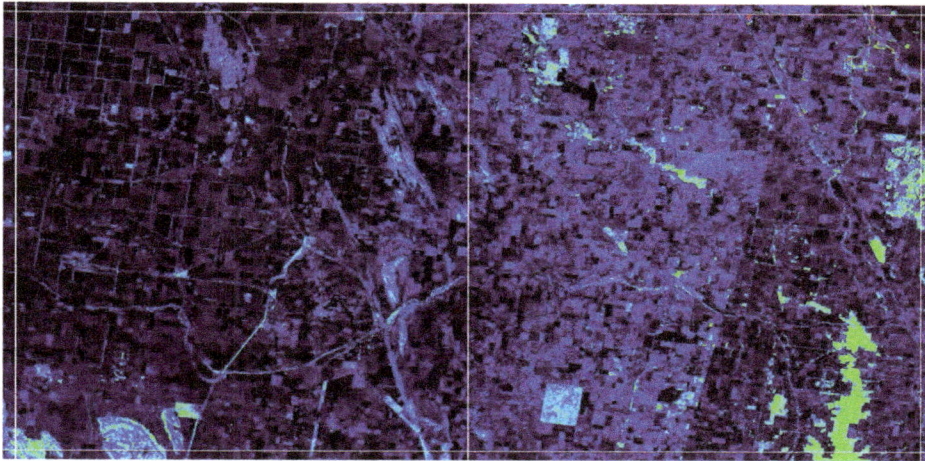

Figure 3. CSIRO-GA ASTER Green Vegetation product where blue is low, green is moderate and red is high green vegetation content. See Figure 2 for coordinate extents.

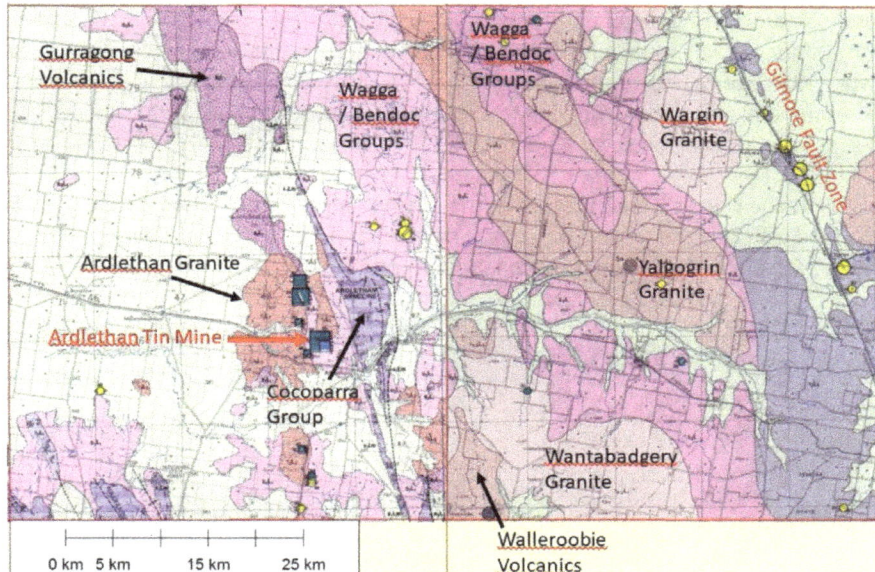

Figure 4. Past published geological mapping within the Ardlethan and Barmedman 1:100 K map sheet areas (Fitzpatrick, 1979; Heugh, 1979).

Notes: Round yellow symbols signify vein hosted gold deposits, square blue/green symbols signify disseminated tungsten/tin deposits. See Figure 2 for coordinate extents.

samples. Weathering was also found to range from up to 60 m but generally below 10 m depth and exhibited decreasing haematite/goethite and kaolinite development with depth (Scott & Rampe, 1984).

Sub-surface structures within the study area include palaeochannels within the Quaternary flood-plain deposits of the Lachlan River and its tributaries (Colquhoun et al., 2005). Considerable interest has been directed to such palaeochannels for their potential hosting of alluvial gold or tin deposits and also for dryland salinity and hydrological importance (Lawrie, Chan, Gibson, & de Souza Kovacs, 1999; Mackey et al., 2000). In particular, the Bland Creek Palaeochannel east of West Wyalong is an example of an alluviated valley hosting the Gibsonvale alluvial tin workings (Gibson & Chan, 1998). The present relief is generally low but it's suggested that the valley was originally 1 km wide and 35 m deep, dated to possibly the Oligocene (Gibson & Chan, 1998). Aero magnetics from airborne geophysical surveys has been found useful to delineate such palaeochannels, particularly where maghemite- and magnetite-bearing pisoliths and alluvium generates high spatial frequency magnetic anomalies (Gibson & Chan, 1998; Lawrie et al., 1999; Mackey et al., 2000). Magnetic model-ling of such palaeochannels has been attempted by Mackey et al. (2000). Alternatively, first vertical derivative (1VD) filtering of the measured Total Magnetic Intensity (TMI) has been demonstrated to be a useful qualitative technique to map near surface sourced magnetic anomalies associated with palaeochannels (Mackey et al., 2000).

3. Methodology

3.1. Data-sets

ASTER measures radiance from fourteen VNIR, SWIR and TIR bands at spatial pixel dimensions of 15 m (VNIR), 30 m (SWIR) and 90 m (TIR) (Table 1) (Fujisada, Sakuma, Ono, & Kudoh, 1998; Yamaguchi et al., 2001). In this study, both the processed ASTER Australian Geoscience products, as well as ASTER day-time (VNIR, SWIR and TIR bands) and ASTER night-time (TIR bands) image acquisitions were utilised. The ASTER Australian Geoscience map products were sourced for the past and current studies directly or indirectly from CSIRO and Geoscience Australia in either GeoTiff or ENVI™ (http://www.exelisvis.com/)/ER Mapper™ (http://www.hexagongeospatial.com/) software compatible for-mats. The Australian ASTER geoscience maps are available from: (i) the AuScope Discovery Portal (http://portal.auscope.org/portal/gmap.html); or (ii) via CSIRO (http://c3dmm.csiro.au/Australia_ ASTER/index.html). The products were generated from traditional band (ratio) parameters (Crowley, Brickey, & Rowan, 1989), targeting mineral related spectral absorption features (albeit at multi-spectral ASTER resolution), followed by masking to exclude areas of cloud cover, deep shadow, wa-ter bodies and significant vegetation cover (Cudahy, 2012).

Individual ASTER image scenes were accessed at the time via Japan's METI Ground Data System (GDS) portal. NASA's Earth Explorer Reverb web portal now handles this ASTER data access (https://reverb.echo.nasa.gov/). Each ASTER image scene encompasses a 60 × 60 km area. Level 1b

Table 1. Spectral bandwidths and central band wavelengths of the ASTER sensor (Fujisada et al., 1998)

Module	VNIR	SWIR	TIR
Spectral bandwidth (µm) [Centre λ (µm)]	Band 1 0.52–0.60 [0.556]	Band 4 1.600–1.700 [1.656]	Band 10 8.125–8.475 [8.291]
	Band 2 0.63–0.69 [0.661]	Band 5 2.145–2.185 [2.167]	Band 11 8.475–8.825 [8.634]
	Band 3 N 0.78–0.86 [0.807]	Band 6 2.185–2.225 [2.209]	Band 12 8.925–9.275 [9.075]
	Band 3B 0.78–0.86 [0.804] (Backward looking)	Band 7 2.235–2.285 [2.262]	Band 13 10.25–10.95 [10.657]
		Band 8 2.295–2.395 [2.336]	Band 14 10.95–11.65 [11.318]
		Band 9 2.360–2.430 [2.400]	
Spatial resolution (m)	15	30	90

Note: 1 µm ≡ 1,000 nm.

Table 2. Acquisition dates of day- and night-time ASTER imagery used in study

Day-time surface temperature & VNIR-SWIR surface reflectance	21/3/2012	18/10/2013	
Night-time surface temperature	11/1/2012	29/11/2013	21/4/2008
Day-time VNIR-SWIR surface reflectance & TIR emissivity	13/1/2005		
GA-CSIRO ASTER inputs: Day-time VNIR-SWIR	16/11/2006 & 15/10/2006		

(radiance at sensor) and Level 2 (e.g. surface reflectance) products were used in this study. Table 2 lists the various day- and night-time ASTER acquisitions used in this particular study.

3.2. Day-time processing of ASTER imagery

Similar terrain issues were encountered within the Ardlethan and Barmedman 1:100 K map sheet areas, to those encountered in previous studies in the Wagga Wagga study, 100 km to the south east of (Hewson, Robson et al., 2015). It was also found that the Ardlethan and Barmedman study area had significantly less topographic relief and associated outcrop than the Wagga Wagga area. The nature of this terrain precluded the application of DEM derived slope masks to extract coherent geological information from the Australian ASTER products, as demonstrated in the Wagga Wagga area (Hewson, Robson et al., 2015). As a consequence, the following approaches were undertaken with the day-time ASTER acquisitions:

(1) Process specific individual subsets of ASTER surface reflectance (VNIR-SWIR) and emissivity (TIR) imagery of the Ardlethan–Barmedman area using band parameter processing, MNF style classification and spectral unmixing (Crowley et al., 1989; Green et al., 1988; Kruse et al., 1993).

(2) Field observations and sampling of rock outcrop and soils collected within the Ardlethan and Barmedman 1:100 K map sheet areas.

(3) Compare the spatial resolution of the ASTER products/imagery and the Airborne Digital Sensor 40 (ADS40) photogrammetry (Sandau et al., 2000) currently used for surface mapping by the GSNSW.

Processed subsets of day-time ASTER imagery within the Ardlethan–Barmedman area focused on key areas such as the Gurragongs Volcanics and the Wagga/Bendoc Group (Figure 4). Topographic slope masking was also incorporated with the MNF and spectral unmixing methodologies (Green et al., 1988; Kruse et al., 1993) applied to those areas within subsets, to target possible geological outcrops. ENVI™ software (http://www.harrisgeospatial.com/) was applied for this purpose Hewson, Robson, et al. (2015).

3.3. Night-time processing ASTER imagery

The basis of using night-time imagery, either processed to surface temperature, or as Thermal Inertia, as a geological mapping tool is explained in Sabins (1997) and Watson (1975). Thermal inertia (P) is a scalar volumetric physical property of a material, that cannot be measured directly but inferred from diurnal variations of temperature at the Earth's surface and related to the following other properties:

$$P = (K\rho c)^{1/2}$$

where K = thermal conductivity, ρ = density and c = specific heat.

Generally geological materials, particularly denser and non-porous, retain the heat acquired during the sunlight exposure more effectively (e.g. higher thermal inertia), and radiates its heat overnight. Unconsolidated and less dense materials however heat up and cool down more quickly (e.g. lower thermal inertia) in particular lower density regolith such as soils have lower thermal inertia than rocks and their constituent minerals (Watson, 1975). The addition of moisture to dry soil can also increase the thermal inertia due to both an effective increase in density as well as reduction in porosity affecting bulk thermal properties (Watson, 1975). Ideally therefore, night-time thermal imagery provides better results under dry summer conditions, or at least undertaken with a knowledge of local rainfall and moisture levels. Rainfall records from the Australian Bureau of Meteorology (http://www.bom.gov.au/climate/data/) were sourced for the nearby Tallimba weather station and precluded moisture as a significant influence on the ASTER acquisitions.

Night-time NOAA satellite thermal imagery has been previously used for geological mapping in Western Queensland (Russell & Lappi, 1988). Russell and Lappi (1988) noted that rock outcrops absorb and retain significantly more day-time thermal energy than the surrounding soil and alluvium, generating night-time temperature anomalies. They estimated that this solar heating effect could penetrate into unconsolidated material of 4 m, potentially indicating the presence of rock outcrops to this depth. Field observations in this study also identified areas with anomalously high night-time surface temperatures associated with dense forested woodland with no apparent geological outcrops. Although there is also the possibility of near-surface outcrop, the association of high temperature anomalies with dense vegetated areas acting as a thermal blanket, has been recognised previously by Fabris (2002).

Several different night-time ASTER acquisitions were obtained for this study to observe the possible effects of seasonal ground temperature and rainfall variations on the imagery (Table 2). Various histogram stretches were applied to the ASTER night-time surface temperature imagery expressed in Kelvin $\times$ 10 values to obtain the best qualitative information. The surface temperature ASTER products were based on the Temperature Emissivity Separation algorithm described by Gillespie et al. (1998) and validated by Hook, Vaughan, Tonooka, and Schladow (2007).

The derivation of Thermal Inertia is not trivial and an approximation is calculated as Apparent Thermal Inertia (ATI) (Kahle,1987; Price, 1977). The generation of ATI imagery requires both day-time T_1 (e.g. maximum) and night-time T_2 (e.g. minimum) time surface temperatures, and an estimate of albedo, ρ_{av} (e.g. average VNIR surface reflectance):

$$ATI \approx \frac{(1 - \rho_{av})}{\Delta T}$$
where $\Delta T = T_1 - T_2$

This approximation by Kahle (1987) and Price (1977) ignored for simplicity, topographic and atmospheric affects. The ATI approximation precludes the comparison of one ATI image values to another. Also, although the dates of ASTER day- and night-time acquisitions were not ideal for the Ardlethan–Barmedman areas, Kahle and Alley (1985) demonstrated that relative and apparent estimates of thermal inertia are still useful for the differentiation of outcrops and alluvium, even if several weeks' separate day- and night-time acquisitions. Another limitation in this particular application using ASTER TIR data is that the acquisition times are not likely to observe the maximum day-time and minimum night-time ground surface temperatures, at the 10:30 am and 10:30 pm acquisition times, rather than at approximately 1–3 pm and 3–5 am suggested by Price (1977).

The derivation of ATI products required sub-setting of all the input ASTER data into coincident overlapping imagery. Different spatial coverage of the ASTER data occurred from different acquisition dates, as well as from changes between their descending and ascending/day- and night-time orbits. Consequently the resulting image subset ATI product was less than 60 $\times$ 60 km coverage. The calculation derivation of the ATI and their image sub-setting was performed using ER Mapper™ software (http://www.hexagongeospatial.com/). An additional potential QC issue of these products was the 90 m spatial resolution of the ASTER TIR imagery, increasing the effective mixed observation of thermal properties between the vegetation, soil and outcrop radiant components.

4. Results

4.1. National Australia ASTER Geoscience maps
In this study both the ASTER Australian Geoscience products and independently processed ASTER day-time (VNIR, SWIR and TIR bands) and night-time (TIR bands) acquisitions were used in an attempt to extract geological mapping information. However, the ASTER False Colour (Figure 2(a)) and Green Vegetation (Figure 3) ASTER products for the Ardlethan and Barmedman 1:100 K areas highlight the issues of agricultural cropland and vegetation cover. Likewise, the ASTER Regolith product (Figure 5(a)) highlights the difficulty in its geological mapping application for this area and mostly highlights crop and paddock patterns and boundaries to variable vegetation and soil

(a)

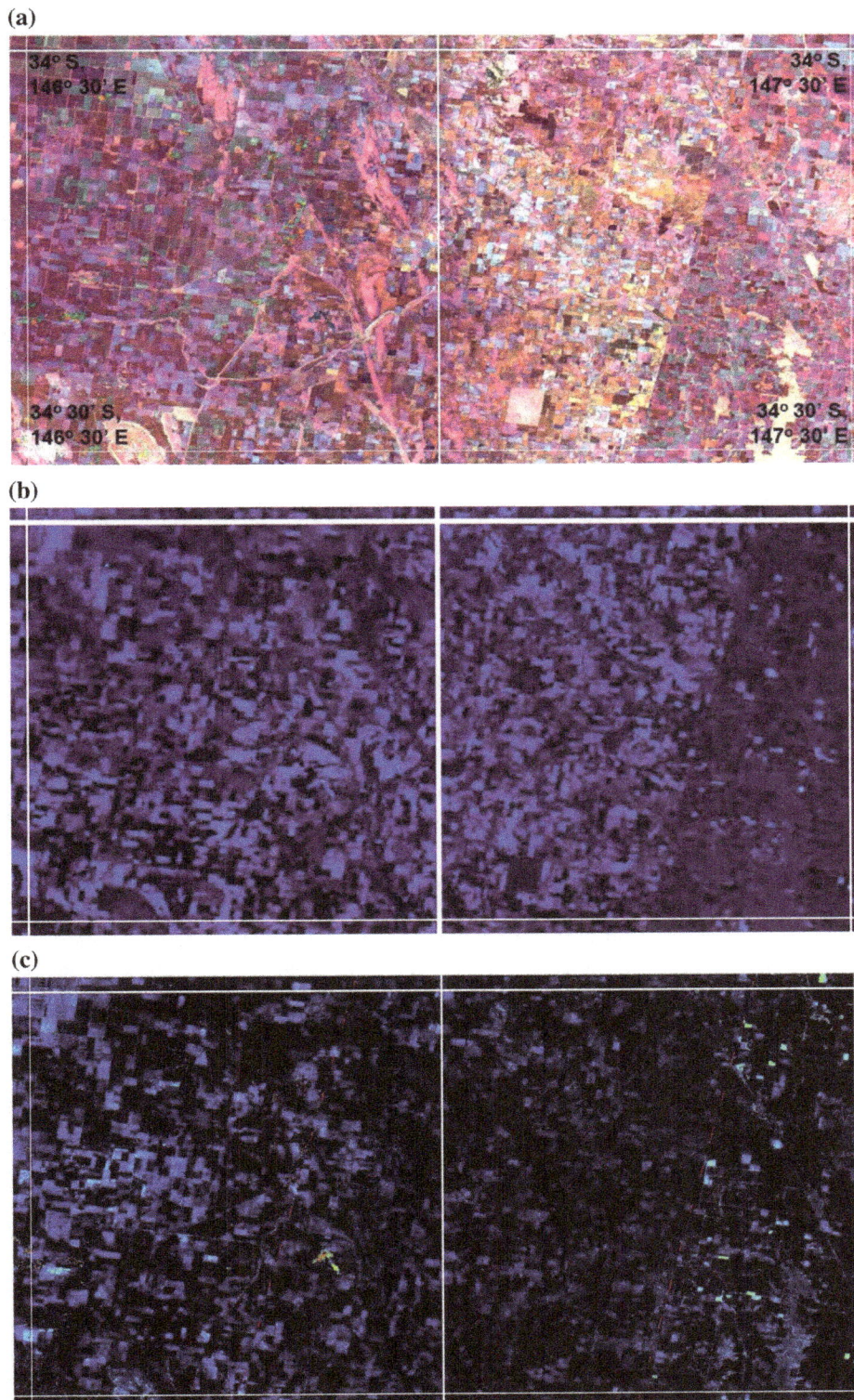

(b)

(c)

Figure 5. Ardlethan and Barmedman 1:100 K sheets:
(a) National Australia ASTER Regolith product; (b) National Australia ASTER Silica Index; (c)
National Australia ASTER AlOH. Derived map product values are represented by blue = low,
green = moderate and red = high content. See Figure 2 for coordinate extents.

exposure. The ASTER Silica Index and AlOH Content products also show cropland patterns (Figure 5(b) and (c), respectively). Some variation in the silica/quartz content appears within the crop fields and possibly related to variable quartz sand content within the soils. Unlike previous studies within the nearby Wagga area, no lineaments or anomalies were apparent and suggestive of sub-surface geology from the ASTER ferric iron product within the soils and regolith within the cultivated areas (Hewson, Robson et al., 2015).

4.2. Spectral unmixing of individual ASTER surface reflectance and emissivity image imagery

The day-time ASTER VNIR-SWIR image acquisition obtained for the individual scene based analysis, straddled both the Ardlethan and Barmedman 1:100 K map sheets (Figure 6(a)). The chosen summer ASTER scene (13/1/2005) indicated the cropland showed high reflectance/albedo due to its dry and predominantly exposed soil/regolith (Figure 6(a)). MNF processing of the ASTER imager of the nine ASTER VNIR-SWIR bands for this summer acquisition generated several classified images combining landscape features of vegetation, soil and potentially outcrop. MNF band 5 generated from the whole scene appeared to highlight Quaternary alluvial cover although most MNF results were dominated by soil/regolith types and vegetated landforms. An image subset for the Wagga/Bendoc Group (blue box, Figure 6(b)) was also processed for its MNF results. A comparison of its MNF's 4, 5 and 6 composite imagery (Figure 6(c)) with an albedo image (greyscale background image of Figure 6(d)) highlighted subtle discrimination of the various beds and north westerly structural trends (493000 mE, 6220000 mN). Some of the apparent bedding features observed may also relate to topographic aspect illumination effects and associated vegetation variation and warrant further examination. MTMF spectral unmixing classification appears to highlight soil/regolith and/or crop vegetation anomalies rather than discriminate the actual outcropping Wagga/Bendoc beds (Figure 6(d)). However, the spectral signatures associated with each unmixed endmember predominantly mapped variety of vegetation and clay/soil types by their 0.8 µm and 2.2 µm spectral features, respectively.

A similar approach was applied at the Gurragong Volcanics subset (red box, Figure 6(b)) for both the day-time ASTER VNIR-SWIR data (Figure 7(a)–(d)). The ASTER vegetation map product (Figure 7(b)) and MNF processing of the subset (Figure 7(c)) again indicated the dominance of vegetation and cropping patterns. A more targeted approach to MNF processing was also attempted by further subsetting the imagery for elevated areas using a 5% slope mask, derived from the SRTM DEM, (Figure 7(d)). This decreased slope threshold, compared to the 10% used in the Wagga study, was assumed given the terrains more weathered and colluvial dominated outcropping slopes. A trend in the MNF spectral response suggests a change of composition and/or vegetation from the northerly Gurragong units (blue-green) to the yellow and red southerly occurrences (Figure 7(d)). However there is no current understanding to this apparent trend in processed ASTER spectral image response.

MNF and MTMF processed results of the ASTER TIR surface emissivity imagery proved noisy and again the results were dominated by vegetation and cropping patterns/soils. The variation of quartz sand within exposed or fallow soils would likely to be a strong determinant in its response.

4.3. Effects of image spatial resolution on mapping

The effect of the spatial resolution of the ASTER sensor was examined in the study area within an area of a mixture of vegetation and exposed geology. A reduction in the surface feature spatial detectability was observed from the comparison of the 0.5 m resolution ADS40 photogrammetry (Sandau et al., 2000) and the 30 m ASTER products and imagery (VNIR-SWIR) within the study area (Figure 8(a) and (b)). Currently ADS40 is commonly used to assist surface mapping by the GSNSW, particularly for identifying the limited surface outcrops. Generally there are sparse and limited outcrops within the Ardlethan– Barmedman area, sometimes located within exposed cropland or on ridges and topographic highs with open woodland or scrubby vegetation. The differences between the ADS40 and ASTER imagery in their ability to highlight vegetation and outcrops is shown in Figure 8(a) and (b). The coarser 30 m ASTER False Colour imagery (Figure 8(a)) discriminates the predominantly woodland areas but not individual trees as discriminated by the ADS40 (Figure 8(b)).

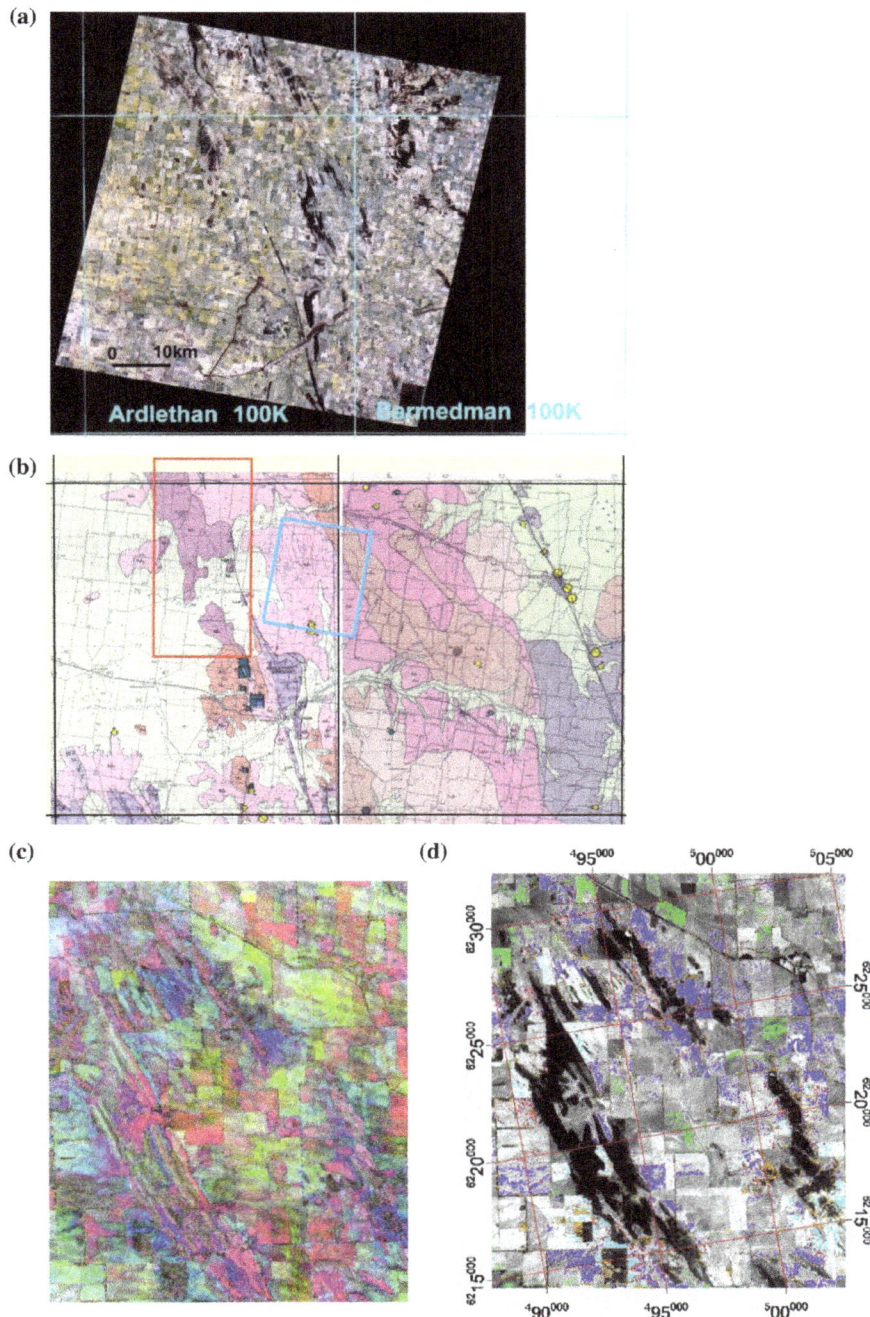

Figure 6. (a) ASTER False Colour imagery (13/1/2005) acquisition within the Ardlethan and Barmedman 1:100 K sheets; (b) Ardlethan–Barmedman 1:100 K sheet published geology showing the Gurragong Volcanics (red box) and Wagga/Bendoc Group (blue box) image subsets. See Figure 4 for definition of main geological units; (c) Wagga/Bendoc Group subset RGB composite of MNF bands 4, 5 and 6; (d) Wagga/Bendoc Group subset MTMF spectral unmixed results, based on 8 spectral endmembers. Greyscale albedo image of ASTER band 4 overlain by examples of various subtle spectral endmembers dominated by clays in soil (blue, green, orange and cyan). See Figure 2 for coordinate extents.

Differences in the cropland soil colour and the presence of likely north-westerly trending Wagga/Bendoc Group outcrops suggest that the ADS40 is more discriminatory for structural interpretation and outcrop delineation than the day-time ASTER products in this terrain (Figure 8(a) and (b)).

(a)   (b)   (c)   (d)

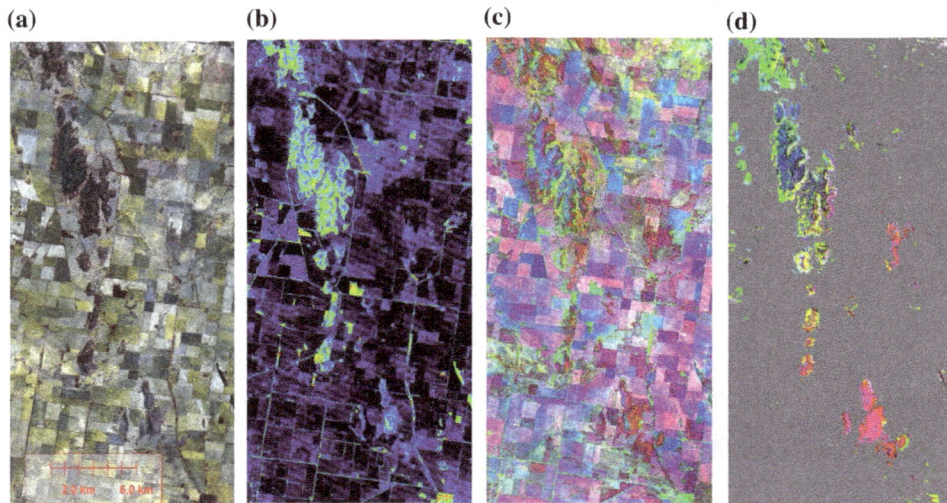

Figure 7. Gurragong Volcanics (a) subset of ASTER False Colour; (b) subset of ASTER Green Vegetation; (c) subset RGB composite of MNF bands 2, 5 and 6; (d) subset of RGB composite of MNF bands 5, 4, and 3 processed with DEM Slope greater than 5% mask. See Figure 6 for subset location.

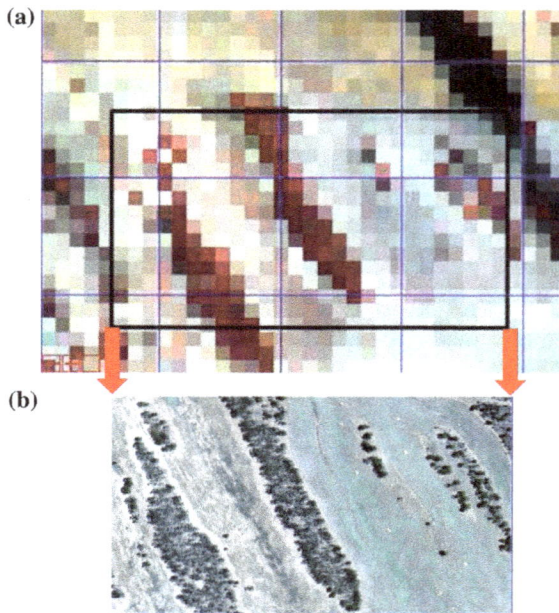

(a)

(b)

Figure 8. (a) Example of the 30 m ASTER False Colour product (~34°6′S, 146°57′) within the Wagga/ Bendoc Group geological unit. The blue gridlines are 250 × 250 m; (b) ADS40 digital photogrammetry for the zoomed in area of (a) (black box).

4.4. ASTER night-time surface temperature imagery

The apparent limitation of using VNIR–SWIR and TIR day-time ASTER remote sensing for geological mapping for this study area's terrain suggests that alternative less surface affected techniques based more on physical properties rather than mineralogical be trialled. In particular this study examined the application of ASTER's surface temperature product, as supplied in values of Kelvin multiplied by 10. Such satellite surface temperature estimates are variable according to times, dates, winds, etc. so the displayed images, showing relative differences from low to high temperatures (e.g. blue to red), are useful more for their qualitative trends. A first pass examination of a night-time ASTER acquisition as surface temperature shows it typically has higher values on ridges and topographic rises between watercourses (Figure 9(a)). Past published 1:250 000 scale geological mapping also indicates many of the temperature anomalies are associated with the boundary of

Quaternary cover and contained within some of the mapped outcropping geological units (Figure 9(b)). The relationship between the night-time temperature anomalies and the topography is clearly illustrated in Figure 10.

The RGB composite imagery of thermal radiance bands 13, 12 and 10, for the night-time ASTER acquisition, also highlighted similar anomalous topographic relief and/or outcropping landforms (Figure 11). It appears that by using a combination of radiance bands, combining both the temperature and emissivity surface properties, the thermal and compositional nature was mapped (Figure 11). The red apron surrounding the anomalously high radiance also could be associated with the relatively higher quartz band 13 emissivity (e.g. red) in the sand colluvium/regolith/soil surrounding the outcrop associated topographic rises (Figure 11). This is consistent with the combined emissivity and temperature nature of radiance imagery where the spectral emissivity of surface materials is related to its compositional nature.

(a)   **(b)**

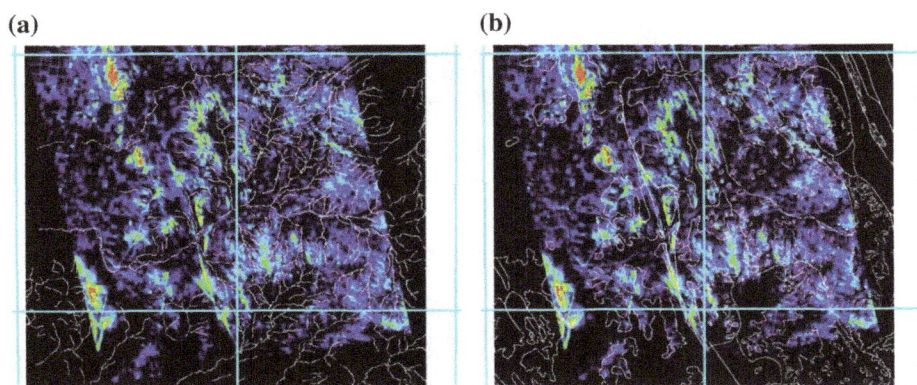

Figure 9. Ardlethan–Barmedman 1:100 K study area (cyan) (a) night-time ASTER surface temperature (99% linear stretch) in relation to 1:250 K watercourses (white), blue: low temperature, red: higher temperature; (b) night-time ASTER surface temperature (99.9% linear stretch) in relation to 1:250 K published geology (white), blue: low temperature, red: higher temperature.

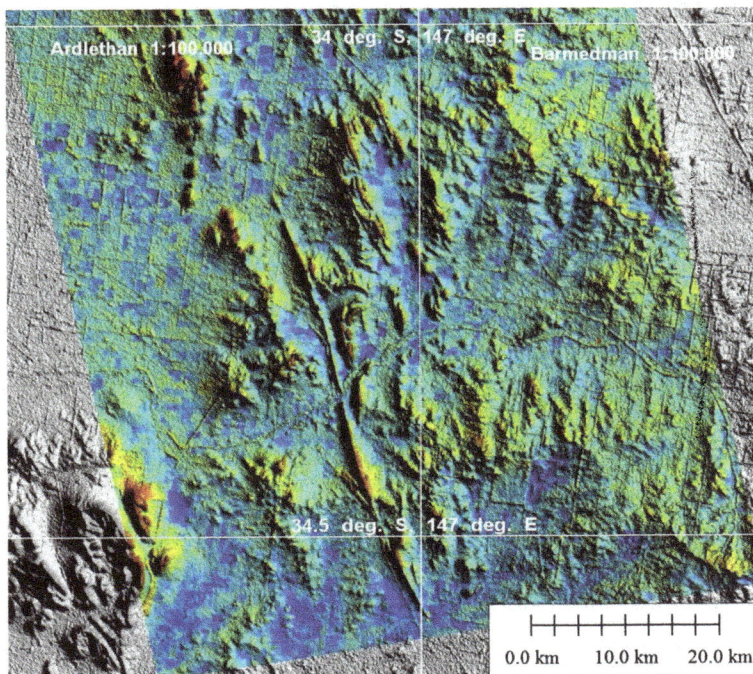

Figure 10. Shaded DEM relief overlain by night-time ASTER surface temperature within the Ardlethan–Barmedman 1:100 K study area, blue: low temperature, red: higher temperature.

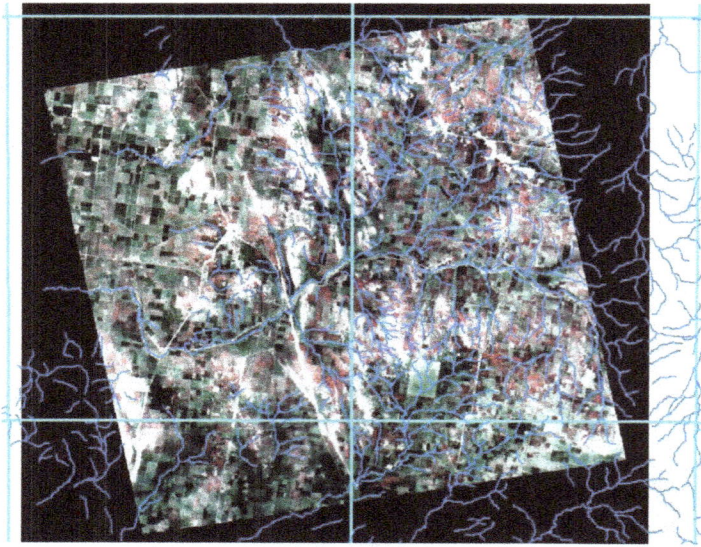

Figure 11. RGB composite image of night-time ASTER radiance bands 13, 12 and 10 with 1:250 K watercourses (blue) and Ardlethan–Barmedman 1:100 K sheet boundaries (cyan) as an overlay.

(a)

(b)

(c)

Figure 12. Site ERIVPJG0033: (a) ADS40 imagery of site; (b) Night-time ASTER Surface Temperature, blue: low temperature, red: higher temperature; (c) Panoramic photo of Ardlethan (?) Granite.

(a)

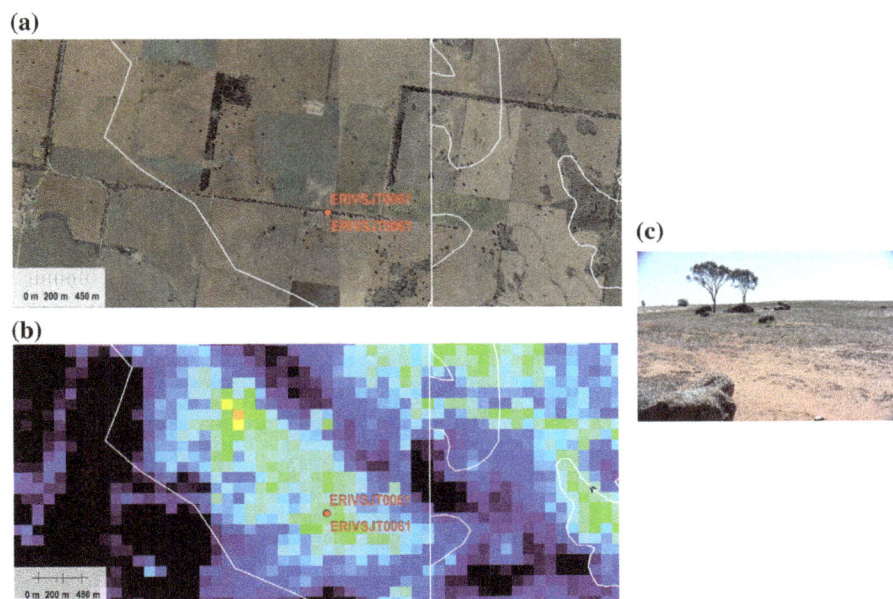

(b)

(c)

Figure 13. Site ERIVSJT0061: (a) ADS40 imagery of site; (b) Night-time ASTER Surface temperature with 1:250 K published geology (white boundary), blue: low temperature, red: higher temperature; (c) Photos of Ardlethan Granite outcrops.

Several field sites were examined for their ADS40 and night-time surface temperature ASTER imagery to investigate its ability to map outcrops. Site ERIVPJG0033 was located at an isolated possible outcrop of Ardlethan Granite. The ADS40 imagery (Figure 12(a)) did not appear to identify any granitic outcrop while the ASTER's surface temperature showed a subtle anomaly (Figure 12(b)). Panoramic photos of ERIVPJG0033 highlighted the isolated limited nature of this outcrop (Figure 12(c)). The night-time surface temperature imagery also appeared to highlight the nearby dams and a linear feature, possibly associated with a canal or road.

Site ERIVSJT0061 with some large outcrops were subtle within the ADSA40 imagery (Figure 13(a)) but part of a pronounced larger surface temperature anomaly striking to the northwest (Figure 13(b)). The outcrops on this hilltop are limited (Figure 13(c)) but were described as Ardlethan Granite within the previous 1:250 K geological mapping (Figure 13(b)).

4.5. ASTER derived apparent thermal inertia imagery

Two pairs of day- and night-time ASTER acquisitions were used to calculate Apparent Thermal Inertia (ATI) as described in Section 3.2. There was a limited choice of available pairs and day- and night-time ASTER acquisitions. However pairs of ASTER imagery were acquired on 18/10/2013 (Figure 14(a)) and on 21/3/2012 (Figure 14(b)) while night-time surface temperature imagery was obtained for 29/11/2013 and 11/1/2012. There was noticeably more vegetation present for the 2013 spring pair (e.g. increased red within the day-time false colour imagery, Figure 14(a)). The ATI was calculated for each pair using day- and night-time temperature products and estimate of the day-time albedo (Figure 14(c)).

Ambiguities due to areas associated with high vegetation cover, particularly woodland, were minimised by masking the ATI with an estimate of green vegetation cover derived by the Normalized Vegetation Difference Index (NDVI) from the ASTER VNIR imagery. The issue of increased green vegetation and spring soil moisture limited the usefulness of the 2013 ATI imagery which showed greater crop and paddock affected anomalies (Figure 14(c)).

The approach by Mackey et al. (2000) and Lawrie et al. (1999) for the detection of subsurface palaeochannels using aero magnetics anomalies was attempted to compare the results of surface

(a)   (b)

(c)   (d)

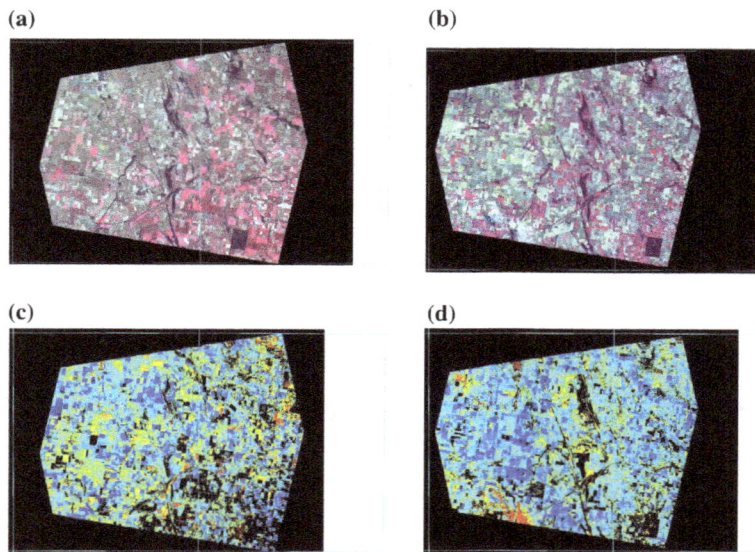

Figure 14. (a) False Colour imagery of day-time ASTER 18/10/2013 acquisition; **(b)** False Colour imagery of day-time ASTER 21/3/2012 acquisition; **(c)** derived ATI using ASTER night-time 29/11/2013 with NDVI mask; **(d)** derived ATI using ASTER night-time 11/1/2012 with NDVI mask.

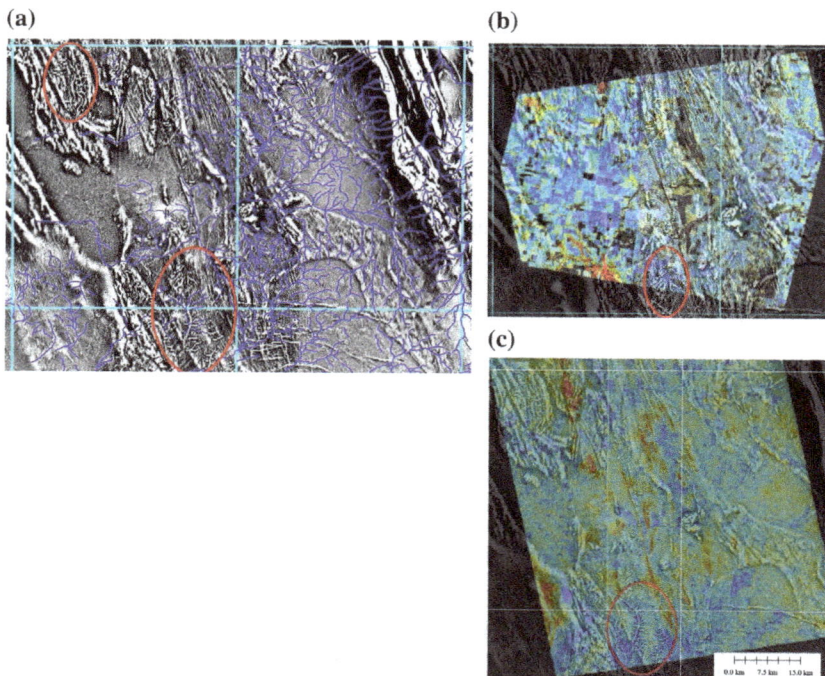

(a)   (b)

(c)

Figure 15. (a) 1VD TMI with 1:250 K watercourses within Ardlethan–Barmedman 1:100 K areas; **(b)** Combined ATI (ASTER night-time 29/11/2013) with NDVI mask overlain with 1VD TMI.

Notes: The dendritic 1VD TMI anomalies (red) in (a), and the corresponding ATI low thermal inertia anomaly (red); (c) 1VD TMI overlain by the night-time ASTER surface temperature.

temperature and ATI anomalies in this study. Several dendritic 1VD TMI anomalies were identified within the Ardlethan–Barmedman study area (red ellipses, Figure 15(a)). Comparison of these 1VD TMI anomalies with present-day watercourses showed a discrepancy and it is viable that this is a result of palaeochannel accumulations of maghemite (Figure 15(a)) as described elsewhere by Mackey et al. (2000) in the nearby area of West Wyalong. There was some limited correspondence to low ATI values in the area shown (blue, Figure 15(b)) however there was insufficient ATI coverage to fully confirm and completely map these features. A low surface temperature (Figure 15(c)) in the

Correlation matrix	SRTM slope%	ASTER (2008-04-21, pm) surf. temp. (Kelvin × 10)	NDVI (2012-03-21, am)	ASTER (2012-03-21, am) surf. temp. Kelvin × 10	ASTER (2012-01-11, pm) surf. temp. Kelvin × 10	ATI (2012)
Table 3. Correlation statistics of ASTER thermal products with topographic and vegetation information						
SRTM slope%	1.00	0.50	0.28	−0.29	0.18	0.36
ASTER (2008-04-21, pm) surf. temp. (Kelvin × 10)	0.50	1.00	0.24	−0.24	0.30	0.36
NDVI (2012-03-21, am)	0.28	0.24	1.00	−0.66	0.22	0.63
ASTER (2012-03-21, am) Surf. Temp. Kelvin × 10	−0.29	−0.24	−0.66	1.00	−0.26	−0.87
ASTER (2012-01-11, pm) surf. temp. Kelvin × 10	0.18	0.30	0.22	−0.26	1.00	0.60
ATI (2012)	0.36	0.36	0.63	−0.87	0.60	1.00

vicinity of these palaeochannels also suggested a cooler hydrological influence, although again, there is limited coverage in this study to be definitive. Although these preliminary results are qualitatively suggestive of indicating sub surface structures, further studies over much larger areas is required to establish the benefits of a combined use of ATI and aero magnetics.

In order examine the likely effects of high relief associated outcrops (e.g. high slope %) and vegetation (NDVI) on the calculated ASTER ATI products, correlation statistics were calculated for spatially coincident data-sets (Table 3). The slope % showed a slightly higher correlation of R = 0.5 to the 2008 night-time surface temperature compared to the ATI (R = 0.36). The ATI seemed more affected by the vegetation (R = 0.63) although coincident 2008 NDVI imagery wasn't available. An interesting result was the cooling effect of the vegetation on day-time surface temperature (R = −0.66). Further work is required and recommended to more clearly distinguish the effects of vegetation and rocky outcrops on such thermal image products.

5. Conclusions

This study tested the ability of ASTER remote sensing imagery to assist the geological mapping of a vegetated and cultivated temperate environment with limited outcrop exposure. Limited information was extracted in the study area using the large-scale regional ASTER compositional maps that used band parameter style processing and uniform histogram methods and thresholds. There appeared improved qualitative results for discriminating subtle geological/regolith variations using spectral processing and the application of masks, generated from higher topographic slopes, on individual day-time ASTER acquired scenes. Overall however, it appeared that the coarse spatial resolution of satellite ASTER (e.g. 30 m) compared to airborne ADS40 imagery (e.g. 0.5 m) significantly limited the delineation of surface feature boundaries and the separation from vegetation cover. Night-time ASTER thermal imagery appeared useful for locating sparse rock outcrops in areas of low or high relief although anomalous "warm" forested areas require the use of day-time imagery to mask their effects. Further studies into the ability of ASTER or other night-time thermal imagery and its integration with geophysics (e.g. 1VD TMI) would be useful to assess its potential for the delineation of sub-surface geological structures in such areas.

In summary, the study's evaluation of the application of ASTER for using day- and night-time acquisitions, showed limited to modest success for routine geological mapping within an area of low relief, sparse outcrops and cultivated terrain. Although a freely available satellite image source imagery, ASTER's spatial resolution limit its application in this such an example of a non-arid environment. Also ASTER's limited day- and night-time paired acquisitions handicap its potential for the reliable interpretation of thermal inertial properties. The availability of the recently launched higher spatial resolution WorldView-3 sensor (Kruse, Baugh, & Perry, 2015) and future NASA HySpiri mission thermal capability (https://hyspiri.jpl.nasa.gov/) offers a possible partial solution to the geological remote sensing technical issues in such a challenging terrain.

Acknowledgements

This study has been supported by the Geological Survey of New South Wales, in part, for the purpose of contributing to the knowledge of the Ardlethan and Barmedman 100 000 map sheet areas geology for mineral exploration. ASTER GeoTiffs map products were also gratefully obtained from Geoscience Australia and CSIRO-Earth Science & Resource Engineering, whilst the original ASTER imagery was made available by NASA and METI of Japan. GIS shapefiles for the Narranadera and Cootamundra 1:250 000 map sheet areas were also gratefully obtained from Geoscience Australia. Assistance from Simon Jones and Laurie Buxton of RMIT University was also forthcoming and appreciated during this study.

Funding

This study was initially instigated and funded by the Geological Survey of New South Wales, NSW Trade & Investment, Division of Resources & Energy.

Author details

R. Hewson[1]
E-mail: r.d.hewson@utwente.nl
D. Robson[2]
E-mail: Robodavidf@gmail.com
A. Carlton[2]
E-mail: Astrid.carlton@trade.nsw.gov.au
P. Gilmore[2]
E-mail: Phil.gilmore@trade.nsw.gov.au
[1] Faculty of Geo-Information Science & Earth Observation (ITC), University of Twente, P.O. Box 217, 7500 AE, Enschede, The Netherlands.
[2] Geological Survey of NSW, NSW Trade & Investment, Division of Resources & Energy, P.O. Box 344, Hunter Region Mail Centre, 2310, Sydney, NSW, Australia.

References

Adams, J. B., & Filice, A. L. (1967). Spectral reflectance 0.4 to 2.0 microns of silicate rock powders. *Journal of Geophysical Research, 72*, 5705–5715. https://doi.org/10.1029/JZ072i022p05705

Caccetta, M., Collings, S., & Cudahy, T. (2013). A calibration methodology for continental scale mapping using ASTER imagery. *Remote Sensing of Environment, 139*, 306–317. https://doi.org/10.1016/j.rse.2013.08.011

Chen, X. Y. (1997). Quaternary sedimentation, parna, landforms, and soil landscapes of the Wagga Wagga 1:1,00,000 map sheet area, south-eastern Australia. *Australian Journal of Soil Research, 35*, 643–668. https://doi.org/10.1071/S96071

Clare, A. P., Fleming, G. D., & Glen, R. A. (1997, May). *Geology of the Cargelligo and Narrandera map sheet areas: Removing the cover using Discovery 2000 Geophysics* (No 104, 22 pp.). Quarterly Notes–Geological Survey of New South Wales.

Colquhoun, G. P., Meakin, N. S., & Cameron, R. G. (2005). *Explanatory notes Cargelligo 1:250 000 geological sheet* (3rd ed., SI/55-6). Geological Survey of NSW.

Crowley, J. K., Brickey, D. W., & Rowan, L. C. (1989). Airborne imaging spectrometer data of the Ruby Mountains, Montana: Mineral discrimination using relative absorption band-depth images. *Remote Sensing of Environment, 29*, 121–134. https://doi.org/10.1016/0034-4257(89)90021-7

Cudahy, T. J. (2012). Australian ASTER Geoscience Product Notes, (Version 1), 7th August, 2012–CSIRO, ePublish No. EP-30-07-12-44.

Downes, P. M., McEvilly, R., & Raphael, N. M. (2004, February). Mineral deposits and models, Cootamundra 1:250,000 map sheet area. *Quarterly Notes, Geological Survey of New South Wales, No, 116*, 37.

Fabris, A. (2002). Thermal satellite imagery–An aid to heavy mineral sand discoveries. *MESA Journal, 24*, 24–26.

Fitzpatrick, K. R. (1979). *Cootamundra 1:2,50,000 metallogenic Map explanatory notes*. Sydney: Geological Survey of New South Wales.

Fujisada, H., Sakuma, F., Ono, A., & Kudoh, M. (1998). Design and preflight performance of ASTER instrument protoflight model. *IEEE Transactions on Geoscience and Remote Sensing, 36*, 1152–1160. https://doi.org/10.1109/36.701022

Gibson, D. L., & Chan, R. A. (1998). Aspects of palaeodrainage in the North Lachlan Fold Belt Region, cooperative research centre for landscape environments and mineral exploration, In G. Taylor, & C. Pain (Eds), *Regolith 98 Proceedings volume, Program and Abstracts volume* (pp. 23–37). Kalgoorlie: Field Trip Guide.

Gillespie, A., Rokugawa, S., Matsunaga, T., Cothern, J. S., Hook, S., & Kahle, A. B. (1998). A temperature and emissivity separation algorithm for Advanced Spaceborne Thermal Emission and Reflection Radiometer (ASTER) images. *IEEE Transactions on Geoscience and Remote Sensing, 36*, 1113–1126. https://doi.org/10.1109/36.700995

Green, A. A., Berman, M., Switzer, P., & Craig, M. D. (1988). A transformation for ordering multispectral data in terms of image quality with implications for noise removal. *IEEE Transactions on Geoscience and Remote Sensing, 26*, 65–74. https://doi.org/10.1109/36.3001

Heugh, J. P. (1979). *Cargelligo-Narrandera 1:250 000 Metallogenic Map* (1st ed.). Sydney: Geological Survey of New South Wales.

Hewson, R. D. (2015). *Using remote sensing, spectral and geophysical information to assist exploration within the Ardlethan and Barmedman 1:100,000 map sheet areas* (49 pp.). New South Wales Geological Survey (Report GS2015/0185). Retrieved from http://digsopen.minerals.nsw.gov.au/

Hewson, R., Carlton, A., Gilmore, P., Jones, S., & Robson, D. (2015, July 7–10). *Geological mapping within NSW using remotely sensed and proximal spectral data*. In proceedings near surface geophysics conference, Society of Exploration Geophysicists, Waikoloa, Hawaii, 4 pp.

Hewson, R. D., & Cudahy, T. J. (2011). Issues affecting geological mapping with ASTER data: A case study of the Mt Fitton Area, South Australia. In B. Ramachandran (Ed.), *Land remote sensing and global environmental change: NASA's earth observing system and the science of ASTER and MODIS: Applications in ASTER*. New York, NY: Springer-Verlag. ISBN: 978-1-4419-6748-0.

Hewson, R. D., Cudahy, T. J., Huntington, J. F. (2001, July 9–13). *Geologic and alteration mapping at Mt Fitton, South*

Australia, using ASTER satellite-borne data. In IEEE 2001 international geoscience and remote sensing symposium (IGARSS), p. 3.

Hewson, R. D., Cudahy, T., Mizuhiko, S., Ueda, K., & Mauger, A. J. (2005). Seamless geological map generation using ASTER in the Broken Hill-Curnamona province of Australia. *Remote Sensing of Environment, 99*, 159–172. https://doi.org/10.1016/j.rse.2005.04.025

Hewson, R., Koch, C., Buchanan, A., & Sanders, A. (2002). *Detailed geological and regolith mapping in the Bangemall Basin, WA, using ASTER multi-spectral satellite-borne data.*. Eleventh Australasian Remote Sensing and Photogrammetric Conference, Brisbane, 2–5th September, pp. 110–125.

Hewson, R., & Robson, D. (2014). *Applying ASTER and geophysical mapping for mineral exploration in the Wagga Wagga and Cobar areas.* Quarterly Notes, Geological Survey of New South Wales, No 140. Retrieved from http://www.resourcesandenergy.nsw.gov.au/about-us/quarterly-notes

Hewson, R., Robson, D., Mauger, A., Cudahy, T., Thomas, M., & Jones, S. (2015). Using the Geoscience Australia-CSIRO ASTER maps and airborne geophysics to explore Australian geoscience. *Journal of Spatial Science.* doi:10.1080/14498596.2015.979891

Hook, S. J., Vaughan, R. G., Tonooka, H., & Schladow, S. G. (2007). Absolute radiometric in-flight validation of mid infrared and thermal infrared data from ASTER and MODIS on the terra spacecraft using the Lake Tahoe, CA/NV, USA, automated validation site. *IEEE Transactions on Geoscience and Remote Sensing, 45*, 1798–1807. https://doi.org/10.1109/TGRS.2007.894564

Hunt, G. R., & Ashley, R. P. (1979). Spectra of altered rocks in the visible and near infrared'. *Economic Geology, 74*, 1613–1629. https://doi.org/10.2113/gsecongeo.74.7.1613

Iwasaki, A., & Tonooka, H. (2005). Validation of a crosstalk correction algorithm for ASTER/SWIR. *IEEE Transactions on Geoscience and Remote Sensing, 43*, 2747–2751. https://doi.org/10.1109/TGRS.2005.855066

Iwasaki, A., Fujisada, H., Akao, H., Shindou, O., & Akagi, S. (2005). Enhancement of spectral separation performance for ASTER/SWIR. *Proceedings of SPIE-The International Society for Optical Engineering, 4486*, 42–50.

Kahle, A. B. (1987). Surface emittance, temperature, and thermal inertia derived from thermal infrared multispectral scanner (TIMS) for death valley. *Geophysics, 52*, 858–874. https://doi.org/10.1190/1.1442357

Kahle, A. B., & Alley, R. E. (1985). Calculation of thermal interia from day-night measurements separated by day or weeks. *Photogrammetric Engineering and Remote Sensing, 51*, 73–75.

Kruse, F. A. (2000). *The effects of spatial resolution, spectral resolution, and signal-to-noise on geologic mapping using hyperspectral data, Northern Grapevine Mountains, Nevada.* Proceedings 9th JPL Airborne Earth Science workshop (pp. 261–269). JPL Publications.

Kruse, F. A., Baugh, W. M., & Perry, S. L. (2015). *Validation of DigitalGlobe WorldView-3 Earth imaging satellite shortwave infrared bands for mineral mapping*, (Vol. 9, pp. 17).

Kruse, F. A., Lefkoff, A. B., Boardman, J. B., Heidebrecht, K. B., Shapiro, A. T., & Barloon, P. J. (1993). The spectral image processing system SIPS–Interactive visualization and analysis of imaging spectrometer data. *Remote Sensing of Environment, 44*, 145–163. https://doi.org/10.1016/0034-4257(93)90013-N

Lawrie, K. C., Chan, R. A., Gibson, D. L., & de Souza Kovacs, N. (1999). *Alluvial gold potential in buried palaeochannels in the Wyalong district, Lachlan Fold Belt*, (Vol. 30, 5 pp.). New South Wales: AGSO Research Newsletter.

Lyon, R. J. P., & Burns, E. A. (1963). Analysis of rocks and minerals by reflected infrared radiation. *Economic Geology, 58*, 274–284. https://doi.org/10.2113/gsecongeo.58.2.274

Mackey, T., Lawrie, K., Wilkes, P., Munday, T., Kovacs, N., Chan, R., ... Evans, R. (2000). Palaeochannels near West Wyalong, New South Wales: A case study in delineation and modelling using aero magnetics. *Exploration Geophysics, 31*, 1–7. https://doi.org/10.1071/EG00001

Price, J. C. (1977). Thermal inertia mapping: A new view of the Earth. *Journal of Geophysical Research, 82*, 2582–2590. https://doi.org/10.1029/JC082i018p02582

Ren, S. K., Walshe, J. L., Paterson, R. G., Both, R. A., & Andrew, A. (1995). Magmatic and hydrothermal history of the porphyry-style deposits of the Ardlethan tin field, New South Wales, Australia. *Economic Geology, 90*, 1620–1645. https://doi.org/10.2113/gsecongeo.90.6.1620

Rowan, L. C., & Mars, J. C. (2003). Lithologic mapping in the Mountain Pass, California area using advanced spaceborne thermal emission and reflection radiometer (ASTER) data. *Remote Sensing of Environment, 84*, 350–366. https://doi.org/10.1016/S0034-4257(02)00127-X

Russell, R., & Lappi, D. W. (1988, February 14–21). *NOAA satellite thermal interpretation of ATP 354P, western Queensland*, Proceedings ASEG/SEG conference, Adelaide, pp. 141–147.

Sabins, F. F. (1997). *Remote sensing–Principles and interpretation* (3rd ed., p. 432). Long Grove, IL: Waveland Press.

Sandau, R., Braunecker, B., Driescher, H., Eckardt, A., Hilbert, S., Hutton, J., Kirchhofer, W., ... Wicki, S. (2000). Design principles of the LH systems ADS40 airborne digital sensor. *International Archives of Photogrammetry and Remote Sensing, XXXIII*, Part B1, 258–265.

Scott, K. M., & Rampe, M. (1984). Integrated mineralogical and geochemical exploration for tin in the bygoo region of the ardlethan tin field, Southern N.S.W., Australia. *Journal of Geochemical Exploration, 20*, 337–354. https://doi.org/10.1016/0375-6742(84)90075-X

Vincent, R. K., Rowan, L. C., Gillespie, R. E., & Knapp, C. (1975). Thermal-infrared spectra and chemical analyses of twenty-six igneous rock samples. *Remote Sensing of Environment, 4*, 199–209. https://doi.org/10.1016/0034-4257(75)90016-4

Vincent, R. K., & Thomson, F. (1972). Spectral compositional imaging of silicate rocks. *Journal of Geophysical Research, 77*, 2465–2472. https://doi.org/10.1029/JB077i014p02465

Warren, A. Y. F., Gilligan, L. B., & Raphael, N. M. (1996). *Cootamundra 1:2,50,000 Geological Sheet SI/55-11* (2nd ed.). Sydney: Geological Survey of New South Wales.

Watson, K. (1975). Geologic applications of thermal infrared images. *Proceedings of the IEEE, 63*, 128–137. https://doi.org/10.1109/PROC.1975.9712

Wynn, D. W. (1977). *Narrandera 1:2,50,000 Geological Sheet SI/55-10* (2nd ed.). Sydney: Geological Survey of New South Wales.

Yamaguchi, Y., Fujisada, H., Kahle, A., Tsu, H., Kato, M., Watanabe, H., ... Kudoh, M., (2001, July 9–13), *ASTER instrument performance, operation status, and application to earth sciences.* Proceedings IEEE 2001 International geoscience and remote sensing symposium.

13

Dynamic response based empirical liquefaction model

Snehal Rajeev Pathak[1] and Asita Nilesh Dalvi[2]*

*Corresponding author: Dalvi Asita Nilesh, Department of Civil Engineering, SITS, Pune, Maharashtra, India
E-mail: asitadalvi@gmail.com
Reviewing editor:
Craig O'Neill, Macquarie University, Australia

Abstract: Dynamic response-based methodology, wherein integrated effect of dynamic soil properties and ground motion parameters proposed by authors, has been found to detect liquefaction susceptibility. The present work necessarily deals with the formulation of a comprehensive empirical liquefaction model (ELM) using this methodology. The absolute form of the ELM is dimensionally homogeneous and yields a correlation between proposed "liquefaction potential term" and "normalized standard penetration blow count corrected for fines content, $(N_1)_{60cs}$." The developed ELM demonstrates unbiased performance when verified over a wide range of significant parameters. One of the prominent features of the present ELM is accurate prediction of possibility of liquefaction. The proposed ELM has proven to work well on varied data-sets of more than 1000 case records within the given range of model parameters. Moreover, the dynamic response-based ELM proves its ability when compared with other liquefaction evaluation procedures. Thus, a generalized and optimistic ELM simulating realistic field conditions is formulated. It is anticipated that for accurate prediction of liquefaction occurrence, it would be more appropriate to employ the proposed ELM which will minimize the enormous losses caused due to liquefaction.

Subjects: Civil, Environmental and Geotechnical Engineering; Georisk & Hazards; Soil Mechanics

Keywords: seismic soil liquefaction; liquefaction susceptibility; empirical approach

ABOUT THE AUTHOR

Pathak is working in this field for more than a decade; the various areas dealt so far are 1-g shake table testing, Static and Cyclic Triaxial testing, and analytical modeling to assess susceptibility of liquefaction by varying soil and seismic parameters. The publications include 20 plus international and national journal papers in the field of seismic soil liquefaction; to name a few: ASCE, Natural Hazards, Geomechanics and Geoengineering, and equivalent papers presented and published in international and national conferences on the research theme.

PUBLIC INTEREST STATEMENT

Liquefaction is a phenomenon wherein a loose sandy soil loses its strength and flows like a fluid and is a topic of interest around the world. It was more thoroughly brought to the attention of engineers after 1964 Niigata and 1964 Alaska earthquakes which have witnessed major damage due to seismic soil liquefaction. Accordingly, prediction of liquefaction potential is the key step which will minimize the enormous losses caused due to liquefaction. The focus of the present work is thus formulation of a comprehensive and cogent empirical liquefaction model (ELM) and validating the same to ensure realistic dynamic soil response to earthquake-induced liquefaction. The final form of the ELM thus developed, engenders realistic yet optimistic prediction of liquefaction at a particular site within a given range of model parameters. ELM shows its versatility by performing on over 1000 case records and accurately predicts liquefaction occurrence.

1. Introduction

The use of empirical relationships based on correlation of observed field behavior with various *in situ* "index" tests is the dominant approach in common engineering practice to assess liquefaction susceptibility. Based on extensive literature review of empirical liquefaction procedures, Pathak and Dalvi (2011) inferred that inclusion of dynamic soil properties in conjunction with ground motion parameters would enhance the predictability of the model. Furthermore, it is observed that although there exist several studies to assess the liquefaction potential (LP), it is well understood that a large number of seismic as well as soil parameters affect the occurrence of liquefaction during an earthquake, making it necessary to identify the significant parameters based on their contribution to liquefaction phenomenon to expedite the assessment of liquefaction. Accordingly, Dalvi, Pathak, and Rajhans (2014) extracted significant parameters responsible for the phenomenon of liquefaction by employing multi-criteria decision-making tools (MCDM), namely: AHP and Entropy analysis. Such an attempt of applying MCDM tools in the field of earthquake-induced liquefaction has been firstly introduced by the authors.

Subsequently, Pathak and Dalvi (2012) presented a preliminary approach referred as, "dynamic response based methodology" to evaluate LP by incorporating thus extracted significant parameters to represent realistic dynamic response of soils. Based on this dynamic response-based methodology, Pathak and Dalvi (2013) established an elementary empirical liquefaction model (ELM) to identify susceptibility of liquefaction which is given by the following Equation. 1 as

$$(N_1)_{60} = 1.24(LPterm)^{0.31} \tag{1}$$

Where $(N_1)_{60}$ = normalized SPT blow count and,

$$LP\ term = \left[\frac{v_{max} * G_{max} * dur}{\sigma'_v} \right]$$

where, "v_{max}" is the peak ground velocity (m/s),

"G_{max}" is the small strain shear modulus (kPa),

"dur" is duration of strong ground motion (sec), and

"σ'_v" is effective overburden pressure (kPa).

The performance of thus developed elementary model ascertained the potential of using the functional form LP term in assessing liquefaction susceptibility (Pathak & Dalvi, 2013). However, in order to predict liquefaction in practical situations, a dimensionless relationship might be required. Moreover, it is also required to take into account all types of soils ranging from sandy soils to silty sands and also site conditions to simulate actual field conditions. Thus, in the present work, an endeavor has been made to establish an exhaustive generalized ELM. The main motive behind the present work is to detect the occurrence of liquefaction at a particular site based on dynamic response. A sequential procedure as elaborated in preceding sections is thus adopted to arrive at the final form of ELM.

2. Development of ELM

In order to develop an exhaustive ELM to replicate pragmatic dynamic response which will take into account these issues to be applicable in realistic situations, the fundamental relationship as in Equation. 1 is employed. Hence, initially, the elementary model is transformed into a dimensionally homogeneous form using Fourier's principle of dimensional homogeneity, expressing the dimensions of the LP term in terms of primary quantities M–L–T, i.e. mass, length and time as

$$LPterm\ dim\ ensions = \frac{\frac{m}{s} * \frac{kN}{m^2} * s}{\frac{kN}{m^2}} = m,$$

Thus, the term remains with the units of length; accordingly, the epicentral distance is chosen as the dimensional homogeneity factor to convert it into a homogeneous form. Now, in order to formulate the model, a total of 314 case records (Appendix I) have been utilized which are extracted from Boulanger, Wilson, and Idriss (2012), Cetin et al. (2004), Davis and Berrill (1982), Hamada and Wakamatsu (1996), Hanna, Ural, and Saygili (2007), and Zhang (1998). Out of the total parameters involved in formulating the present model, the values of earthquake magnitude (M_w), peak ground acceleration (a_{max}), epicentral distance (r), normalized SPT blow count ($(N_1)_{60}$, and effective overburden pressure (σ_v) are used directly as available in the database, whereas the parameters, namely, peak ground velocity(v_{max}),small strain shear modulus (G_{max}), and duration of strong ground motion (*dur*) have been computed as detailed in Pathak and Dalvi (2013). Sampling bias and class balance have been maintained to ensure realistic predictability as illustrated by Oommen, Baise, and Vogel (2011) and accordingly, 220 cases have been employed for formulating the model and 94 cases have been used to validate the ELM. Thus, regression analysis has been performed in MATLAB using the training data-set of 220 cases out of the total 314 cases using the functional form as in Equation 1 and regression coefficients are obtained after applying dimensional homogeneity factor which resulted in Equation 2 as:

$$(N_1)_{60} = 7.47(LP_m term)^{0.28} \tag{2}$$

where LP_m term = modified LP term which is dimensionally homogeneous

The boundary curve representing Equation 2 thus obtained is found to accurately discriminate the cases of liquefaction and non-liquefaction when validated on remaining 94 cases. All the 47 liquefaction cases are correctly predicted by Equation 2, hence signifies the theme underlying the present work. However, this proposed model is observed to predict "no" liquefaction conservatively, as some of the non-liquefied cases have been identified as liquefied by the model. Few such misclassified non-liquefaction case records indicate the normalized SPT values < 5 which actually illustrate high risk of liquefaction. Further, it is observed that for many of the remaining wrongly predicted "NO" cases, the value of normalized SPT blow count is within the range of 10–20, indicating intermediate risk of liquefaction. Thus, though the cases are of "no" liquefaction, the model rightly predicts them as "yes" cases. The model thereby achieves an overall success rate of 78% with accurate prediction for liquefied cases. Such a dimensionally homogeneous proposed model ensures it with physical significance in liquefaction assessment studies.

Further to simulate different soil conditions on site, range of soils is then incorporated through the compositional factor "fines content" which is known to affect both cyclic shear strength and penetration resistance of soils. To do so, initially, the effect of fines content on the model performance is verified by categorizing the whole data-set of 314 case records as: Class (A) Low FC, clean sand (FC ≤ 5%), Class (B) Intermediate FC, silty sand (6 ≤ FC ≤ 35%), and Class (C) High FC, silty sand to sandy silt (FC > 35%), to distinguish the soil types. Consequently, three separate equations each representing a particular type of soil have been obtained. The regression coefficients as evaluated for Classes A, B, and C are as tabulated below.

These three categories are also associated with corresponding relative risk of liquefaction (Dalvi et al., 2014) as stated in Table 1. It can also be observed from Table 1 that the regression coefficients corresponding to intermediate fines content are the same as those proposed through Equation 2. Interestingly, the other two ranges of fines content define the upper and lower bounds of boundary curves as illustrated in Figure 1.

It is worth mentioning that despite considering a different form to represent LP, the trend of boundary curves as proposed in present work (Figure 1) indicates similarity with those established by previous researchers such as (Andrus & Stokoe II, 2000; Bolton Seed, Tokimatsu, Harder, & Chung, 1985; Rezania, Javadi, & Giustolisi, 2010). The trend of boundary curves as depicted in Figure 1 clearly indicates that fines content is an important parameter which affects the performance of the

Table 1. Summary of model parameters for range of fines content

Class	Fines content	Soil type	Relative risk of liquefaction	α_1	β_1
A	Low (FC ≤ 5%)	Clean sand	High	8.53	0.26
B	Intermediate (6 ≤ FC ≤ 35%)	Silty sand	Intermediate	7.47	0.28
C	High (FC > 35%)	Silty sand to sandy silt	Low	5.39	0.34

Note: *The algebraic form of the ELM is* $(N_1)_{60} = \alpha_1(LP_m term)^{\beta_1}$.

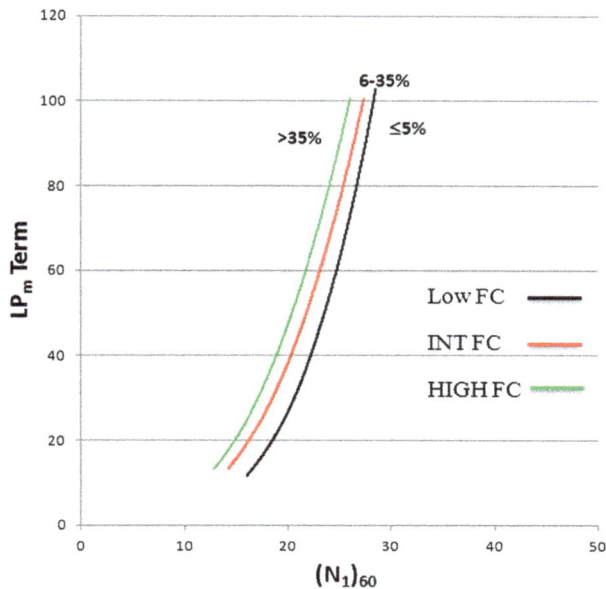

Figure 1. Fines content-based curves.

proposed model (Equation 2). Subsequently, the normalized SPT blow count is corrected for fines content presented as $(N_1)_{60cs}$ through Equation 3 to include its effect directly into the model,

$$(N_1)_{60cs} = (N_1)_{60} * C_{FINES} \tag{3}$$

where C_{FINES} is as given by Seed et al. (2003); thus, the final form of the model culminates into Equation 4:

$$(N_1)_{60cs} = 5.22(LP_m term)^{0.37} \tag{4}$$

This liquefaction triggering correlation obtained is in terms of dimensionally homogeneous "liquefaction potential term" (LP_m term) and the normalized SPT blow count corrected for fines content $(N_1)_{60cs}$. After verifying the predictive performance of the developed Equation. 4, on remaining 94 cases, it is found that by inclusion of fines content as model parameter, the predictive performance improves by over 4%, giving an overall success rate of 82%. Most importantly, all the 47 liquefaction cases are correctly predicted and 30 cases out of the 47 non-liquefaction cases have been identified correctly as shown in Figure 2.

From Figure 2, it can be seen that the scatter of non-liquefaction cases lying above the boundary curve is lesser when correction for fines content is applied. It is also noticed that the wrongly predicted "no" liquefaction cases having $(N_1)_{60cs}$ between 15 and 20, lie in the vicinity of the proposed curve. Thus, the transformed form of the empirical model stated via Equation. 4 certainly signifies remarkable potential in correctly identifying liquefaction-prone sites. *Thus, Equation. 4 has been proposed as the dynamic response-based ELM.*

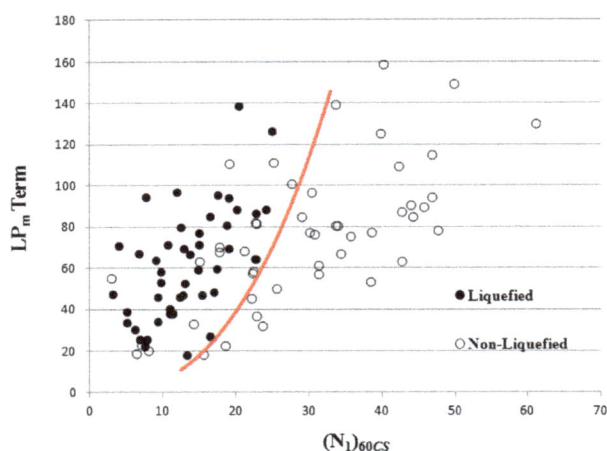

Figure 2. Validation of the proposed model with inclusion of fines content.

3. Performance of ELM

Now in order to justify the comprehensiveness of the proposed ELM, it is projected to assess its performance over diversified applications such as:

(3.1) Performance of ELM over the range of significant parameters,

(3.2) Predictability of ELM over an entirely new data-set, and

(3.3) Comparison of ELM with similar empirical procedures.

3.1. Performance of ELM over the range of significant parameters

Based on MCDM analysis, the effect of range of significant parameters, namely: a_{max}, M_w, and effective overburden pressure, is studied in present work. It has been demonstrated that these parameters are not only statistically significant but also possess physical relevance with the actual phenomenon (Dalvi et al., 2014). As a wide range of these parameters is included in the proposed ELM, the ranges are classified such that a particular range of respective parameter reflects the actual field behavior, indicating possible relative risk of liquefaction. Accordingly, the entire data-set of 314 cases is classified into three categories, namely: Class (A) High risk of liquefaction, Class (B) Intermediate risk of liquefaction, and Class (C) Low risk of liquefaction. These ranges of the respective parameters and their corresponding potential risk of liquefaction are summarized in Table 2.

3.1.1. Effect of peak ground acceleration (a_{max})

a_{max} has been commonly used to describe the ground motion because of its inherent relationship with inertial forces; indeed, the largest dynamic forces are assumed to be closely related to the a_{max} (Kramer, 1996). The scatter of data points relative to each class corresponding to ranges of acceleration as stated in Table 2 is represented graphically through Figure 3(a–c). Figure 3(a) is indicating low range of a_{max} (a_{max} < 0. 2 g) categorized as Class C. Figure 3(b) demonstrates variation of ELM over

Table 2. Range and degree of risk of liquefaction			
Parameter	Peak ground acceleration (a_{max}) (g)	Magnitude (M_w)	Effective overburden pressure (σ_v') (kPa)
Relative risk of liquefaction			
High risk of liquefaction, class A	> 0.5	< 6.5	< 61
Intermediate risk of liquefaction, class B	0.2–0.5	6.5–7.5	61–81
Low risk of liquefaction, class C	< 0.2	> 7.5	> 81

case records falling under Class B, i.e. intermediate acceleration level (a_{max} = 0.2 to 0.5 g). Figure 3(c) depicts the variation of ELM for Class A, i.e. high risk of liquefaction (a_{max} > 0.5 g).

From these Figures (3 (a–c), it can be noted that the overall distribution of liquefaction and non-liquefaction points relative to the proposed ELM across these ranges as specified above appears to be fairly balanced, indicating efficiency of the developed model. Moreover, for intermediate and high a_{max} ranges, (Figure 3(b) and (c), respectively), more than 60% of correct prediction of non-liquefied sites is observed. It is worth mentioning that the positive predictability of the proposed ELM through Equation (4) remains unaltered, although a particular range of a_{max} is considered.

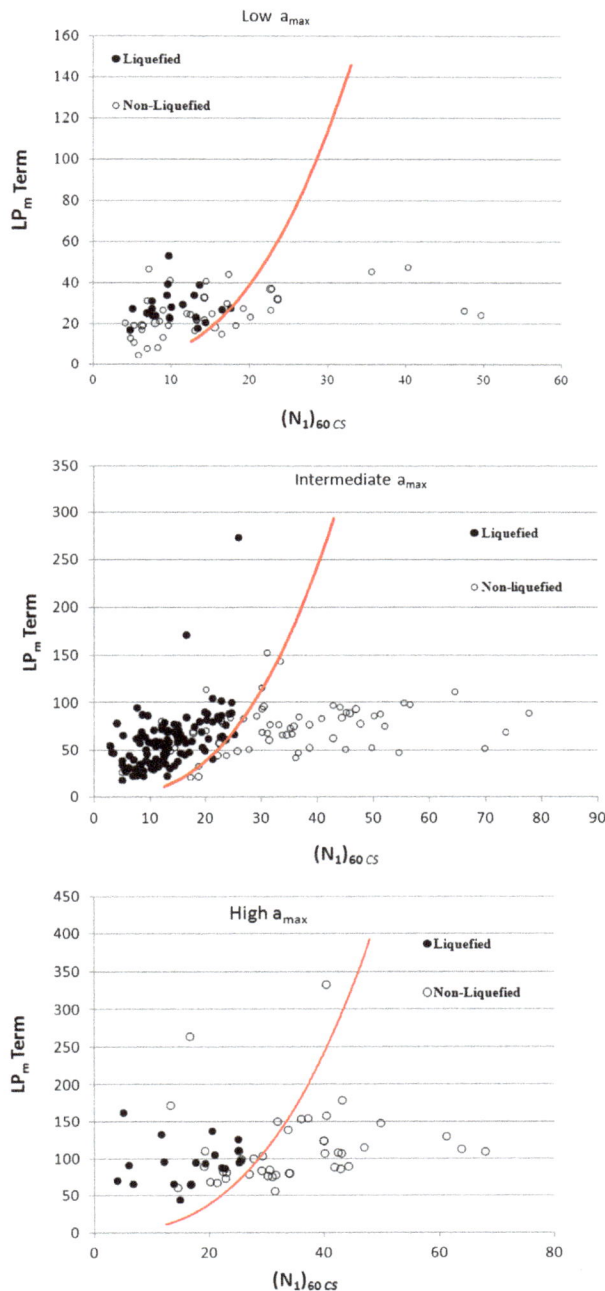

Figure 3. Variation of ELM with peak ground acceleration (a_{max}).

3.1.2. Effect of earthquake magnitude (M_w)

It is a known fact that for liquefaction to occur, there must be ground shaking and the potential for liquefaction increases with the increase in earthquake shaking represented by M_w. A similar observation as above is found when the model is verified relative to ranges of M_w as: Class C ($M_w < 6.5$), Class B ($M_w = 6.5$ to 7.5), and Class A ($M_w > 7.5$) as depicted in Figure 4(a–c), respectively.

The scatter of data points shown in Figure 4 (a) indicates that most of the misclassified non-liquefaction cases bear the $(N_1)_{60cs}$ value less than 10 which is indicative of high risk of liquefaction. Figure 4(b) represents Class B data points having intermediate risk of liquefaction and few of the liquefied cases could be seen in the vicinity of demarcation of curve. Figure 4 (c) clearly indicates that all the liquefaction cases of Class A are correctly predicted by the ELM. It can be stated that the model performance remains impartial at defined ranges of magnitude as well as acceleration of an earthquake as the overall success rate of accurate prediction remains unchanged. The functional form of LP term as employed in the present study includes these two parameters through v_{max}. This ensures the realistic seismic soil response as frequency content gets included.

3.1.3. Effect of vertical effective overburden pressure

Further, representing geological setting of soil strata, overburden stress effects for liquefaction analysis procedures have been investigated by Boulanger (2003), Boulanger and Idriss (2012) and Seed et al. (2003). Moreover, it is known that soil behaves as a deformable body with increasing depth; hence, depth is a site condition parameter and has a huge impact on soil response against its capacity to resist liquefaction. In addition to this, for liquefaction to occur, groundwater table should be at sufficient depth to create saturated soil conditions. Thus, the depth to groundwater table is an another important consideration in identifying soils that are susceptible to liquefaction. The vertical effective overburden pressure obviously takes into account the effect of depth as well as groundwater table level. For this purpose, data have been categorized w.r.t σ_v' in three different classes to represent the pertinent degree of risk of liquefaction (Table 2). The variation of ELM w.r.t these classes is illustrated through Figure 5 (a) to (c).

Figure 5 (a) to (c) implies that the proposed ELM presented through Equation (4) is unbiased, relative to the variation in vertical effective overburden pressure. Moreover, it rightly indicates that the susceptibility of liquefaction decreases (LP_m term) with increase in vertical effective overburden pressure. Thus, it is inferred that the observed behavior on field is replicated through the proposed model. In summary, the deterministic ELM not only yields accurate prediction of liquefied sites within a given range of model parameters but also replicates actual field behavior, thus an optimistic yet realistic ELM is developed.

3.2. Performance of proposed ELM using varied data-sets

Based on the investigation of predictive performance of recently developed empirical models, it is inferred that the success rate of accurate prediction reduces if the developed model is tested upon a totally different data-set and thus in general, the models are data specific (Pathak & Dalvi, 2011).

In this work, the predictability of the present ELM is ascertained on a varied data-set of around reported 740 cases. Among these, 386 case records are obtained from Hanna et al. (2007) and Boulanger et al. (2012). Based on verification over these data points, it is prominently observed that the developed ELM succeeds to accurately detect the occurrence of liquefaction for all the 227 "yes" cases out of the total 386 cases under consideration.

As the remaining 354 case records extracted from Kayen et al. (2013) include database of corrected shear wave velocity (V_{s1}), the $(N_1)_{60}$ value has been computed using the established correlation of Andrus and Stokoe II (2000) as given by:

$$V_{s1} = B_1[(N_1)_{60}]^{B_2} \tag{5}$$

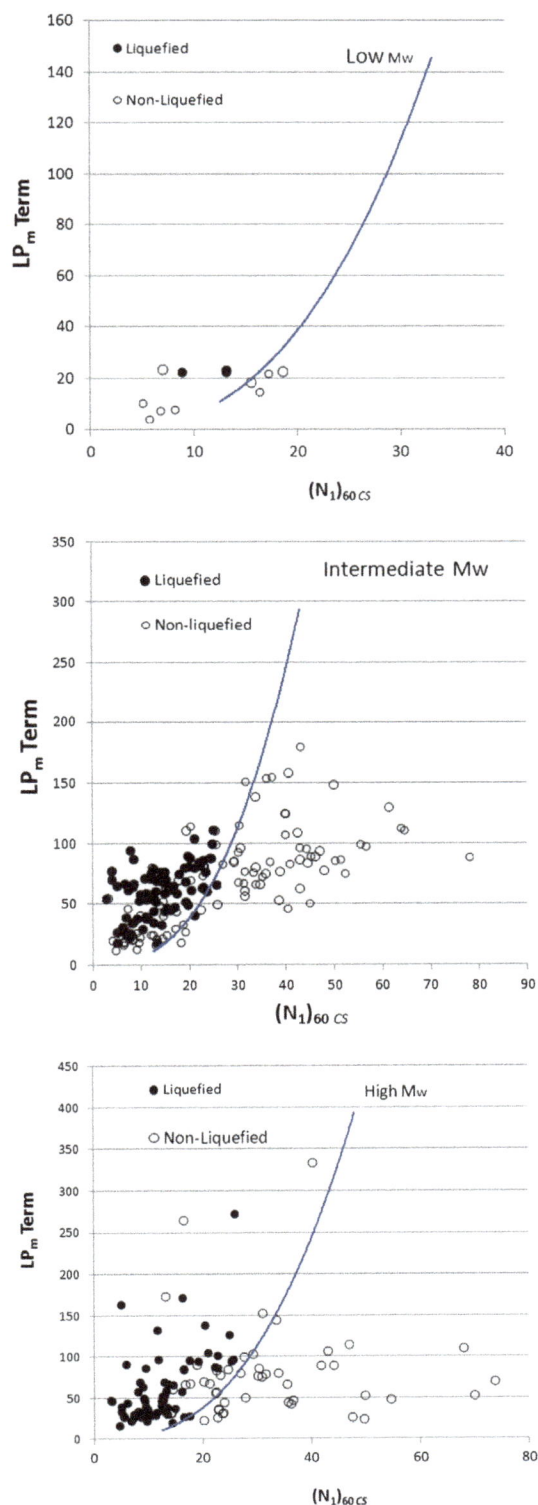

Figure 4. Variation of ELM with earthquake magnitude (M_w).

where V_{s1} = overburden stress-corrected shear wave velocity, B_1 = 93.2 ± 6.5 and B_2 = 0.231 ± 0.022. For this data-set, the predictability of ELM is verified against Equation (2) which gives the relation between LP_m term and $(N_1)_{60}$. The scatter of liquefied cases is as depicted in Figure 6 below.

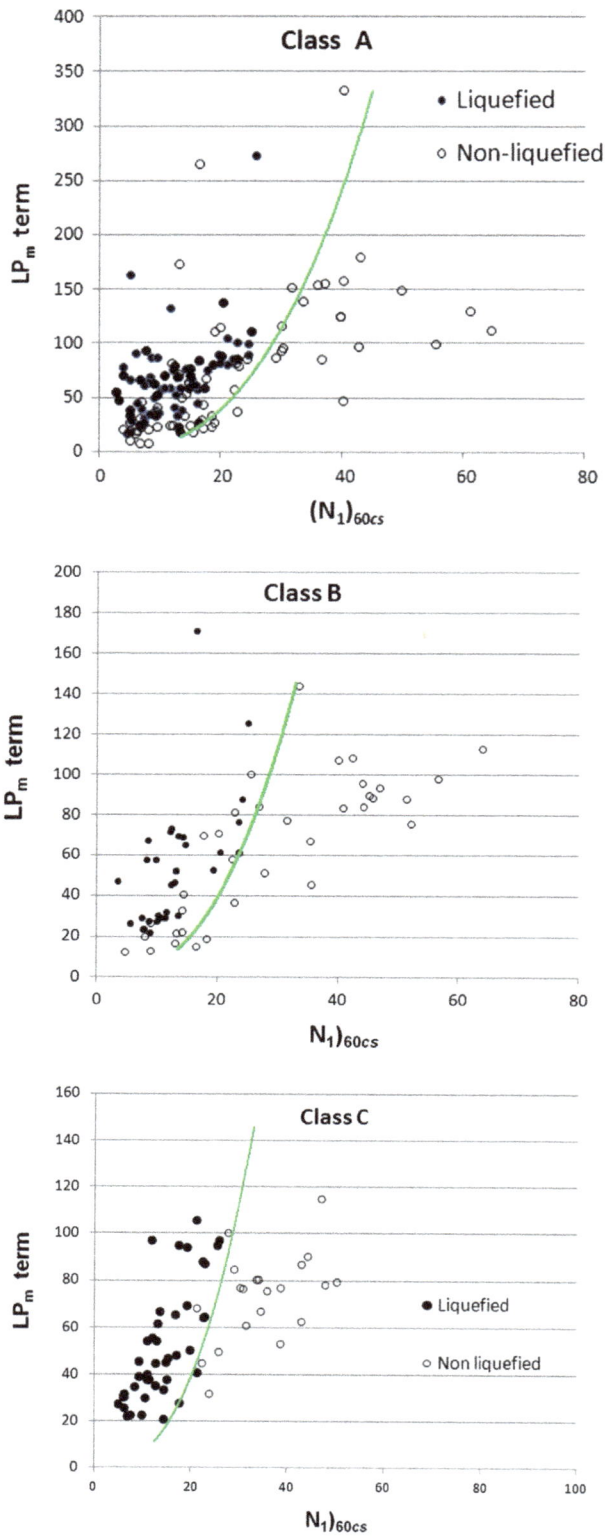

Figure 5. Variation of ELM with vertical effective pressure.

It is interesting to note that most of the 238 "liquefied" cases are accurately predicted as could be seen lying within the zone of liquefaction as shown in Figure 6. Although few of the data points could

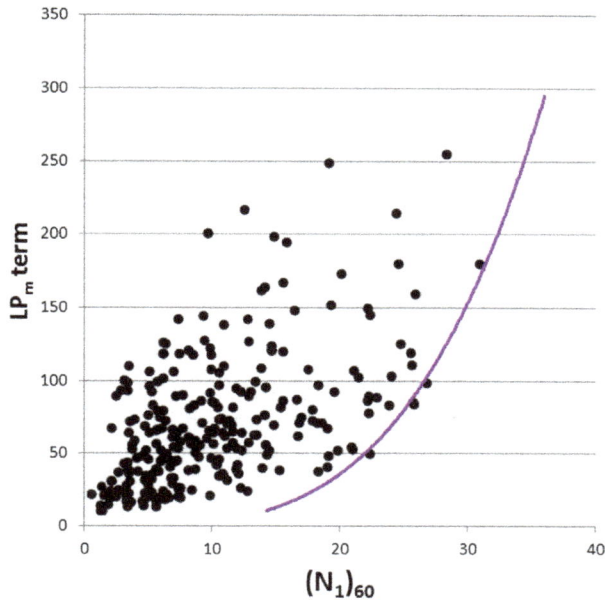

Figure 6. Scatter of liquefied cases in LP$_m$ term-$(N_1)_{60}$ space.

be seen lying just on the boundary curve, but these are the records with $(N_1)_{60}$ value more than 23 and thus indicative of equal risk of liquefaction.

Thus, overall, the ELM performance is verified on 740 case records apart from the 314 cases employed in formulating and validating the model. It is prominently observed that the developed ELM succeeds to accurately predict the occurrence of liquefaction for all the 465 "yes" cases out of the total 740 cases under consideration. Although conservative results are observed for non-liquefaction cases, it is to be noted that such wrongly predicted "no" liquefaction cases bear the properties which are prone to liquefaction. Finally, the ELM has performed well on over 1,054 case history records within the given range of model parameters, covering worldwide 46 number of earthquakes occurred during 1906–2011. Thus, the proposed final form ELM (Equation 4) certainly bears remarkable potential in detecting liquefaction by correlating the field conditions with the dynamic soil properties and ground motion parameters. The performance of the developed ELM to predict liquefaction is then compared with existing empirical models to demonstrate its applicability in the field as discussed in the next section. The comparison is purely meant to focus the inclusion of dynamic response in liquefaction evaluation procedures as proposed by the authors.

3.3. Comparison of proposed ELM with other approaches

The proposed ELM being fundamental in its nature has to be verified for its performance with SPT- and V_s-based simplified procedures. The SPT-based approach as originally proposed by Bolton Seed et al. (1985) is based on empirical evaluations of field observations. Although this method has been modified and improved periodically since that time, it presents the fundamental concept of inclusion of fines content into liquefaction evaluation procedures. Thus, the model performance when

Table 3. Comparison of predictive performance of ELM with other empirical approaches		
Approach	OA (%)	SRL (%)
I	82.6	81.6
II	83.1	72.4
III	85.2	89.9
IV	66.8	71.8
Present study	85.6	100

compared with these SPT-based approaches indicates that the developed ELM is efficient in correctly identifying the liquefied sites which are misclassified by these approaches. Further, it is known that Andrus and Stokoe (1997, 2000) pioneered in the use of V_s, a dynamic soil property as a field index of liquefaction resistance. Similar to the simplified procedure, they proposed CRR-V_{s1} curves corresponding to the range of fines content using compiled case histories. As stated in Andrus and Stokoe II (2000), only two evidences from 1989, Loma Prieta event ($M_w = 6.93$), lie incorrectly in "No" liquefaction zone; however, these two misclassifications lie very near to the demarcation curve; interestingly, the proposed ELM rightly predicts liquefaction occurrence for these two evidences which were actually reported as liquefied. This also indicates that success rate of correct prediction is more when dynamic soil properties are taken into account.

The performance of ELM is verified using the approaches by Youd et al. (2001): (Approach I) and Cetin et al. (2004): (Approach II) and also two recently developed approaches, namely: SPT-based approach by Oommen, Baise, and Vogel (2010): (Approach III) and V_s-based approach by Zhang (2010): (Approach IV). The research carried out by various researchers is undeniably exemplary. It is to be noted that the basis for this comparison is to verify the performance of various ELMs over a common data-set which in present work is as extracted from Cetin et al. (2004). The values of overall accuracy (OA) and success rate for liquefaction (SRL) by approaches I, II, and III have been mentioned as reported in Oommen et al. (2010), whereas the values of the same for approach IV and ELM are computed. The predictive performance in terms of OA and SRL is then summarized in Table 3.

From Table 3 it is observed that the OA in correct prediction can be seen to be at par with other four approaches. Moreover, the proposed ELM succeeds to detect liquefaction occurrence more accurately. It is to be noted that the comparison of ELM with other models is cited purely to symbolize the effectiveness of using the proposed methodology, although the basic framework of liquefaction assessment for each of these models differs significantly. Overall, the above discussion justifies the extensiveness of the developed ELM and thus for accurate prediction of liquefaction occurrence, it would be more appropriate to employ the proposed ELM which will minimize the enormous losses caused due to liquefaction.

4. Conclusions

Formulation, validation, and verification of a fundamental yet comprehensive ELM for detection of liquefaction occurrence is presented. Although it follows the general format of simplified procedure based on SPT, the liquefaction potential is characterized by integrating the effect of dynamic soil properties and ground motion parameters. The final form of the model is not only dimensionally homogeneous but also takes into account the fines content correction. Realistic predictability is achieved as sampling bias and class balance is maintained while formulating the ELM. Based on the performance of the ELM relative to soil and seismic parameters, it bears the potential to replicate the actual field behavior. A final correlation thus established is found to perform well for the diversified 1,000 case history records. Accurate prediction of liquefied cases for the specified range of model parameters is the prominent feature of the developed ELM.

Funding
The authors received no direct funding for this research.

Author details
Snehal Rajeev Pathak[1]
E-mail: srp.civil@coep.ac.in
Asita Nilesh Dalvi[2]
E-mail: asitadalvi@gmail.com
ORCID ID: http://orcid.org/0000-0003-2812-0242
[1] Department of Civil Engineering, College of Engineering Pune, Pune, Maharashtra, India.
[2] Department of Civil Engineering, SITS, Pune, Maharashtra, India.

Cover image
Source: Authors.

References
Andrus, R., & Stokoe, K. H. (1997). Liquefaction resistance based on shear wave velocity. *NCEER workshop on evaluation of liquefaction resistance of soils, national center for Earthquake Engineering Research* (pp. 89–128). Buffalo, NY: State University of New York.

Andrus, R. D., & Stokoe II, K. H. (2000). Liquefaction resistance of soils from shear-wave velocity. *Journal of Geotechnical and Geoenvironmental Engineering, 126*, 1015–1025. http://dx.doi.org/10.1061/(ASCE)1090-0241(2000)126:11(1015)

Bolton Seed, H., Tokimatsu, K., Harder, L. F., & Chung, R. M. (1985). Influence of SPT procedures in soil liquefaction resistance evaluations. *Journal of Geotechnical Engineering, 111*, 1425–1445. http://dx.doi.org/10.1061/(ASCE)0733-9410(1985)111:12(1425)

Boulanger, R. W. (2003). High overburden stress effects in liquefaction analyses. *Journal of Geotechnical and Geoenvironmental Engineering, 129*, 1071–1082. http://dx.doi.org/10.1061/(ASCE)1090-0241(2003)129:12(1071)

Boulanger, R. W., & Idriss, I. M. (2012). Probabilistic standard penetration test–based liquefaction–triggering procedure. *International Journal of Geotechnical and Geoenvironmental Engineering, 138*, 1185–1195. http://dx.doi.org/10.1061/(ASCE)GT.1943-5606.0000700

Boulanger, R. W., Wilson, D. W., & Idriss, I. M. (2012). Examination and reevalaution of spt-based liquefaction triggering case histories. *Journal of Geotechnical and Geoenvironmental Engineering, 138*, 898–909. http://dx.doi.org/10.1061/(ASCE)GT.1943-5606.0000668

Cetin, K. O., Seed, R. B., Der Kiureghian, A., Tokimatsu, K., Harder, L. F., Jr, Kayen, R. E., ... Moss, R. E. (2004). Standard penetration test-based probabilistic and deterministic assessment of seismic soil liquefaction potential. *Journal of Geotechnical and Geoenvironmental Engineering, 130*, 1314–1340. http://dx.doi.org/10.1061/(ASCE)1090-0241(2004)130:12(1314)

Dalvi, A. N., Pathak, S. R., & Rajhans, N. R. (2014). Entropy analysis for identifying significant parameters for seismic soil liquefaction. *Geomechanics and Geoengineering, 9*(1), 1–8. doi:10.1080/17486025.2013.805255

Davis, R. O., & Berrill, J. B. (1982). Energy dissipation and seismic liquefaction in sands. *Earthquake Engineering Structural Dynamics, 10*, 51–68.

Hamada, M., & Wakamatsu, K. (1996). Liquefaction, ground deformation and their caused damage to structures. *The 1995 Hyogoken-Nanbu earthquake. Investigation into damage to Civil Engineering structures, committee of earthquake Engineering.* Japan Society of Civil Engineering, pp. 45–99.

Hanna, A. M., Ural, D., & Saygili, G. (2007). Neural network model for liquefaction potential in soil deposits using Turkey and Taiwan earthquake data. *Soil Dynamics and Earthquake Engineering, 27*, 521–540. http://dx.doi.org/10.1016/j.soildyn.2006.11.001

Kayen, R., Moss, R. E. S., Thompson, E. M., Seed, R. B., Cetin, K. O., Kiureghian, D., ... Tokimatsu, K. (2013). Shear-wave velocity-based probabilistic and deterministic assessment of seismic soil liquefaction potential. *Journal of Geotechnical and Geoenvironmental Engineering, 139*, 407–419. http://dx.doi.org/10.1061/(ASCE)GT.1943-5606.0000743

Kramer, S. L. (1996). *Geotechnical earthquake Engineering.* Upper Saddle River, NJ: Prentice Hall .

Oommen, T., Baise, L. G., & Vogel, R. (2010). Validation and application of empirical liquefaction models. *Journal of Geotechnical and Geoenvironmental Engineering, 136*, 1618–1633.

Oommen, T., Baise, L. G., & Vogel, R. M. (2011). Sampling bias and class imbalance in maximum-likelihood logistic regression. *Mathematical Geosciences, 43*, 99–120. http://dx.doi.org/10.1007/s11004-010-9311-8

Pathak, S. R. & Dalvi, A. N. (2011). Performance of empirical models for assessment of seismic soil liquefaction. *International Journal of Earth Sciences and Engineering, 4*, 83–86.

Pathak, S. R., & Dalvi, A. N. (2012). Liquefaction potential assessment: An elementary approach. *International Journal of Innovative Research in Science, Engineering and Technology, 1*, 253–255.

Pathak, S. R., & Dalvi, A. N. (2013). Elementary empirical liquefaction model. *Natural HAZARDS, Journal of the International Society for the Prevention and Mitigation of Natural Hazards, 69*, 425–440. doi:10.1007/s11069-013-0723-x.

Rezania, M., Javadi, A. A., & Giustolisi, O. (2010). Evaluation of liquefaction potential based on CPT results using evolutionary polynomial regression. *Computers and Geotechnics, 37*, 82–92. http://dx.doi.org/10.1016/j.compgeo.2009.07.006

Seed, R. B, Cetin, K. O, Moss R, E, Kammerer A, M, Wu J, Pestana J. M, ... Faris, A. (2003). *Recent advances in soil liquefaction engineering: A unified and consistent framework.* 26th Annual Los Angeles Geotechnical Spring Seminar. Keynote Presentation, California.

Youd, T. L., Idriss, I. M., Andrus, R. D., Arango, I., Castro, G., Christian, J. T., ... Stokoe, II, K. H. (2001). Liquefaction resistance of soils: Summary report from the 1996 NCEER/NSF workshops on evaluation of liquefaction resistance of soils. *Journal of Geotechnical and Geoenvironmental Engineering, 127*, 817–833.

Zhang, L. (1998). Assessment of liquefaction potential using optimum seeking method. *Journal of Geotechnical and Geoenvironmental Engineering, 124*, 739–748. http://dx.doi.org/10.1061/(ASCE)1090-0241(1998)124:8(739)

Zhang, L. (2010). A simple method for evaluating liquefaction potential from shear wave velocity. *Frontiers of Architecture and Civil Engineering in China, 4*, 178–195.

Appendix I

314 case records used for ELM development

No	Site	Mag.	σ'_v (kPa)	FC(%)	$(N_1)_{60}$	$(N_1)_{60cs}$	G_{max} (MPa)	v_{max} (m/s)	dur (s)	LP_m term	Obs. Liq?
		Extracted parameters				Computed parameters					
1	Ienaga	8.10	68.00	10.00	8.20	9.03	58.98	0.67	19.88	58.12	Y
2	Komei	8.10	61.00	30.00	3.40	5.31	42.87	0.67	19.88	47.09	Y
3	Shonenji temple	7.00	48.00	0.00	11.80	11.80	54.99	1.14	8.97	58.74	Y
4	Takaya 45	7.00	104.00	4.00	21.10	21.10	94.73	1.00	8.97	40.87	Y
5	Arayamotomachi	7.60	41.00	5.00	4.70	4.70	38.81	0.28	12.54	16.70	Y
6	Cc17-1	7.60	72.00	2.00	9.90	9.90	64.08	0.50	12.54	27.92	Y
7	Rail road-2	7.60	100.00	2.00	17.50	17.50	88.43	0.50	12.54	27.74	Y/N
8	Cc17-2	7.60	43.00	8.00	12.70	13.51	53.13	0.50	12.54	38.76	Y
9	Old town-1	7.60	81.00	2.00	22.70	22.70	85.20	0.56	12.54	37.12	N
10	Old town-2	7.60	109.00	2.00	23.50	23.50	99.72	0.56	12.54	32.29	N
11	Road site	7.60	79.00	0.00	14.10	14.10	74.12	0.56	12.54	33.12	N
12	Showa Br 4	7.60	67.00	0.00	35.50	35.50	86.54	0.56	12.54	45.59	N
13	Aomori station	8.30	38.00	3.00	16.50	16.50	53.66	0.74	16.24	84.80	Y
14	Nanaehama1-2-3	8.30	45.00	20.00	7.60	9.21	46.93	0.69	13.97	50.61	Y
15	Hachinohe- 2	8.30	76.00	5.00	35.30	35.30	92.04	0.80	13.97	67.59	N
16	Hachinohe- 4	8.30	45.00	5.00	23.00	23.00	63.72	0.80	13.97	79.02	N
17	Juvenilehall	6.61	96.00	55.00	3.90	6.20	56.10	1.21	8.56	30.33	Y
18	Panjin Ch. F. P.	7.00	89.00	67.00	7.60	10.41	66.00	0.57	14.06	29.79	Y
19	Amatitlan B-1	7.50	86.00	3.00	5.00	5.00	57.27	0.42	19.76	27.35	Y
20	Luan Nan L2	7.60	32.00	3.00	8.50	8.50	40.89	0.69	15.89	69.84	Y
21	Coastal region	7.60	54.00	12.00	11.70	12.86	58.19	0.41	15.58	34.13	Y
22	Le Ting L8-14	7.60	53.00	12.00	11.50	12.65	57.37	0.63	15.58	52.74	Y
23	Qing Jia Ying	7.60	59.00	20.00	20.10	22.71	70.46	1.09	15.58	101.82	Y
24	Amatitlan B-3&4	7.50	71.00	3.00	14.30	14.30	70.54	0.42	19.76	40.81	N
25	Luan Nan L1	7.60	38.00	5.00	24.40	24.40	59.44	0.69	15.89	85.50	N
26	Tangshan City	7.60	75.00	10.00	31.60	33.36	89.04	1.56	15.58	144.61	N
27	San Juan B-6	7.50	56.00	50.00	5.80	8.36	48.32	0.62	12.99	34.52	Y
28	Amatitlan B-2	7.50	55.00	3.00	9.70	9.70	55.68	0.42	19.76	41.58	Y/N
29	San Juan B-4	7.50	39.00	4.00	14.30	14.30	52.28	0.62	12.99	53.62	N
30	San Juan B-5	7.50	44.00	3.00	13.60	13.60	54.77	0.62	12.99	49.79	N
31	Nakamura 4	6.50	30.00	5.00	6.90	6.90	37.24	0.32	12.82	25.32	Y
32	Nakamura 5	7.70	42.00	4.00	9.60	9.60	48.51	1.02	14.83	87.00	Y
33	Ishinomaki-2	7.70	45.00	10.00	5.50	6.22	42.63	0.63	14.83	44.60	Y
34	Arahama	6.50	67.00	0.00	12.80	12.80	66.46	0.27	12.82	16.86	N
35	Kitawabuchi-2	6.50	59.00	5.00	12.30	12.30	61.68	0.37	12.82	24.88	N
36	Nakajima-18	6.50	79.00	3.00	14.10	14.10	74.12	0.37	12.82	22.33	N
37	Nakamura 5	6.50	42.00	4.00	9.60	9.60	48.51	0.32	12.82	23.56	N
38	Oiiri-1	6.50	85.00	5.00	9.40	9.40	68.60	0.37	12.82	19.21	N
39	Yuriage Br-3	6.50	42.00	2.00	11.80	11.80	51.44	0.32	12.82	24.98	N
40	Yuriagekami-2	6.50	47.00	0.00	15.10	15.10	58.26	0.32	12.82	25.28	N
41	Ishinomaki-2	6.50	45.00	10.00	5.50	6.22	42.63	0.32	12.82	19.32	N

Appendix (Continued)

No	Site	Mag.	σ'_v (kPa)	FC(%)	$(N_1)_{60}$	$(N_1)_{60cs}$	G_{max} (MPa)	v_{max} (m/s)	dur (s)	LP_m term	Obs. Liq?
		Extracted parameters				**Computed parameters**					
42	Shiomi-6	6.50	60.00	10.00	7.50	8.30	53.98	0.37	12.82	21.41	N
43	Yuriage Br-1	6.50	56.00	10.00	5.40	6.12	47.30	0.32	12.82	17.23	N
44	Yuriage Br-2	6.50	34.00	7.00	16.20	17.00	50.50	0.32	12.82	30.30	N
45	Hiyori-1 B	6.50	71.00	20.00	11.10	12.99	65.74	0.37	12.82	22.03	N
46	Yuriagekami-1	6.50	63.00	60.00	2.50	4.60	39.59	0.32	12.82	12.82	N
47	Radio tower b1	6.53	50.00	64.00	2.90	5.06	36.94	0.53	9.24	18.18	Y
48	Riverpark a	6.53	20.00	80.00	4.60	6.99	26.93	0.64	9.24	39.76	Y
49	Radio tower b2	6.53	38.00	30.00	15.20	18.52	52.48	0.53	9.24	33.98	N
50	Wildlifeb	6.53	54.00	30.00	10.30	13.04	56.13	0.45	9.24	21.74	N
51	Kombloom b	6.53	62.00	92.00	6.20	8.82	51.87	0.35	9.24	13.38	N
52	Owi 1	6.00	57.00	13.00	7.10	8.12	51.77	0.23	7.94	8.43	N
53	Owi 2	6.00	123.00	27.00	3.90	5.67	63.50	0.23	7.94	4.79	N
54	Wildlifeb	5.90	54.00	30.00	10.30	13.04	56.13	0.63	6.84	22.40	Y
55	Kombloom b	5.90	62.00	92.00	6.20	8.82	51.87	0.78	6.84	22.19	Y
56	Mckim ranch a	5.90	32.00	31.00	4.60	6.72	34.06	0.22	6.84	7.94	N
57	Riverparkc	5.90	45.00	18.00	15.20	17.19	57.11	0.51	6.84	22.09	N
58	Gaiko wharf B2	7.70	53.00	1.00	12.40	12.40	58.59	0.72	10.38	41.32	Y
59	Noshiro section N7	7.70	38.00	1.00	16.20	16.20	53.39	0.79	10.38	57.84	Y
60	Takeda elememtary school	7.70	42.00	0.00	13.30	13.30	53.18	0.90	10.38	59.01	Y
61	Giako 1	7.70	79.00	3.00	8.70	8.70	64.67	0.65	10.38	27.63	Y
62	Gaiko 2	7.70	107.00	4.00	6.90	6.90	70.34	0.65	10.38	22.19	Y
63	Nakajima no 2(1)	7.70	81.00	3.00	13.50	13.50	74.16	0.65	10.38	30.91	Y
64	Nakajima no 3(3)	7.70	71.00	2.00	10.20	10.20	64.18	0.65	10.38	30.51	Y
65	Nakajima no 3(4)	7.70	68.00	2.00	11.50	11.50	64.98	0.65	10.38	32.26	Y
66	Ohama no 2-1	7.70	56.00	2.00	5.40	5.40	47.30	0.65	10.38	28.51	Y
67	Ohama no 3-1	7.70	63.00	2.00	7.40	7.40	55.09	0.65	10.38	29.52	Y
68	Ohama no 3-3	7.70	64.00	2.00	5.60	5.60	51.12	0.65	10.38	26.96	Y
69	Ohama no 3-4	7.70	49.00	2.00	8.10	8.10	49.89	0.65	10.38	34.37	Y
70	Nakajima no 1(5)	7.70	74.00	8.00	9.90	10.62	64.96	0.65	10.38	29.63	Y
71	Nakajima no 2(2)	7.70	48.00	7.00	9.30	9.91	51.39	0.65	10.38	36.14	Y
72	Arayamotomachi	6.80	37.00	5.00	5.10	5.10	37.79	0.42	9.10	19.33	N
73	Arayamotomachi coarse sand	6.80	77.00	0.00	18.10	18.10	78.29	0.42	9.10	19.25	N
74	Heber road A1	6.54	42.00	12.00	37.80	40.21	69.53	0.42	13.77	47.42	N
75	Heber road A2	6.54	50.00	18.00	2.90	4.01	36.94	0.40	13.77	20.35	N
76	Heber road A3	6.54	56.00	25.00	16.20	19.07	64.81	0.35	13.77	27.63	N
77	mckim ranch a	6.54	32.00	31.00	4.60	6.72	34.06	0.43	13.77	31.27	N
78	River park C	6.54	45.00	18.00	15.20	17.19	57.11	0.51	13.77	44.27	N
79	Radio tower B1	6.54	50.00	64.00	2.90	5.06	36.94	0.53	13.77	27.13	N
80	River park A	6.54	20.00	80.00	4.60	6.99	26.93	0.51	13.77	46.98	N
81	Kornbloom B	6.54	62.00	92.00	6.20	8.82	51.87	0.46	13.77	26.73	N
82	Wildlife B	6.20	61.85	75.00	12.80	16.34	63.86	0.23	12.85	15.14	N
83	Radio tower B1	6.22	50.00	64.00	2.90	5.06	36.94	0.23	13.09	11.06	N

Appendix (*Continued*)

No	Site	Mag.	σ'_v (kPa)	FC(%)	$(N_1)_{60}$	$(N_1)_{60cs}$	G_{max} (MPa)	v_{max} (m/s)	*dur* (s)	LP_m term	Obs. Liq?
		Extracted parameters				Computed parameters					
84	Wildlife B	6.22	54.00	30.00	10.30	13.04	56.13	0.34	13.09	22.99	Y
85	Marine laboratory UC B1	6.93	65.00	3.00	13.10	13.10	65.88	0.79	13.02	52.24	Y
86	Sandholdt UC B10	6.93	43.00	2.00	15.30	15.30	55.92	0.79	13.02	67.02	Y
87	State Beach UC B1	6.93	46.00	1.00	10.30	10.30	51.80	0.79	13.02	58.04	Y
88	Marine laboratory UC B2	6.93	44.00	3.00	15.90	15.90	57.16	0.74	13.02	62.17	Y
89	Farris farm	6.93	92.00	8.00	10.20	10.93	73.06	1.05	13.02	54.08	Y
90	SFOBB-1,2	6.93	86.00	8.00	8.60	9.28	67.25	0.76	13.02	38.86	Y
91	Miller Farm CMF3	6.93	101.00	32.00	9.90	12.77	75.90	1.10	13.02	53.94	Y
92	Miller Farm CMF8	6.93	95.00	25.00	9.80	12.03	73.39	1.10	13.02	55.46	Y
93	Treasure Island	6.93	67.00	20.00	6.40	7.91	54.43	0.45	13.02	23.92	Y
94	Miller farm	6.93	66.00	22.00	10.00	11.98	61.53	1.19	13.02	72.07	Y
95	Woodmarine UC B4	6.93	25.00	35.00	9.10	12.12	36.86	0.79	13.02	75.98	Y
96	POR 2,3 4	6.93	73.00	50.00	5.10	7.56	53.08	0.51	13.02	24.09	Y
97	MBARI NO 3 EB5	6.93	47.00	1.00	14.90	14.90	58.04	0.79	13.02	63.65	N
98	Alameda BF dike	6.93	91.00	7.00	43.30	44.86	105.61	0.68	13.02	51.26	N
99	Perez B1v B11	7.70	90.00	19.00	13.00	14.94	77.36	0.79	11.01	37.54	Y
100	Cereenan St.B12	7.70	68.00	19.00	24.90	27.74	79.93	0.79	11.01	51.33	N
101	Kushiro Port site A	7.60	68.00	2.00	16.40	16.40	71.66	1.25	26.04	171.60	Y
102	Kushiro Port Seismo st	7.60	47.00	5.00	25.90	25.90	67.11	1.47	26.04	273.21	Y
103	Kushiro Port site D	7.60	118.00	0.00	30.90	30.90	111.08	1.25	26.04	153.29	N
104	Wynne Ave Unit C1	6.69	105.00	33.00	11.60	14.78	80.94	1.39	8.37	44.89	Y
105	Balboa B1vUnit C	6.69	143.00	50.00	13.10	16.68	97.72	2.29	8.37	65.54	Y
106	Ashiyama CDE (marine sand)	6.90	115.00	2.00	12.50	12.50	86.50	1.13	10.52	44.56	Y
107	Kobe alluvial site number 5	6.90	116.00	1.00	6.10	6.10	70.60	0.99	10.52	31.55	Y
108	Kobe alluvial site number 7	6.90	60.00	0.00	10.90	10.90	60.12	1.13	10.52	59.36	Y
109	Kobe alluvial site number 9	6.90	64.00	2.00	12.20	12.20	64.09	1.41	10.52	74.16	Y
110	Kobe alluvial site number 11	6.90	62.00	5.00	8.50	8.50	56.91	1.41	10.52	67.97	Y
111	Kobe alluvial site number 16	6.90	60.00	5.00	25.00	25.00	75.15	1.69	10.52	111.30	N/Y
112	Kobe alluvial site number 17	6.90	43.00	5.00	21.10	21.10	60.92	1.41	10.52	104.90	Y
113	Kobe alluvial site number 24	6.90	51.00	0.00	24.60	24.60	69.01	1.41	10.52	100.20	Y
114	Kobe alluvial site number 29	6.90	49.00	0.00	17.90	17.90	62.27	1.13	10.52	75.29	Y
115	Kobe alluvial site number 37	6.90	79.00	0.00	19.30	19.30	80.66	0.99	10.52	52.93	Y
116	Kobe alluvial site number 44	6.90	43.00	5.00	8.30	8.30	47.07	1.13	10.52	64.85	Y
117	Kobe alluvial site number 13	6.90	74.00	15.00	12.70	14.21	69.69	1.41	10.52	69.74	Y
118	Kobe alluvial site number 28	6.90	44.00	8.00	21.10	22.18	61.62	1.13	10.52	82.96	Y

Appendix (Continued)

No	Site	Mag.	σ'_v (kPa)	FC(%)	$(N_1)_{60}$	$(N_1)_{60cs}$	G_{max} (MPa)	v_{max} (m/s)	dur (s)	LP_m term	Obs. Liq?
			Extracted parameters				Computed parameters				
119	Kobe alluvial site number 34	6.90	73.00	9.00	24.20	25.52	82.22	1.13	10.52	66.72	Y
120	Kobe alluvial site number 35	6.90	55.00	6.00	18.90	19.65	66.93	1.41	10.52	90.12	Y
121	Kobe alluvial site number 42	6.90	46.00	10.00	12.10	13.08	54.21	1.13	10.52	69.82	Y
122	ashiyama CDE (mountain sand 2)	6.90	80.00	18.00	21.10	23.52	83.09	1.13	10.52	61.53	Y
123	Port island borehole array st	6.90	96.00	20.00	6.80	8.34	66.34	0.96	10.52	34.79	Y
124	Port island site I	6.90	123.00	20.00	10.80	12.66	85.86	0.96	10.52	35.15	Y
125	Rokko Island building D	6.90	107.00	25.00	16.80	19.73	90.47	1.13	10.52	50.09	Y
126	Rokko island site G	6.90	146.00	20.00	12.30	14.28	97.03	0.96	10.52	33.46	Y
127	Torishma dike	6.90	46.00	20.00	14.00	16.12	56.45	0.70	10.52	45.44	Y
128	Kobe alluvial site number 6	6.90	72.00	21.00	17.80	20.35	75.37	1.13	10.52	62.02	Y
129	Kobe alluvial site number 1	6.90	80.00	3.00	52.00	52.00	103.14	1.13	10.52	76.38	N
130	Kobe alluvial site number 40	6.90	59.00	0.00	39.70	39.70	83.36	1.69	10.52	125.55	N
131	Kobe alluvial site number 36	6.90	36.00	3.00	31.60	31.60	61.69	1.69	10.52	152.27	N
132	Kobe alluvial site number 4	6.90	54.00	1.00	36.60	36.60	78.25	1.13	10.52	85.85	N
133	Kobe alluvial site number 25	6.90	50.00	3.00	35.80	35.80	74.91	1.97	10.52	155.31	N
134	Kobe alluvial site number 26	6.90	37.00	0.00	37.00	37.00	64.94	1.69	10.52	155.96	N
135	Kobe alluvial site number 20	6.90	75.00	0.00	63.70	63.70	104.30	1.55	10.52	113.28	N
136	Kobe alluvial site number 22	6.90	79.00	6.00	38.60	39.83	95.83	1.69	10.52	107.80	N
137	Kobe alluvial site number 12	6.90	72.00	14.00	24.70	26.78	82.08	1.41	10.52	84.41	N
138	Kobe alluvial site number 19	6.90	124.00	10.00	21.30	22.65	103.70	1.69	10.52	74.31	N
139	Kobe alluvial site number 23	6.90	72.00	10.00	24.00	25.46	81.48	1.69	10.52	100.56	N
140	Kobe alluvial site number 27	6.90	29.00	10.00	40.80	42.93	58.81	1.69	10.52	180.21	N
141	Kobe alluvial site number 32	6.90	41.00	6.00	29.10	30.10	64.52	1.41	10.52	116.54	N
142	Ashiyama A (mountain sand 1)	6.90	80.00	18.00	21.10	23.52	83.09	1.13	10.52	61.53	N
143	Adapzari	7.40	57.40	4.00	22.00	22.00	71.15	1.21	11.38	85.59	Y
144	Adapzari	7.40	84.10	3.00	13.00	13.00	74.78	1.21	11.38	61.41	Y
145	Adapzari	7.40	25.90	2.00	4.00	4.00	29.37	1.21	11.38	78.30	Y
146	Adapzari	7.40	61.30	14.00	12.00	13.37	62.44	1.21	11.38	70.34	Y
147	Yuanlin	7.60	160.00	13.00	13.00	14.33	103.15	0.56	11.36	20.60	Y
148	Yuanlin	7.60	68.10	15.00	10.00	11.35	62.50	0.56	11.36	29.33	Y
149	Nantou	7.60	73.00	6.00	14.00	14.64	71.11	1.19	11.36	65.72	Y

Appendix (*Continued*)

No	Site	Mag.	σ'_v (kPa)	FC(%)	$(N_1)_{60}$	$(N_1)_{60cs}$	G_{max} (MPa)	v_{max} (m/s)	dur (s)	LP_m term	Obs. Liq?
			Extracted parameters				Computed parameters				
150	Wufeng	7.60	104.10	15.00	19.00	20.89	92.21	2.09	11.36	105.37	Y
151	Wufeng	7.60	145.30	8.00	24.00	25.17	115.75	2.09	11.36	94.76	Y
152	Wufeng	7.60	157.80	9.00	21.00	22.21	116.55	2.09	11.36	87.85	Y
153	Yuanlin	7.60	33.60	18.00	8.00	9.48	41.16	0.56	11.36	39.15	Y
154	Yuanlin	7.60	45.70	19.00	6.00	7.41	44.10	0.56	11.36	30.84	Y
155	Wufeng	7.60	157.50	21.00	20.00	22.73	114.97	2.09	11.36	86.83	Y
156	Wufeng	7.60	43.10	31.00	9.00	11.67	48.24	2.09	11.36	133.14	Y
157	Wufeng	7.60	56.90	31.00	4.00	6.05	43.53	2.09	11.36	90.99	Y
158	Wufeng	7.60	132.50	26.00	22.00	25.59	108.09	2.09	11.36	97.04	Y
159	Adapzari	7.40	65.60	70.00	19.00	23.41	73.20	1.21	11.38	77.06	Y
160	Adapzari	7.40	97.30	75.00	23.00	27.97	93.69	1.21	11.38	66.50	Y
161	Adapzari	7.40	21.50	87.00	1.00	2.89	17.26	1.21	11.38	55.45	Y
162	Adapzari	7.40	50.30	39.00	20.00	24.55	64.97	1.21	11.38	89.20	Y
163	Adapzari	7.40	56.40	54.00	17.00	21.13	65.89	1.21	11.38	80.68	Y
164	Adapzari	7.40	16.80	51.00	2.00	4.03	19.05	1.21	11.38	78.32	Y
165	Adapzari	7.40	26.50	97.00	6.00	8.59	33.58	1.21	11.38	87.51	Y
166	Adapzari	7.40	53.10	99.00	16.00	19.99	62.90	1.21	11.38	81.81	Y
167	Adapzari	7.40	30.30	57.00	3.00	5.17	29.06	1.21	11.38	66.24	Y
168	Adapzari	7.40	48.00	91.00	11.00	14.29	53.91	1.21	11.38	77.56	Y
169	Adapzari	7.40	47.00	65.00	5.00	7.45	42.34	1.21	11.38	62.21	Y
170	Yuanlin	7.60	51.80	90.00	5.00	7.45	44.45	0.56	11.36	27.42	Y
171	Yuanlin	7.60	92.60	94.00	7.00	9.73	65.71	0.56	11.36	22.68	Y
172	Wufeng	7.60	14.80	65.00	3.00	5.17	20.31	2.09	11.36	163.24	Y
173	Adapzari	7.40	85.40	5.00	33.00	33.00	96.01	1.21	11.38	77.64	N
174	Adapzari	7.40	106.20	5.00	31.00	31.00	105.46	1.21	11.38	68.58	N
175	Adapzari	7.40	96.60	4.00	35.00	35.00	103.56	1.21	11.38	74.03	N
176	Adapzari	7.40	79.10	1.00	20.00	20.00	81.47	1.21	11.38	71.13	N
177	Adapzari	7.40	63.40	4.00	44.00	44.00	88.47	1.21	11.38	96.37	N
178	Adapzari	7.40	55.40	4.00	30.00	30.00	75.57	1.21	11.38	94.19	N
179	Adapzari	7.40	82.50	4.00	50.00	50.00	103.84	1.21	11.38	86.92	N
180	Adapzari	7.40	79.00	8.00	39.00	40.65	96.06	1.21	11.38	83.97	N
181	Adapzari	7.40	79.20	9.00	49.00	51.21	101.29	1.21	11.38	88.32	N
182	Adapzari	7.40	92.50	8.00	75.00	77.80	119.81	1.21	11.38	89.45	N
183	Yuanlin	7.60	139.80	8.00	19.00	20.01	106.86	0.56	10.98	23.62	N
184	Adapzari	7.40	97.00	19.00	27.00	30.00	97.42	1.21	11.38	69.36	N
185	Yuanlin	7.60	113.10	20.00	20.00	22.60	97.42	0.56	10.98	26.62	N
186	Yuanlin	7.60	195.30	29.00	43.00	49.44	154.47	0.56	10.98	24.44	N
187	Nantou	7.60	158.90	35.00	42.00	49.63	138.59	1.19	10.36	53.65	N
188	Nantou	7.60	160.10	20.00	21.00	23.68	117.40	1.19	10.36	45.11	N
189	Nantou	7.60	177.20	19.00	33.00	36.46	138.30	1.19	10.36	48.01	N
190	Nantou	7.60	195.30	26.00	62.00	69.75	167.36	1.19	10.36	52.72	N
191	Nantou	7.60	210.50	22.00	49.00	54.41	165.14	1.19	10.36	48.26	N
192	Nantou	7.60	217.00	29.00	31.00	36.05	150.75	1.19	10.36	42.74	N

Appendix (*Continued*)

No	Site	Mag.	σ'_v (kPa)	FC(%)	$(N_1)_{60}$	$(N_1)_{60cs}$	G_{max} (MPa)	v_{max} (m/s)	*dur* (s)	LP_m term	Obs. Liq?
			Extracted parameters				Computed parameters				
193	Wufeng	7.60	129.70	35.00	36.00	42.79	120.80	2.09	11.02	107.53	N
194	Wufeng	7.60	13.10	23.00	14.00	16.44	30.13	2.09	11.02	265.51	N
195	Wufeng	7.60	13.10	33.00	34.00	40.14	37.87	2.09	11.02	333.80	N
196	Adapzari	7.40	75.20	58.00	26.00	31.39	84.97	1.21	11.38	78.03	N
197	Adapzari	7.40	62.80	53.00	48.00	56.47	89.79	1.21	11.38	98.73	N
198	Adapzari	7.40	52.60	87.00	10.00	13.15	54.93	1.21	11.38	72.12	N
199	Adapzari	7.40	60.40	71.00	47.00	55.33	87.64	1.21	11.38	100.20	N
200	Adapzari	7.40	57.70	97.00	24.00	29.11	72.94	1.21	11.38	87.30	N
201	Adapzari	7.40	51.70	99.00	55.00	64.45	83.93	1.21	11.38	112.11	N
202	Adapzari	7.40	26.90	99.00	16.00	19.99	44.77	1.21	11.38	114.94	N
203	Adapzari	7.40	56.60	88.00	36.00	42.79	79.80	1.21	11.38	97.37	N
204	Adapzari	7.40	104.80	97.00	28.00	33.67	102.18	1.21	11.38	67.33	N
205	Adapzari	7.40	67.60	41.00	38.00	45.07	88.33	1.21	11.38	90.23	N
206	Adapzari	7.40	38.40	99.00	9.00	12.01	45.53	1.21	11.38	81.89	N
207	Yuanlin	7.60	162.40	47.00	40.00	47.35	138.54	0.56	10.98	26.36	N
208	Nantou	7.60	111.70	75.00	63.00	73.57	126.99	1.19	10.36	69.94	N
209	Wufeng	7.60	153.40	38.00	58.00	67.87	146.24	2.09	11.02	110.06	N
210	Wufeng	7.60	25.40	58.00	10.00	13.15	38.17	2.09	11.02	173.50	N
211	Wufeng	7.60	169.20	67.00	25.00	30.25	126.20	2.09	11.02	86.12	N
212	Wufeng	7.60	180.60	52.00	22.00	26.83	126.20	2.09	11.02	80.68	N
213	Wufeng	7.60	116.90	72.00	15.00	18.85	91.71	2.09	11.02	90.58	N
214	Wufeng	7.60	183.10	49.00	35.00	41.65	142.58	2.09	11.02	89.90	N
215	Wufeng	7.60	113.40	51.00	24.00	29.11	102.26	2.09	11.02	104.11	N
216	Wufeng	7.60	197.70	76.00	13.00	16.57	114.66	2.09	11.02	66.96	N
217	Wufeng	7.60	213.30	82.00	11.00	14.29	113.65	2.09	11.02	61.52	N
218	Wufeng	7.60	150.60	74.00	18.00	22.27	109.33	2.09	11.02	83.82	N
219	Wufeng	7.60	202.00	67.00	16.00	19.99	122.69	2.09	11.02	70.12	N
220	Wufeng	7.60	202.00	47.00	26.00	31.39	139.26	2.09	11.02	79.60	N
221	Meiko	8.10	39.00	27.00	1.70	3.23	27.57	0.67	19.88	47.37	Y
222	River Site	7.60	47.00	0.00	9.40	9.40	51.01	0.50	12.54	34.05	Y
223	Old Town -1	7.60	81.00	2.00	22.70	22.70	85.20	0.56	12.54	37.12	N
224	Old Town -2	7.60	109.00	2.00	23.50	23.50	99.72	0.56	12.54	32.29	N
225	Road Site	7.60	79.00	0.00	14.10	14.10	74.12	0.56	12.54	33.12	N
226	Aomori station	8.30	38.00	3.00	16.50	16.50	53.66	0.74	16.24	84.80	Y
227	Hachinohe-6	8.30	42.00	5.00	9.10	9.10	47.77	0.80	13.97	63.48	Y
228	Juvenile Hall	6.61	96.00	55.00	3.90	6.20	56.10	1.21	8.56	30.33	Y
229	Van Norman	6.61	96.00	50.00	8.10	10.98	69.83	1.21	8.56	37.75	Y
230	Ying Kou P. P.	7.00	92.00	5.00	11.00	11.00	74.64	0.86	11.50	39.99	Y
231	Amatitlan B-2	7.50	34.00	3.00	9.70	9.70	43.78	0.42	19.76	52.89	Y/N
232	Yao Yuan village	7.60	67.00	20.00	10.50	12.34	62.86	0.63	15.58	45.71	Y
233	San Juan B-3	7.50	156.00	5.00	7.60	7.60	87.37	0.62	12.99	22.40	Y
234	San Juan B-1	7.50	106.00	20.00	6.30	7.80	68.14	0.62	12.99	25.71	Y
235	Nakamura 4	6.50	30.00	5.00	6.90	6.90	37.24	0.32	12.82	25.32	Y
236	Arahama	7.70	67.00	0.00	12.80	12.80	66.46	0.63	14.83	46.69	Y

Appendix (Continued)

No	Site	Mag.	σ'_v (kPa)	FC(%)	$(N_1)_{60}$	$(N_1)_{60cs}$	G_{max} (MPa)	v_{max} (m/s)	dur (s)	LP_m term	Obs. Liq?
			Extracted parameters				**Computed parameters**				
237	Oiiri-1	7.70	85.00	5.00	9.40	9.40	68.60	0.76	14.83	45.59	Y
238	Hiyori-18	7.70	71.00	20.00	11.10	12.99	65.74	0.76	14.83	52.30	Y
239	yuriagebr-3	7.70	42.00	12.00	11.80	12.97	51.44	0.76	14.83	69.18	Y
240	Kitawabuchi-3	7.70	73.00	0.00	17.60	17.60	75.67	0.89	14.83	68.31	N
241	Ishinomaki-2	6.50	45.00	10.00	5.50	6.22	42.63	0.32	12.82	19.32	N
242	Ishinomaki-4	7.70	57.00	10.00	20.90	22.24	69.96	0.63	14.83	57.78	N
243	yuriagebr-5	7.70	78.00	17.00	20.10	22.32	81.01	0.76	14.83	58.67	N
244	heber road a2	6.53	50.00	18.00	2.90	4.01	36.94	2.08	9.24	70.91	Y
245	mckim ranch a	6.53	32.00	31.00	4.60	6.72	34.06	1.36	9.24	66.80	Y
246	heber road a1	6.53	42.00	12.00	37.80	40.21	69.53	2.08	9.24	158.89	N
247	heber road a3	6.53	56.00	25.00	16.20	19.07	64.81	2.08	9.24	111.08	N
248	Radiotowerb1	5.90	5.00	64.00	2.90	5.06	11.68	0.48	6.84	38.73	Y
249	Radiotowerb2	5.90	38.00	30.00	15.20	18.52	52.48	0.48	6.84	22.89	N
250	Riverpark a	5.90	20.00	80.00	4.60	6.99	26.93	0.51	6.84	23.43	N
251	Takeda elememtary school	6.80	42.00	0.00	13.30	13.30	53.18	0.31	9.10	17.74	Y
252	Aomori station	7.70	38.00	3.00	16.50	16.50	53.66	0.37	10.38	26.97	Y
253	Arayamotomachi	7.70	37.00	5.00	5.10	5.10	37.79	0.63	10.38	33.64	Y
254	Radio tower B1	6.20	39.79	75.00	12.00	15.43	50.30	0.23	12.85	18.53	N
255	Marine laboratory UC B2	6.93	55.00	3.00	14.90	14.90	62.79	0.79	13.02	58.83	Y
256	POO7-1	6.93	89.00	3.00	15.40	15.40	80.60	0.79	13.02	46.67	Y
257	POO7-3	6.93	89.00	3.00	17.00	17.00	82.78	0.79	13.02	47.93	Y/N
258	Miller Farm CMF5	6.93	108.00	13.00	20.90	22.64	96.30	1.10	13.02	64.01	Y
259	Miller Farm CMF10	6.93	105.00	20.00	20.20	22.82	94.11	1.10	13.02	64.34	Y
260	MBARI NO 3 EB1	6.93	35.00	1.00	22.60	22.60	55.94	0.79	13.02	82.37	N
261	MBARI NO 3 EB5	6.93	47.00	1.00	14.90	14.90	58.04	0.79	13.02	63.65	N
262	Hall Avenue	6.93	64.00	30.00	5.70	7.88	51.39	0.40	13.02	20.69	N
263	Potrero Canyon C1	6.69	88.00	64.00	8.50	11.44	67.80	1.17	8.37	37.82	Y
264	Malden street Unit D	6.69	101.00	25.00	27.20	31.17	99.59	1.39	8.37	57.42	N
265	Kobe alluvial site number 8	6.90	65.00	0.00	24.10	24.10	77.50	1.41	10.52	88.29	Y
266	Kobe alluvial site number 38	6.90	94.00	5.00	19.10	19.10	87.75	1.41	10.52	69.13	Y
267	Kobe alluvial site number 41	6.90	50.00	0.00	15.00	15.00	59.98	1.13	10.52	71.06	Y
268	Kobe alluvial site number 43	6.90	55.00	20.00	15.20	17.42	63.13	0.99	10.52	59.50	Y
269	Ashiyama A (marine sand)	6.90	97.99	2.00	31.30	31.30	101.54	1.13	10.52	61.39	N
270	Kobe alluvial site number 18	6.90	171.00	0.00	42.60	42.60	144.23	1.97	10.52	87.44	N
271	Kobe alluvial site number 31	6.90	46.00	0.00	49.70	49.70	77.44	1.69	10.52	149.59	N
272	Kobe alluvial site number 39	6.90	66.00	0.00	61.00	61.00	96.95	1.69	10.52	130.54	N

Appendix (*Continued*)

No	Site	Mag.	σ'_v (kPa)	FC(%)	$(N_1)_{60}$	$(N_1)_{60cs}$	G_{max} (MPa)	v_{max} (m/s)	dur (s)	LP_m term	Obs. Liq?
			Extracted parameters				Computed parameters				
273	Kobe alluvial site number 40	6.90	59.00	0.00	39.70	39.70	83.36	1.69	10.52	125.55	N
274	Kobe alluvial site number 21	6.90	44.00	0.00	33.50	33.50	69.17	1.69	10.52	139.69	N
275	Kobe alluvial site number 3	6.90	77.00	3.00	49.80	49.80	100.23	1.13	10.52	111.30	N
276	Port island improved site (Ikegaya)	6.90	125.00	20.00	22.70	25.52	105.84	1.13	10.52	50.16	N
277	Port island improved site (Tanahashi)	6.90	140.00	20.00	19.50	22.06	107.67	1.13	10.52	45.56	N
278	Port island improved site (Watanabe)	6.90	135.00	20.00	34.60	38.37	122.09	1.13	10.52	53.58	N
279	Kobe alluvial site number 2	6.90	103.00	15.00	39.50	42.62	110.01	1.13	10.52	63.27	N
280	Kobe alluvial site number 10	6.90	107.00	9.00	27.40	28.84	102.69	1.69	10.52	85.29	N
281	Kobe alluvial site number 14	6.90	69.00	19.00	20.30	22.79	76.39	1.41	10.52	81.98	N
282	Kobe alluvial site number 30	6.90	78.00	10.00	40.10	42.20	96.07	1.69	10.52	109.45	N
283	Kobe alluvial site number 33	6.90	83.00	50.00	27.90	33.56	90.85	1.41	10.52	81.06	N
284	Adapzari city	7.40	43.20	28.00	10.00	12.52	49.78	1.21	11.38	79.58	Y
285	Adapzari city	7.40	48.50	18.00	18.00	20.20	62.05	1.21	11.38	88.35	Y
286	Adapzari city	7.40	55.00	13.00	21.00	22.74	68.81	1.21	11.38	86.39	Y
287	Adapzari city	7.40	47.40	33.00	8.00	10.71	48.89	1.21	11.38	71.23	Y
288	Adapzari city	7.40	55.90	9.00	14.00	14.95	62.23	1.21	11.38	76.88	Y
289	Adapzari city	7.40	25.10	9.00	7.00	7.70	34.21	1.21	11.38	94.12	Y
290	Adapzari city	7.40	76.40	9.00	9.00	9.77	64.23	1.21	11.38	58.05	Y
291	Wufeng	7.60	80.10	14.00	23.00	24.99	85.01	2.09	11.36	126.24	Y
292	Wufeng	7.60	91.50	10.00	11.00	11.94	74.44	2.09	11.36	96.77	Y
293	Wufeng	7.60	112.80	23.00	15.00	17.53	90.09	2.09	11.36	95.00	Y
294	Wufeng	7.60	123.90	18.00	17.00	19.12	97.67	2.09	11.36	93.76	Y
295	Wufeng	7.60	202.40	17.00	12.00	13.67	113.45	2.09	11.36	66.68	Y
296	Wufeng	7.60	58.70	20.00	18.00	20.44	68.26	2.09	11.36	138.32	Y
297	Adapzari city	7.40	53.00	35.00	15.00	18.85	61.75	1.21	11.38	80.46	Y
298	Adapzari	7.40	66.20	34.00	14.00	17.60	67.72	1.21	11.38	70.64	N
299	Adapzari	7.40	65.40	17.00	43.00	46.77	89.39	1.21	11.38	94.39	N
300	Adapzari	7.40	71.70	17.00	42.00	45.71	93.09	1.21	11.38	89.66	N
301	Adapzari	7.40	88.00	20.00	32.00	35.56	96.74	1.21	11.38	75.92	N
302	Adapzari	7.40	96.30	11.00	45.00	47.53	109.59	1.21	11.38	78.59	N
303	Adapzari	7.40	77.80	20.00	40.00	44.20	95.89	1.21	11.38	85.12	N
304	Adapzari	7.40	86.90	24.00	34.00	38.46	97.55	1.21	11.38	77.52	N
305	Adapzari	7.40	106.80	32.00	29.00	34.31	104.05	1.21	11.38	67.28	N
306	Wufeng	7.60	118.00	32.00	23.00	27.54	103.18	2.09	11.02	100.96	N
307	Wufeng	7.60	216.60	10.00	32.00	33.78	151.78	2.09	11.02	80.90	N

Appendix *(Continued)*

No	Site	Mag.	σ'_v (kPa)	FC(%)	$(N_1)_{60}$	$(N_1)_{60cs}$	G_{max} (MPa)	v_{max} (m/s)	*dur* (s)	LP_m term	Obs. Liq?
		Extracted parameters				**Computed parameters**					
308	Wufeng	7.60	120.00	27.00	41.00	46.78	119.77	2.09	11.02	115.23	N
309	Wufeng	7.60	213.50	26.00	26.00	30.00	143.17	2.09	11.02	77.42	N
310	Wufeng	7.60	224.70	17.00	28.00	30.75	149.62	2.09	11.02	76.88	N
311	Adapzari	7.40	48.00	71.00	25.00	30.25	67.22	1.21	11.38	96.71	N
312	Adapzari	7.40	21.60	81.00	1.00	2.89	17.30	1.21	11.38	55.32	N
313	Wufeng	7.60	217.20	67.00	17.00	21.13	129.31	2.09	11.02	68.74	N
314	Wufeng	7.60	183.50	35.00	37.00	43.93	144.62	2.09	11.02	90.99	N

Sr. No. 1–220 used in Formulation; Sr. No. 221–314 used in Validation.

14

Review of preprocessing techniques used in soil property prediction from hyperspectral data

S. Minu[1*], Amba Shetty[1] and Binny Gopal[2]

*Corresponding author: S. Minu, Department of Applied Mechanics and Hydraulics, National Institute of Technology Karnataka, Surathkal, Mangalore 575025, Karnataka, India
E-mail: minu.s88@gmail.com
Reviewing editor:
Lachezar Hristov Filchev, Space Research and Technology Institute - Bulgarian Academy of Sciences (SRTI-BAS), Bulgaria

Abstract: Soil properties are neither static nor homogenous with space and time. Capturing the spatial variation of soil properties through conventional methods is a difficult task. Hyperspectral remote sensing data provide rich source of information produced in the form of spectrum at each pixel which can be used to identify surface materials. Airborne and spaceborne narrowband hyperspectral sensors have come to the fore which provides spectral information across large area. Thus, it is a promising tool for studying soil properties and can be used as an alternative to conventional method. But atmospheric attenuation and low signal to noise ratio are major problems with this type of data. Preprocessing of hyperspectral airborne/spaceborne data is required to extract soil properties. This paper reviews previous studies on prediction of soil properties from hyperspectral airborne and satellite data during the past years and the preprocessing techniques used in these predictions.

Subjects: Earth Sciences; Engineering & Technology; Environment Agriculture

Keywords: airborne; hyperspectral; prediction model; preprocessing techniques; soil properties; spaceborne

1. Introduction

Remotely sensed hyperspectral satellite data have great potential for quantitative assessment of soil and vegetation parameter at spatial scale. The development of methods to map soil properties using optical remote sensing data in combination with field measurements has been the objective of several studies during the last decade (Ben-Dor et al., 2009). Also it has been a challenge to find the most appropriate technique for studying soil properties from optical data and thus reducing the time and effort involved in field sampling and laboratory analysis.

Soil reflectance in the visible near-infrared and mid-infrared regions has been widely used in many studies. Some of the soil properties predicted from reflectance data were organic matter (OM), soil

ABOUT THE AUTHORS

Our group works on application of Hyperspectral data for soil and vegetation discrimination applications. In the process of applying this data to any application, major issue to be addressed is to account for the effects of atmosphere on the hyperspectral data and to account for it appropriately. Though there are several algorithms are available to address this, there is no guideline on the application of them. One of the issues that we wish to address is to how best to account for the effect of atmosphere so that proper signal of the targets is extracted for further analysis.

PUBLIC INTEREST STATEMENT

The present review paper would be very useful in the process of digital soil mapping mission from satellite data. As soil is a precious non-renewable resource, it has to be examined periodically. Prediction from satellite data provides a continuous method of monitoring soil quality. The accuracy of prediction depends on the quality of satellite data. The methods to improve quality of data are reviewed in this paper.

organic carbon (SOC), total nitrogen (TN), pH, moisture content (MC), electrical conductivity (EC), phosphorous (P), potassium (K), calcium (Ca), magnesium (Mg), sodium (Na), manganese (Mn), zinc (Zn), and iron (Fe) with various levels of prediction accuracy. Various prediction models such as multiple linear regression (MLR), principal components regression (PCR), stepwise multiple linear regression (SMLR), partial least squares regression (PLSR), artificial neural networks (ANN), etc. were used. These models work well with signals obtained under laboratory conditions, with minimal source of noise. Thus, performance of these models on remotely sensed airborne or spaceborne data is influenced by atmospheric interference and the occurrence of spectral noises. At this juncture, the role of preprocessing techniques on the prediction accuracy of soil properties from remotely sensed data needs to be studied.

Preprocessing techniques consist of atmospheric correction algorithms as well as spectral pretreatment and smoothening methods. Over the years, atmospheric correction algorithms have evolved from applied math approach to ways supported on rigorous radiative transfer (RT) modeling (Minu & Shetty, 2015). Noise and unwanted spectral signals are removed by spectral pretreatment and smoothening methods. Only good-quality data with better signal-to-noise ratios can be conveniently used for the purpose.

Minu and Shetty (2015) review different hyperspectral atmospheric correction algorithms developed during the past years. Internal average reflectance approach (Kruse, Raines, & Watson, 1985), flat field approach (Roberts, Yamaguchi, & Lyon, 1986), empirical line (EL) method (Roberts, Yamaguchi, & Lyon, 1985), QUick atmospheric correction (Bernstein et al., 2005) etc. are empirical or semi-empirical atmospheric correction methods. RT codes try to simulate the transfer process of an electromagnetic wave in the atmosphere. The normally used RT codes are LOWTRAN (Kneizys et al., 1988), MODTRAN (Berk, Bernstein, & Robertson, 1989), 5S (Tanré, Deroo, Duhaut, Herman, & Morcrette, 1990), and 6S (Vermote et al., 1997). There are a range of software programs available to model the atmosphere including ATmospheric REMoval algorithm (ATREM) (Gao, Heidebrecht, & Goetz, 1993), ATmospheric CORrection (ATCOR) (Richter, 1996), Fast Line-of-sight Atmospheric Analysis of Spectral Hypercubes (FLAASH) (Adler-Golden et al., 1998), Imaging Spectrometer Data Analysis System (ISDAS) (Staenz, Szeredi, & Schwarz, 1998), High-accuracy ATmosphere Correction for Hyperspectral data (HATCH) (Qu, Goetz, & Heidbrecht, 2001), Atmospheric CORrection Now (ACORN) (ACORN 4.0, 2002) etc. Hybrid methods include combinations of empirical approaches and radiative modeling for the derivation of surface reflectance from hyperspectral imaging data. Each preprocessing technique is made of its own assumptions. So there is a need to analyze limitations of different preprocessing techniques and to come up with a universal method.

2. Prediction of soil properties from airborne/spaceborne hyperspectral data

Hyperspectral sensors operate with more than hundreds of bands with good spatial and spectral resolution producing continuous spectra. With the progress and maturity of technology, hyperspectral remote sensing has found a wide range of applications in mapping soil types and quantifying soil constituents. Review papers by Ben-Dor et al. (2009); Ge, Thomasson, and Sui (2011); Mulder, de Bruin, Schaepman, and Mayr (2011), etc. point toward it. Airborne sensors provide high spatial resolution (2–20 m), high spectral resolution (10–20 nm), and high SNR (>500:1) data. Even though satellite hyperspectral imageries have become available since 2000, only few attempts have been made to use them for mapping soil properties. This may be due to their low signal to noise ratio. Tables 1 and 2 summarize previous studies carried out using airborne and satellite hyperspectral imageries to predict soil properties. The preprocessing techniques used are also mentioned in the table.

It is seen that RT models are mainly used in preprocessing of airborne imagery. It may be due to the fact that more information on atmospheric conditions are available in the case of airborne sensors, so that modeling of atmosphere can be done precisely and it can be removed to obtain pure signal. Whereas semi-empirical models like FLAASH are mainly used in hyperspectral imageries. Comparison of different models are still lacking in this field. Also EL method which also requires ground information gives good results. But it is limited only to the areas where ground information

Table 1. Summary of soil properties prediction, using airborne hyperspectral imagery							
Soil property	Platform/ spectral range/spatial resolution	Field nature	Country	Preprocessing method	Prediction tech.	R^2 value	Author
Fe	AVIRIS (400–2,500 nm) (20 m)	Pasture and seasonal crops	Brazil	MODTRAN-based (Green, Conel, & Roberts, 1993)	Regression equations	0.83	Galvão, Pizarro, and Epiphanio (2001)
TiO$_2$						0.74	
Al$_2$O$_3$						0.68	
OM	DAIS-7915 (400–2,500 nm) (5 m)	Agriculture fields	Israel	Minimum noise fraction (MNF) (Green, Berman, Switzer, & Craig 1988) for noise reduction; EL technique	Visible and NIR analysis	0.827	Ben-Dor, Patkin, Banin, and Karnieli (2002)
MC						0.647	
EC						0.665	
EC	RDACS/H3 (471–828 nm) (1 m)	Bare soil of corn–soy-bean rotation field	Missouri	Calibrated with chemically treated reference traps with known reflectance	SMLR	0.66	Hong et al. (2002)
pH						0.68	
Mg						0.67	
K						0.59	
OM						0.55	
OM	CASI (408.73–947.07 nm) (2 m)	Corn field with clay-loam soil	Canada	CAM5S model O'Neill et al. (1997)	SMLR	0.49	Uno et al. (2005)
					ANN	0.592	
Iron oxide	CASI-A (400–1,000 nm) (3 m)	Sand dunes	Israel	EL technique	Spectral indices based model	0.59	Ben-Dor et al. (2006)
Gravel coverage %	DAIS-7915 (400–2,500 nm) (5 m)	Alluvial fan	Negev desert, Israel	MNF technique for noise reduction and EL technique	Ferric absorption feature depth(AFD) model	0.83	Crouvi, Ben-Dor, Beyth, Avigad, and Amit (2006)
					Al–OH AFD model	0.67	
					Carbonate AFD model	0.57	
SOC	HyMap (450–2,500 nm) (3.5 m)	Agriculture fields	Germany	ATCOR Richter & Schläpfer, (2002; Schläpfer & Richter, 2002)	MLR	0.9	Selige, Böhner, and Schmidhalter (2006)
TN						0.92	
Sand						0.95	
Clay						0.71	
SOC					PLSR	0.86	
TN						0.87	
Sand						0.87	
Clay						0.65	
EC	HyMap (420–2,480 nm) (6 m)	Wetland	Western Australia	Corrected for atmospheric effects and multiplicative signal correction (MSC) techniques	PLSR	0.86	Farifteh, Van der Meer, Atzberger, and Carranza (2007)
					ANN	0.86	
EC	AVNIR (429–1,010 nm) (1.2 m)	Cotton field	California	Atmospheric calibrated with black and gray reference panels	MLR	0.6696	De Tar, Chesson, Penner, and Ojala (2008)
Ca						0.6188	
Mg						0.582	
Na						0.6224	
Cl						0.7376	
Clay	HYMAP (400–2,500 nm) (5 m)	Area is mainly devoted to vineyards	France	ATCOR4, Savitzky–Golay filter	PLSR	0.64	Gomez, Lagache-rie, and Coulouma (2008)
CaCO$_3$						0.77	
Clay					Continuum removal	0.58	
CaCO$_3$						0.47	

(Continued)

Table 1. (Continued)

Soil property	Platform/ spectral range/spatial resolution	Field nature	Country	Preprocessing method	Prediction tech.	R^2 value	Author
MC	HyMap (440–2,470 nm) (4 m)	Sandy substrates and low vegetation cover area	Germany	MODTRAN4 based ACUM algorithm	Normalized soil moisture index (NSMI) model	0.819	Haubrock, Chabrillat, Kuhnert, Hostert, and Kaufmann (2008)
Clay	HYMAP (400–2,500 nm) (5 m)	Area is devoted to vineyards	France	ATCOR4 code for airborne sensors	Continuum removal analysis	0.58	Lagacherie, Baret, Feret, Madeira Netto, and Robbez-Masson (2008)
CaCO$_3$						0.47	
SOC	AHS-160 sensor (430 nm–2540 nm) (2.6 m)	Agriculture fields	Belgium	MODTRAN-4 embedded with ATCOR-4 (Richter, Schläpfer, & Müller, 2006)	PLSR	RPD = 1.47	Stevens et al. (2008
Clay	HYMAP (400–2,500 nm) (5 m)	Area is devoted to vineyards	France	ATCOR4 code for airborne sensors	PLSR	0.64	Lagacherie, Gomez, Bailly, Baret, and Coulouma (2010)
CaCO$_3$						0.77	
SOC	AHS-160 sensor (430 nm–2,540 nm) (2.6 m)	Cropland	Luxembourg	MODTRAN4-based algorithm; (Richter, 2005; Rodger & Lynch, 2001)	PLSR	0.71	Stevens et al. (2010)
					PSR	0.75	
					SVMR	0.69	
C	HyperSpecTIR (400–2,450 nm) (2.5 m)	Tilled agricultural fields	MD, USA	Imagery processing by ENVI 4.7; & different signal smoothening methods	PLSR	0.65	Hively et al. (2011)
Al						0.76	
Fe						0.75	
Silt						0.79	
Clay	MIVIS (430–1,270 nm) (4.8 m)	Maize field, but the crop had not emerged	Central Italy	MODTRAN4-based model (Vermote, Tanre, Deuze, Herman, & Morcette, 1997)	PLSR	0.78	Casa, Castaldi, Pascucci, Palombo, and Pignatti (2013)
Silt						0.56	
Sand						0.81	
SOC	CASI 1500 (380–1,050 nm) (0.2 m)	Compost added soil	Italy	EL calibration with asphalt spectral signatures	Correlation between the second derivative value and SOC	0.85	Matarrese et al. (2014)

is available. Also it is seen that prediction of SOC gives good results compared to other properties. This may be because the soil reflectance curve is affected more by presence of OM.

3. Inference

Several surface soil properties were modeled from remotely sensed hyperspectral imagery. Since soil is a more heterogeneous material, more careful spectral manipulations need to be done in assessing its properties from spectral data. For the best performance of any prediction system, the key influencing factors are to be identified and optimized. Although there are many soil properties prediction models, the prediction accuracy is found to be still very low.

The noises should be removed from the hyperspectral imagery in order to utilize it to the best. The signal to noise ratio should be maximum. Several spectral pre-processing methods are employed in various studies to improve the performance and robustness of the prediction models. Even though the preprocessing techniques affect the prediction model considerably, it was not given that much importance. So to develop a good model there is a need to perform a better preprocessing. In this percept, different preprocessing techniques used in various studies are listed in this review paper. Hybrid methods which combine physical model and image statistics need to be promoted. There is a need to give guidelines on selection of suitable preprocessing technique for the prediction of soil chemical properties.

Table 2. Summary of soil properties prediction using satellite remote sensing techniques

Soil Prop	Platform	Field characteristics	Country	Preprocessing method	Prediction tech.	R^2 values	Author
SOC	EO1 Hyperion (400–2,500 nm) (30 m)	Cotton crops and pasture. Field size = 100 × 500 m²	Australia	Algorithm based on ATREM and 5S code.	PLSR	0.5	Gomez, Viscarra Rossel, and Mc-Bratney (2008))
OM	EO1 Hyperion (400–2,500 nm) (30 m)	Row-crop agriculture field	Central Indiana, USA	ENVI FLAASH module	PLSR	0.74	Zheng (2008)
TN						0.72	
TP						0.67	
TN	EO1 Hyperion (400–2,500 nm) (30 m)	Arid regions; 4,332 km².	Shanxi, China	EL atmospheric correction	Linear regression model	0.84	Wu, Liu, Chen, Wang, and Chai (2009)
MC	EO1 Hyperion (400–2,500 nm) (30 m)	Bare field	Central Indiana, USA	ACORN	PLSR	0.79	Zhang, Li, and Zheng (2009)
OM						0.89	
TN						0.7	
TP						0.69	
TC						0.86	
Clay						0.49	
OM	EO1 Hyperion (400–2,500 nm) (30 m)	Agriculture–pasture mixed area.	Hengshan County, China	Internal average relative reflectance	Land degradation spectral response units (DSRU) model	0.722	Wang, He, Lv, Chen, and Jian (2010)
Clay	CHRIS-PROBA (415–1,050 nm) (17 m)	Maize field, but the crop had not emerged, 12 and 17 ha plots	Central Italy	FLAASH	PLSR	0.6	Casa et al. (2013)
Silt						0.3	
Sand						0.62	
OM	EO1 Hyperion (400–2,500 nm) (30 m)	Wheat and potato fields. Field size = 90 × 90 m².	China	FLAASH	PLSR	0.63	Lu, Wang, Niu, Li, and Zhang (2013)
pH						0.68	
P						0.62	
N	EO1 Hyperion (400–2,500 nm) (30 m)	Scattered paddy fields, 47 km²	Karnataka India	FLAASH,	PLSR	0.63	Gopal, Shetty, and Ramya (2014)
				Moving average			
				Savitzky–Golay		0.63	
POM	EO1 Hyperion (400–2,500 nm) (30 m)	Coastal soils densely covered with vegetation	Florida, USA	FLAASH, MNF filter	PLSR	0.67	Anne, Abd-Elrahman, Lewis, and Hewitt (2014)
MAOM						0.74	
labile C						0.93	
labile N						0.96	

Funding
The authors received no direct funding for this research.

Author details
S. Minu[1]
E-mail: minu.s88@gmail.com
Amba Shetty[1]
E-mail: amba_shetty@yahoo.co.in
Binny Gopal[2]
E-mail: binnycoorg07@gmail.com
[1] Department of Applied Mechanics and Hydraulics, National Institute of Technology Karnataka, Surathkal, Mangalore 575025, Karnataka, India.
[2] Department of Agronomy, University of Agricultural and Horticultural Sciences, Navile, Shimoga, Karnataka 577225, India.

References
ACORN 4.0. (2002). *"User's Guide", Analytical Imaging and Geophysics.* Boulder, CO: LLC.
Adler-Golden, S. M., Berk, A., Bernstein, L. S., Richtsmeier, S., Acharya, P. K., Matthew, M. W., ... Chetwynd, J. (1998). FLAASH, A MODTRAN4 atmospheric correction package for hyperspectral data retrievals and simulations. *Proceedings of the 7th Annual JPL Airborne Earth Science Workshop* (Vols. 97–21, pp. 9–14). CA: JPL Publication Pasadena.
Anne, N. J. P., Abd-Elrahman, A. H., Lewis, D. B., & Hewitt, N. A. (2014). Modeling soil parameters using hyperspectral

image reflectance in subtropical coastal wetlands. *International Journal of Applied Earth Observation and Geoinformation, 33,* 47–56. http://dx.doi.org/10.1016/j.jag.2014.04.007

Ben-Dor, E., Patkin, K., Banin, A., & Karnieli, A. (2002). Mapping of several soil properties using DAIS-7915 hyperspectral scanner data—A case study over clayey soils in Israel. *International Journal of Remote Sensing, 23,* 1043–1062. http://dx.doi.org/10.1080/01431160010006962

Ben-Dor, E., Levin, N., Singer, A., Karnieli, A., Braun, O., & Kidron, G. J. (2006). Quantitative mapping of the soil rubification process on sand dunes using an airborne hyperspectral sensor. *Geoderma, 131*(1–2), 1–21. http://dx.doi.org/10.1016/j.geoderma.2005.02.011

Ben-Dor, E., Chabrillat, S., Demattê, J. A. M., Taylor, G. R., Hill, J., Whiting, M. L., & Sommer, S. (2009). Using imaging spectroscopy to study soil properties. *Remote Sensing of Environment, 113,* S38–S55. http://dx.doi.org/10.1016/j.rse.2008.09.019

Berk, A., Bernstein, L. S., & Robertson, D. C. (1989). MODTRAN: A moderate resolution model for LOWTRAN7. Final report, GL-TR-89-0122, AFGL, Hanscom AFB, MA 42 pp.

Bernstein, L. S., Adler-Golden, S. M., Sundberg, R. L., Levine, R. Y., Perkins, T. C., Berk, A., ... Hoke, M. L. (2005). A new method for atmospheric correction and aerosol optical property retrieval for VIS-SWIR multi- and hyperspectral imaging sensors: QUAC (QUick Atmospheric Correction). *Geoscience and Remote Sensing Symposium, IEEE International, 5,* 3552.

Casa, R., Castaldi, F., Pascucci, S., Palombo, A., & Pignatti, S. (2013). A comparison of sensor resolution and calibration strategies for soil texture estimation from hyperspectral remote sensing. *Geoderma, 197–198,* 17–26. http://dx.doi.org/10.1016/j.geoderma.2012.12.016

Crouvi, O., Ben-Dor, E., Beyth, M., Avigad, D., & Amit, R. (2006). Quantitative mapping of arid alluvial fan surfaces using field spectrometer and hyperspectral remote sensing. *Remote Sensing of Environment, 104,* 103–117. http://dx.doi.org/10.1016/j.rse.2006.05.004

De Tar, W. R., Chesson, J. H., Penner, J. V., & Ojala, J. C. (2008). Detection of soil properties with airborne hyperspectral measurements of bare fields. *American Society of Agricultural and Biological Engineers, 51,* 463–470.

Farifteh, J., Van der Meer, F. D., Atzberger, C. G., & Carranza, E. J. M. (2007). Quantitative analysis of salt-affected soil reflectance spectra: A comparison of two adaptive methods (PLSR and ANN). *Remote Sensing of Environment, 110,* 59–78. http://dx.doi.org/10.1016/j.rse.2007.02.005

Galvão, L. S., Pizarro, M. A., & Epiphanio, J. C. N. (2001). Variations in reflectance of tropical soils. *Remote Sensing of Environment, 75,* 245–255. http://dx.doi.org/10.1016/S0034-4257(00)00170-X

Gao, B. C., Heidebrecht, K. B., & Goetz, A. F. H. (1993). Derivation of scaled surface reflectances from AVIRIS data. *Remote Sensing of Environment, 44,* 165–178. http://dx.doi.org/10.1016/0034-4257(93)90014-O

Ge, Y., Thomasson, A., & Sui, R. (2011). Remote sensing of soil properties in precision agriculture: A review. *Frontiers of Earth Science, 5,* 229–238.

Gomez, C., Lagacherie, P., & Coulouma, G. (2008). Continuum removal versus PLSR method for clay and calcium carbonate content estimation from laboratory and airborne hyperspectral measurements. *Geoderma, 148,* 141–148. http://dx.doi.org/10.1016/j.geoderma.2008.09.016

Gomez, C., Viscarra Rossel, R. A., & McBratney, A. B. (2008). Soil organic carbon prediction by hyperspectral remote sensing and field vis-NIR spectroscopy: An Australian case study. *Geoderma, 146,* 403–411. http://dx.doi.org/10.1016/j.geoderma.2008.06.011

Gopal, B., Shetty, A., & Ramya, B. J. (2014). Prediction of topsoil nitrogen from spaceborne hyperspectral data. *Geocato International, 30,* 82–92. doi:10.1080/1010604 9.2014.894585

Green, A. A., Berman, M., Switzer, P., & Craig, M. D. (1988). A transformation for ordering multispectral data in terms of image quality with implications for noise removal. *IEEE Transactions on Geoscience and Remote Sensing, 26,* 65–74. http://dx.doi.org/10.1109/36.3001

Green, R. O., Conel, J. E., & Roberts, D. A. (1993). Estimation of aerosol optical depth and additional atmospheric parameters for the calculation of the reflectance from radiance measured by the Airborne Visible/Infrared Imaging Spectrometer. *In Summaries of the Forth Annual JPL Airborne Geoscience Workshop, JPL Publication, 93–26,* 73–76.

Haubrock, S.-N., Chabrillat, S., Kuhnert, M., Hostert, P., & Kaufmann, H. (2008). Surface soil moisture quantification and validation based on hyperspectral data and field measurements. *Journal of Applied Remote Sensing., 2*(023552), 1–26.

Hively, W. D., McCarty, G. W., Reeves, J. B., Lang, M. W., Oesterling, R. A., & Delwiche, S. R. (2011). Use of airborne hyperspectral imagery to MAP soil properties in tilled agricultural fields. *Applied and Environmental Soil Science,* Article ID 358193, 13.

Hong, S. Y., Sudduth, K. A., Kitchen, N. R, Drummond, S. T., Palm, H. L., & Wiebold, W. J. (2002). Estimating within-field variations in soil properties from airborne hyperspectral images. *In Pecora 15/Land Satellite Information IV/ISPRS Commission I/FIEOS 2002 Conference Proceedings.* Denver, CO.

Kneizys, F. X., Shettle,E .P., Abreau, L. W., Chetwynd, J. H., Anderson, G. P., Gallery, W. O, ... Clough, S. A. (1988). Users guide to LOWTRAN-7. In AFGL-TR-8-0177 Air Force Geophysics Laboratories, Bedford, MA.

Kruse, F. A., Raines, G .I, & Watson, K. (1985). Analytical techniques for extracting geologic information from multichannel airborne spectroradiometer and airborne imaging spectrometer data. In *Proceedings of the 4th thematic conference on remote sensing for exploration geology.* Ann Arbor, MI.

Lagacherie, P., Baret, F., Feret, J.-B., Madeira Netto, J. M., & Robbez-Masson, J. M. (2008). Estimation of soil clay and calcium carbonate using laboratory, field and airborne hyperspectral measurements. *Remote Sensing of Environment, 112,* 825–835. http://dx.doi.org/10.1016/j.rse.2007.06.014

Lagacherie, P., Gomez, C., Bailly, J. S., Baret, F., & Coulouma, G. (2010). The use of hyperspectral imagery for digital soil mapping in Mediterranean areas. *Digital Soil Mapping, Progress in Soil Science, 2,* 93–102. http://dx.doi.org/10.1007/978-90-481-8863-5

Lu, P., Wang, L., Niu, Z., Li, L., & Zhang, W. (2013). Prediction of soil properties using laboratory VIS–NIR spectroscopy and Hyperion imagery. *Journal of Geochemical Exploration, 132,* 26–33. http://dx.doi.org/10.1016/j.gexplo.2013.04.003

Matarrese, R., Ancona, V., Salvatori, R., Muolo, M. R., Uricchio, V. F., & Vurro, M. (2014). Detecting soil organic carbon by CASI hyperspectral images. In *Geoscience and Remote Sensing Symposium* (INSPEC Accession Number: 14716443, pp. 3284–3287). IEEE Conference Publications. doi:10.1109/IGARSS.2014.694718

Minu, S., & Shetty, A. (2015). Atmospheric correction algorithms for hyperspectral imageries: A review. *International Research Journal of Earth Sciences, 3,* 14–18.

Mulder, V. L., de Bruin, S., Schaepman, M. E., & Mayr, T. R. (2011). The use of remote sensing in soil and terrain mapping—A review. *Geoderma, 162,* 1–19. http://dx.doi.org/10.1016/j.geoderma.2010.12.018

O'Neill, N. T., Zagolski, F., Bergeron, M., Royer, A., Miller, J. R., & Freemantle, J. (1997). Atmospheric correction validation of casi images acquired over the Boreas Southern study area. *Canadian Journal of Remote Sensing, 23*, 143–162. http://dx.doi.org/10.1080/07038992.1997.10855196

Qu, Z., Goetz, A. F. H., & Heidbrecht, K. B. (2001). High accuracy atmosphere correction for hyperspectral data (HATCH). *Proceedings of the Ninth JPL Airborne Earth Science Workshop JPL-Pub, 00-18*, 373–381.

Richter, R. (1996). A spatially adaptive fast atmosphere correction algorithm. *International Journal of Remote Sensing, 11*, 159–166.

Richter, R. (2005). *Atmospheric/topographic correction for airborne imagery* (DLR report, DLR-IB 562-02/05, p. 107). Wesseling.

Richter, R., & Schläpfer, D. (2002). Geo-atmospheric processing of airborne imaging spectrometry data. Part 2: Atmospheric/topographic correction. *International Journal of Remote Sensing, 23*, 2631–2649. http://dx.doi.org/10.1080/01431160110115834

Richter, R., Schläpfer, D., & Müller, A. (2006). An automatic atmospheric correction algorithm for visible/NIR imagery. *International Journal of Remote Sensing, 27*, 2077–2085. http://dx.doi.org/10.1080/01431160500486690

Roberts, D. A., Yamaguchi, Y., & Lyon, R. J .P. (1985). Calibration of airborne imaging spectrometer data to percent reflectance using air borne imaging spectrometer data to percent reflectance using field spectral measurements. In Proceedings of the Nineteenth International Symposium on Remote Sensing of the Environment (pp. 21–25). Michigan, 21–25 October 1985.

Roberts, D. A., Yamaguchi, Y., & Lyon, R. (1986). Comparison of various techniques for calibration of AIS data. In *Proceedings of the 2nd Airborne Imaging Spectrometer Data Analysis Workshop* (Vols. 86–35, pp. 21–30). Pasadena, CA: JPL Publication Laboratory.

Rodger, A., & Lynch, M. J. (2001). Determining atmospheric column water vapour in the 0.4–2.5 μm spectral region. In *Proceedings of the AVIRIS Workshop 2001*. Pasadena, CA.

Schläpfer, D., & Richter, R. (2002). Geo-atmospheric processing of airborne imaging spectrometry data: Part 1. Parametric orthorectification. *International Journal of Remote Sensing, 23*, 2609–2630. http://dx.doi.org/10.1080/01431160110115825

Selige, T., Böhner, J., & Schmidhalter, U. (2006). High resolution topsoil mapping using hyperspectral image and field data in multivariate regression modeling procedures. *Geoderma, 136*, 235–244. http://dx.doi.org/10.1016/j.geoderma.2006.03.050

Staenz, K., Szeredi, T., & Schwarz, J. (1998). ISDAS—A system for processing/analyzing hyperspectral data. *Canadian Journal of Remote Sensing, 24*, 99–113. http://dx.doi.org/10.1080/07038992.1998.10855230

Stevens, A., van Wesemael, B., Bartholomeus, H., Rosillon, D., Tychon, B., & Ben-Dor, E. (2008). Laboratory, field and airborne spectroscopy for monitoring organic carbon content in agricultural soils. *Geoderma, 144*, 395–404. http://dx.doi.org/10.1016/j.geoderma.2007.12.009

Stevens, A., Udelhoven, T., Denis, A., Tychon, B., Lioy, R., Hoffmann, L., & van Wesemael, Bv (2010). Measuring soil organic carbon in croplands at regional scale using airborne imaging spectroscopy. *Geoderma, 158*, 32–45. http://dx.doi.org/10.1016/j.geoderma.2009.11.032

Tanré, D., Deroo, C., Duhaut, P., Herman, M., & Morcrette, J. J. (1990). Technical note description of a computer code to simulate the satellite signal in the solar spectrum: The 5S code. *International Journal of Remote Sensing, 11*, 659–668. http://dx.doi.org/10.1080/01431169008955048

Uno, Y., Prasher, S. O., Patel, R. M., Strachan, I. B, Pattey, E., & Karimi, Y. (2005). Development of field-scale soil organic matter content estimation models in Eastern Canada using airborne hyperspectral imagery. *Canadian Biosystems Engineering, 47*, 1.9–1.14.

Vermote, E. F., Tanre, D., Deuze, J. L., Herman, M., & Morcette, J. J. (1997). Second simulation of the satellite signal in the solar spectrum, 6S: An overview. *IEEE Transactions on Geoscience and Remote Sensing, 35*, 675–686. http://dx.doi.org/10.1109/36.581987

Vermote, E. F., El Saleous, N., Justice, C. O., Kaufman, Y. J., Privette, J. L., Remer, L., ... Tanré, D. (1997). Atmospheric correction of visible to middle-infrared EOS-MODIS data over land surfaces: Background, operational algorithm and validation". *Journal of Geophysical Research, 102*, 131–141.

Wang, J., He, T., Lv, C., Chen, Y., & Jian, W. (2010). Mapping soil organic matter based on land degradation spectral response units using Hyperion images. *International Journal of Applied Earth Observation and Geoinformation, 12*, S171–S180. http://dx.doi.org/10.1016/j.jag.2010.01.002

Wu, J., Liu, Y., Chen, D., Wang, J., & Chai, X. (2009). Quantitative mapping of soil nitrogen content using field spectrometer and hyperspectral remote sensing. *IEEE International Conference on Environmental Science and Information Application Technology, 2*, 379–382. doi:10.1109/ESIAT.2009.296

Zhang, T., Li, L., Zheng, B. (2009). Partial least squares modeling of Hyperion image spectra for mapping agricultural soil properties. *Proceedings of SPIE—The International Society for Optical Engineering, 7454*, 74540P-1-74540P-12.

Zheng, B. (2008). *Using satellite hyperspectral imagery to map soil organic matter, total nitrogen and total phosphorus* (MSc thesis, pp. 1–81). Department of Earth Science, Indiana University.

15

Mapping and characterization of salt-affected and waterlogged soils in the Gangetic plain of central Haryana (India) for reclamation and management

A.K. Mandal[1]*

*Corresponding author: A.K. Mandal, Department of Soil and Crop Management, Central Soil Salinity Research Institute, Karnal 132001, Haryana, India
E-mail: arupkmondal@gmail.com
Reviewing editor:
Paolo Paron, UNESCO-IHE Institute for Water Education, Netherlands

Abstract: IRS LISS III Resource SAT data (2005–07) were integrated with ground truth and soil studies for delineation and characterization of salt-affected and waterlogged soils in the Indo-Gangetic plain of central Haryana. The quality appraisal for salty ground water was also conducted prior to its use for irrigation. Such studies are useful for planning reclamation and management of salt-affected soils and poor quality ground water. Strongly sodic soils were easily identified based on the white to yellowish white tones, high spectral and low NDVI values. Waterlogged areas (surface ponding) were detected based on higher absorption in infrared range. Sodic soils with poor quality ground water showed higher reflectance from dry salts during June and freshly precipitated moist salts in March and October. Sodic soils irrigated with normal ground water showed higher cropping density and higher NDVI values. Moderately and slightly sodic soils showed mixed spectral signatures for salt crusts, moderate cropping density and surface wetness. Soil profile studies indicated higher moisture content at sub-surface depths. The presence of iron and manganese mottles indicated the incidences of water stagnation. Soils with high pHs, ESP, and SAR values and showing the dominance of carbonate and bicarbonates of sodium in the saturation extract indicated sodic nature. Significant presence of $CaCO_3$ concretions at 1 m depth, low organic carbon contents, clay illuviation at sub-surface depth are typical features in sodic soil profiles. Water samples with high pH and SAR values and at places high RSC (Residual Sodium Carbonate) content indicated their sodic nature. Gypsum application is recommended for the reclamation of sodic soils and sodic water.

ABOUT THE AUTHOR

A.K. Mandal has been working in the central Soil Salinity research Institute Karnal India 132001 for the last 25 years as Scientist and currently working in the position of Principal Scientist (Soil Science—Pedology) in the Soil and Crop Management Division of the institute. He has been engaged in the resource inventory of salt-affected soils using remote sensing and geo-informatics and specialized in the soil survey, characterization and mapping for soil salinity classification, land reclamation, soil and water management, and land use planning.

PUBLIC INTEREST STATEMENT

The submitted paper highlighted the potential of remote sensing data for natural resource inventory for salt-affected soils and poor quality ground water in the arid and semiarid regions and also used geo-informatics for decision-making and planning in the land reclamation, soil and water management and sustainable land use planning in the Gangetic plain of India.

Subjects: Earth Sciences; Environment & Agriculture; Environmental Studies & Management

Keywords: sodic soil; poor quality water; remote sensing; reclamation; management; gypsum

1. Introduction

1.1. Global and national distribution

Salt-affected soils are commonly distributed in arid and semiarid climatic zones and covered 1,307 M ha at global scale (FAO/IIASA/ISRIC/ISS-CAS/JRC, 2008). The largest areas of salt-affected soils are in Australia followed by North and Central Asia, South America and South and West Asia. An estimated area of 6.73 M ha salt-affected soils are in India, of which 2.5 M ha is in the Indo-Gangetic plain (Mandal, Obi Reddy, & Ravisankar, 2011; Mandal & Sharma, 2006; National Remote Sensing Agency, 2008; Saxena, Sharma, Verma, Pal, & Mandal, 2004). In central Haryana, four districts Karnal, Kurukshetra, Panipat and Sonepat, were worst affected, showing 52% of the geographical area (TGA) under salt-affected soils (Mandal & Sharma, 2005). Interpretation of Landsat images showed old levees, relict flood plain and poorly drained low-lying flats are common topographic zones with salt infestation along the Gangetic alluvial plain (Manchanda & Iyer, 1983). The introduction of canal irrigation from Western Yamuna Canal (WYC) in Haryana during the 1950s accentuated upward movement of salt by rising water table (Singh, Bundela, Sethi, Lal, & Kamra, 2010). Due to the over use of irrigation water in poorly drained areas, waterlogging and secondary salinization appeared and caused losses in productivity for rice (42%), wheat (38%) and sugarcane (61%) crops (Samra, Singh, & Ramakrishna, 2006). Due to the use of salty ground water (60–70% of TGA) for irrigation, secondary salt enrichment in soil profiles occurred along the Ghaggar and Markanda river plains (Gupta, 2010; Manchanda, 1976; Phogat, Satyavan, & Sharma, 2011).

1.2. Modern tools and techniques for diagnosis and assessment of salt-affected soils

Because of the large spectral coverage and discreet bands, remote sensing data have been used for mapping and monitoring salt-affected and waterlogged soils in a time and cost-effective manner (Dwivedi, 2006; Mandal & Sharma, 2013; Rao et al., 1998; Saxena, 2003; Shrestha, 2006). Mougenot, Pouget, and Epema (1993) easily identified barren, salt-affected soils by high reflectance in the visible range, while studies conducted in thermal, infrared and microwaves ranges were used to characterize hygroscopic characteristics of salts and vegetation-covered soils, respectively. Howari (2003) and Howari, Goodell, and Miyamoto (2002) used spectro-radiometry as a remote sensing tool in visible and near infrared bands to quantify spectral ranges for salt-affected soils with variable salt composition. Khan, Rastoskuev, Sato, and Shiozawa (2005) used ratio indices, spectral properties and digital image classification for mapping hydro-saline land degradation in the Indus basin of Pakistan.

1.3. Traditional methods for ground estimation and characterization of degraded soils

Studies conducted by Metternicht and Zinck (1997) showed different approaches for mapping sodium and salt-affected soils, combining digital analysis with field observations and laboratory analysis. They concluded that the main causes of spectral confusions, masking different soil salinity-alkalinity degrees, were the type and abundance of salt-tolerant vegetation cover, topsoil texture and other field properties. Joshi and Sahai (1993), Sharma, Saxena, and Verma (2000) and Verma, Saxena, Barthwal, and Deshmukh (1994) used a similar approach combining remote sensing, ground truth and soil analysis data for mapping coastal salt-affected soils in Saurashtra (Gujarat State) and inland salt-affected soils of Uttar Pradesh State. Such methods are laborious and need concerted efforts for image analysis, collection of ground truth and laboratory analysis of soil and water to integrate for mapping, but produce results and classified outputs of salt-affected soils with higher accuracies. Classification of soils for salinity/alkalinity classes such as slight, moderate and strong, is useful for deciding precise soil reclamation and management options.

1.4. Justification and objectives

The complexity of soil salinity, alkalinity and waterlogging problems in central Haryana and the Gangetic Plain of India were reported by several authors (Gupta, 2010; Mandal & Sharma, 2005; Raj kumar, Ghabru, Singh, Ahuja, & Sharma, 2010; Singh et al., 2010) and is a primary concern for reclamation and management. The complex surface properties of salt-affected and waterlogged soils varied in seasonal imageries causing low mapping accuracies (Sharma, Saxena, Verma, & Mandal, 2008; Verma, Saxena, & Bhargava, 2007). Field validation is therefore necessary for spatial characterization of salinized areas followed by the chemical characterization of soil samples to assess degrees of limitations required for reclamation and management. Keeping in view the use of poor quality ground water for irrigation and its impact on soil degradation (salt enrichment) and reduced crop production (Gupta, 2010), chemical characterization of ground water is also necessary before its use in irrigation. To address these issues, the present study is aimed at the delineation and characterization of degraded (salt-affected and waterlogged) soils, and appraisal of ground water quality in central Haryana useful for planning reclamation and management.

2. Methodology

2.1. Study area

The study area (29°52′58.32″N to 30°15′34.42″N latitude and 76°25′31.31″ to 77°21′19.19″E longitude) covered administrative boundary of Kurukshetra district of central Haryana (1,530 km²) and lies 253 m above mean sea level. The average annual rainfall is 608 mm, mean winter temperature is 12.7°C and mean summer temperature is 38.5°C. The landform is alluvial under the Gangetic alluvium. The area is drained by the Yamuna, Ghaggar and its tributaries Markanda, Saraswati, Chautang, Tangri and other seasonal streams Sahibi, Dohan and Krishnawati that originate from the Aravalli Hills. The primary source of irrigation is the WYC and Bhakra canal. In the absence of canal irrigation supply ground water from tube wells is commonly used for irrigation. Prolonged irrigation altered the moisture regime and chemical characteristics of soils leading to salt infestations, waterlogging and low productivity (Singh, 2009).

2.2. Data, software, tools, and equipment used

(1) IRS 1C LISS III (Resource SAT) data (www.nrsc.gov.in) for March 2005 (pre-monsoon), June 2006 (summer) and October 2007 (post-monsoon) seasons. The specifications are shown in Table 1.

(2) Survey of India topographical maps at 1:50,000 scale (www.surveyofindia.gov.in) showing administrative boundaries, infrastructure (roads/railways), irrigation/drainage (canal/river) and settlements (state/district HQ, villages).

(3) Software: ILWIS (ver. 3.3), MS Office-Excel (2007), ERDAS IMAGINE, ARC GIS.

(4) Legacy data: Salt-affected soil maps at 1:250,000 scale (National Remote Sensing Agency, 1997), water quality and soil mineralogy data (Gupta, 2010; Kapoor et al., 1981; Manchanda, 1976; Verma et al., 2012) and Soil map of Haryana (Sachdev, Lal, Rana, & Sehgal, 1995).

(5) Soil sampling tools: color chart, auger, spade, knife etc.

(6) GPS (Lawrence global) for collecting location-data for soil profiles and soil sampling sites, water samples and tube wells.

Table 1. Particulars of satellite imageries in Kurukshetra district, Central Haryana, India

Sensor	Spectral resolution	Spatial resolution	Image no. and scale	Period
IRS-IB LISS III	B1 0.52–0.59 nm (Green)	23.5 m	53 C/01, 05, 09, 13	FCC
	B2 0.62–0.68 nm (Red)	Swath 140 km	53 B/04, 08, 12, 16	March 2005
	B3 0.77–0.86 nm (NIR)	No. of pixel/ha	43 F/4, 53 G/1	June 2006
	B4 1.55–1.70 nm (SWIR)	18.11	Scale 1: 50,000 scale	October 2007

2.3. Image processing and spatial analysis

The pre-processed IRS images for atmospheric corrections (by NRSC) were geo-referenced using the Survey of India topographical maps at 1:50,000 scale. The data from different bands were integrated to prepare a digital mosaic for the study area, using ERDAS software. Different band combinations, B321 (NIR, R, G) and B432 (SWIR, NIR, R) with histogram equalized (256 intervals) stretches were used to develop False Color Composites (FCC) for visual analysis of degraded soils (National Remote Sensing Agency, 2007). Based on the different manifestations of soil salinity such as tone, texture and patterns, the images (Figures 1 and 2) were visually interpreted to identify degraded soils (Table 2). Spectral reflectance was calculated based on the mean reflectance in bands B2, B3 and B4 (Table 1). A principal component analysis was carried out to prepare homogenous data-sets and filters were used to improve sharpness of the images for visual analysis. The spectral response patterns were analyzed for spatial characterization of image elements such as crop, riverine sand, salt-affected and waterlogged soils (Figure 3). The NDVI values were calculated using the band ratios [(B3 − B2)/ (B3 + B2)] for differentiation of crop and non-crop areas (Figures 1 and 4). A supervised classification of digital data was carried out using a nearest neighborhood operator. An interactive-database was prepared comparing the map units prepared by digital and visual analysis to generate a confusion (error) matrix for accuracy assessment (Table 3). A flow chart showing methodology for mapping salt-affected soils were presented in Figure 5 for clear understanding.

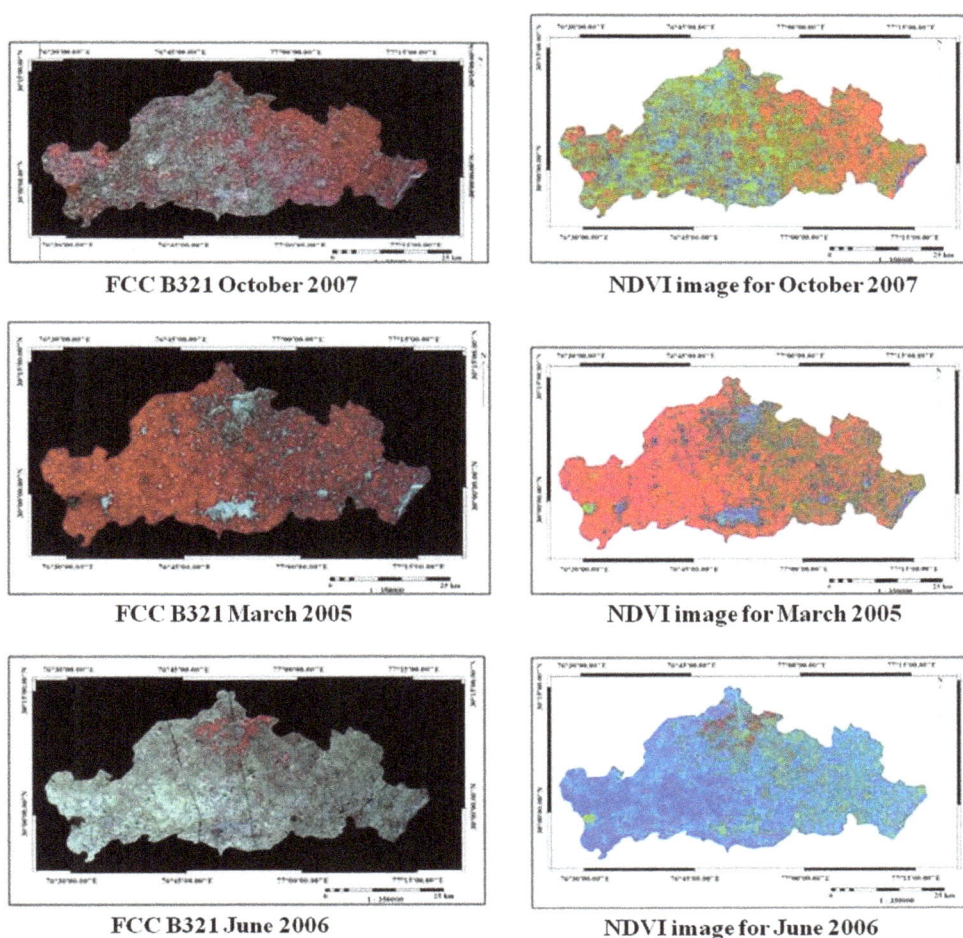

FCC B321 October 2007

NDVI image for October 2007

FCC B321 March 2005

NDVI image for March 2005

FCC B321 June 2006

NDVI image for June 2006

Figure 1. IRS FCC and NDVI images for the study area.

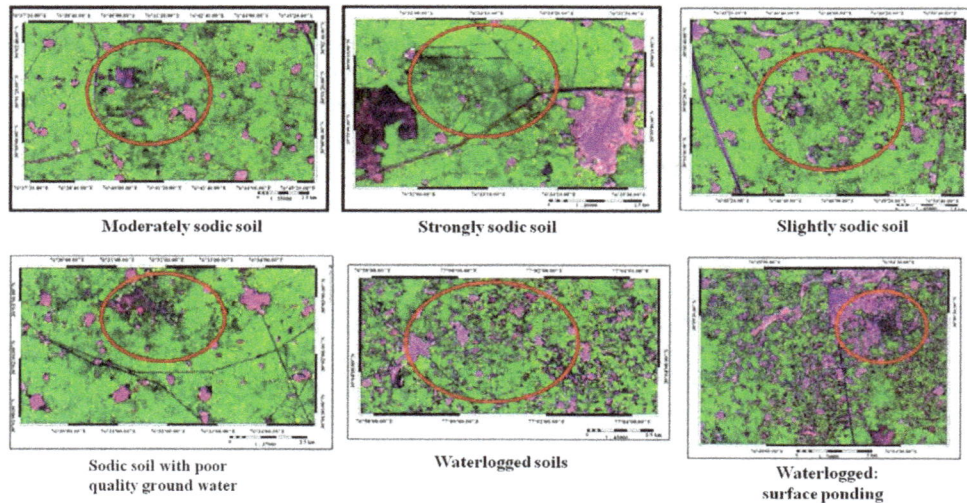

Figure 2. FCC (B432) showing salt-affected soils and waterlogged areas.

2.4. Ground truth studies for soil profile and water quality

A ground truth survey was conducted during March (pre-monsoon) and October (post-monsoon) 2005–07 seasons to authenticate interpreted units in the field and locate salt-affected and water-logged areas. The areas showing salinity emergence in different topographic zones and land uses such as crop and non-crop areas were studied and the data on status, condition and types of vegetation tolerant, partially tolerant and non-tolerant crops were also recorded. The salinity status at surface and sub-surface depths were obtained from soil profile studies. The field salinity/alkalinity status of soil samples was measured by portable pH and EC meters. The ground truth observations sites, soil profiles/soil and water sampling sites and topographical data on slope, aspects, contours, and related ground control points were collected during the ground truth study and were marked on the topographical maps. Water table depths data were also collected in waterlogged areas under canal irrigation to 1.5 m depth below the surface. Ground water samples were collected from tube wells for detailed chemical analysis. Ground water table depths were also recorded to relate with geology data.

Representative soil profiles (1.5 m depth) were studied to assess status and distribution of soil salinity and alkalinity at 24 sites covering the study area. Soil morphological properties such as soil moisture content, texture, color, structure and drainage were recorded from soil profile studies. Soil samples were collected at representative depths up to 1.2 m and properly stored in polythene and cloth bags to minimize moisture loss and changes in salt composition. These were further air-dried, processed to pass through <2 mm sieve and stored for physical and chemical properties (Table 4).

2.5. Studies for physical and chemical properties of soil and water samples

In the laboratory, soil samples were analyzed for physical and chemical properties such as soil reaction (pHs) and electrical conductance (ECe, dS m^{-1}); salt composition for soluble Na$^+$, K$^+$, Ca^{2+}, Mg^{2+}, CO_3^{2-}, HCO_3^- and Cl$^-$, me L^{-1}; $CaCO_3$ (<2 mm size, %) and organic carbon (%); cation exchange capacity (CEC, c mol (p$^+$) kg^{-1}) and Exchangeable Sodium percentage (ESP, %) and soil separates for sand, silt and clay percentages (Jackson, 1986). Based on soil pHs, electrical conductivity (ECe) and ESP values, soils are classified as saline, sodic and saline-sodic with degrees of soil salinity and sodicity classed as slight, moderate and strong (Richards, 1954). Waterlogged areas were classified as permanent waterlogged for surface ponding and sub-surface waterlogging as a result of high water table depth (<1.5 m) close to the surface (National Remote Sensing Agency, 2007).

Table 2. Spatial characteristic of interpreted units in Kurukshetra district

	Interpreted units	Land use/Land cover/ Surface soil moisture/ Wetness and other visible/*In situ* local observations	Image tone	Ranges of soil properties		
				pHs	ECe dS m⁻¹	Depth (m) of WT
1	Strongly sodic soil and sodic GW	Barren salt crust and associated with salt grasses/bushes/pasture/scrub lands, forest covered and higher moisture content at surface	Dark grayish white, spotted red to cark red scattered surface	9.6–10.2	4.5–7.4	100–120
2	Strongly sodic soil and normal (good quality) GW	Barren, surface salt crust, sparse vegetation, forest cover*, higher moisture accumulation at surface during October and March season	Grayish white, red to dark red cover, defined boundary*	9.8–10.7	1.2–8.5	100–120
3	Moderately sodic soil and normal GW	Tiny salt patches with patchy crop stand showing poor germination, cropped areas around the patches showed moderate crop growth	Red to grayish red with white or yellow mottles	9.3–10.2	2.4–6.6	60–100
4	Moderately sodic soil and sodic GW	Scattered cropped areas with very poor vegetative growth, prolonged water stagnation after irrigation with poor quality ground water	Irregular grayish to dark grayish mixed with red to light red color	9.1–9.6	1.2–1.8	100–120
5	Slightly sodic soil and crop covered	Vegetative growth of crops are comparable to normal soils, yield is relatively low	Red to dark red gray patches at some locations	8.5–9.0	<4.0	100–120
6	Surface ponding-slight	Stagnant water, scattered growth of aquatic grasses, canal irrigated, excess irrigation water from the cropped areas accumulated at low-lying flats/depressions	Dark gray/black patches, higher water absorption in post-monsoon (October) data	8.4–8.9	8–12	Surface water
7	Sub-surface water-logging	Partially cropped, low vegetative growth, high moisture content in soil profile, high water table depth	Dark blue/blue-black tone	7.8–8.2	6–10	<1.5
8	Irrigated crop	Normal vegetation	Red to dark red	7.5–8.0	<4	100–120
9	Riverine sand	Barren sandy soil at surface along the river course and often found with natural vegetation viz, grasses, shrubs, bushes	Yellowish white/white with red mottles	7.0–7.5	2–5	40–80

Notes: WT = water table, GW = ground water.

*Forest cover with defined boundaries in the satellite imagery which is different from crop covers with no defined boundary.

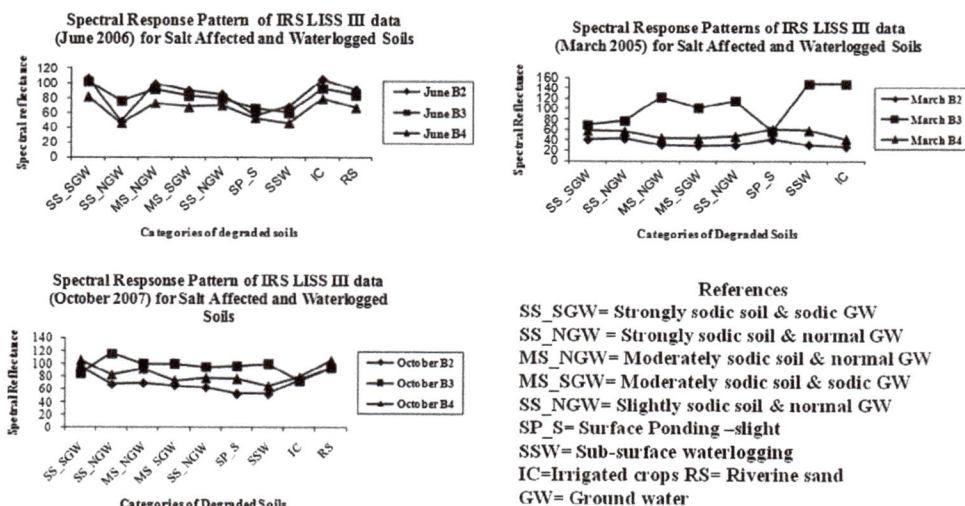

Figure 3. Spectral response patterns of degraded soils.

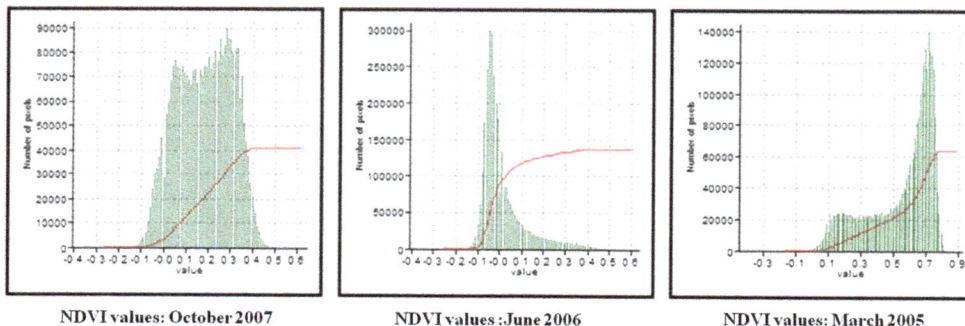

| NDVI values: October 2007 | NDVI values :June 2006 | NDVI values: March 2005 |

Figure 4. Seasonal NDVI values of study area

Table 3. Confusion matrix and accuracy estimates of classified image

	A	B	C	D	E	F	G	H	I	J	K	L	M	N	O	Accuracy
A	0	0	0	0	0	0	0	0	0	0	0	0	0	0	0	0
B	927	559	0	1,117	103	41	715	7	633	277	577	873	0	45	691	0.09
C	0	0	0	0	0	0	0	0	0	0	0	0	0	0	0	0
D	137	79	0	1,019	64	277	248	0	548	106	307	838	0	46	42	0.27
E	0	0	0	0	0	0	0	0	0	0	0	0	0	0	0	0
F	0	0	0	0	0	0	0	0	0	0	0	0	0	0	0	0
G	0	0	0	0	0	0	0	0	0	0	0	0	0	0	0	0
H	0	0	0	0	0	0	0	0	0	0	0	0	0	0	0	0
I	3,975	2,500	0	18,889	56,153	2,575	4,365	76	63,603	3,146	18,286	8,294	0	1,948	4,394	0.34
J	0	0	0	0	0	0	0	0	0	0	0	0	0	0	0	0
K	1,824	1,275	0	9,748	28,589	1,096	1,866	26	33147	1,648	11411	7,976	0	1,762	2,711	0.11
L	0	0	0	0	0	0	0	0	0	0	0	0	0	0	0	0
M	3	9	0	4	0	15	0	13	73	0	430	5	0	56	49	0
N	224	94	0	134	20	109	10	13	318	4	1,162	288	0	1,033	402	0.27
O	0	0	0	0	0	0	0	0	0	0	0	0	0	0	0	0
Reliability	0	0.12	0	0.03	0	0	0	0	0.65	0	0.35	0	0	0.21	0	

Notes: A = canal, B = surface ponding-slight, C = surface ponding-moderate, D = sub-surface waterlogging, E = normal field crop, F = reserve forest, G = road, H = river, I = sodic-slight, J = sodic-slight under forest, K = sodic-moderate, L = sodic-moderate under forest, M = sodic-strong, N = riverine sand, O = urban settlements.

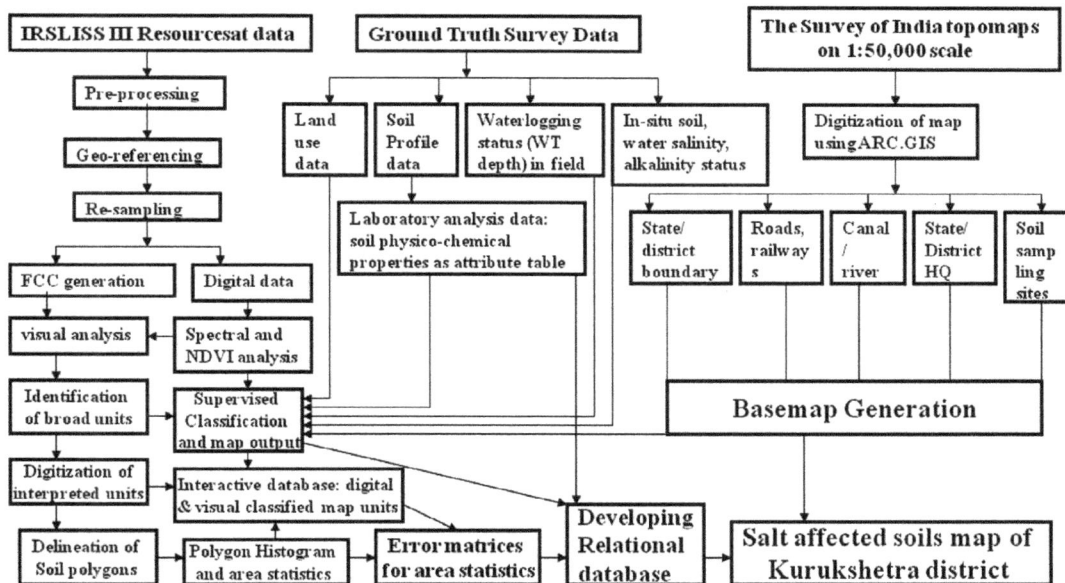

Figure 5. Flow chart showing methodology for mapping degraded soils.

Fifteen ground water samples were collected from different locations to study water quality for agricultural applications (Table 5). These were analyzed for pH_{iw} and EC_{iw} (dS m^{-1}), soluble cations and anions (Na$^+$, K$^+$, Ca^{2+}, Mg^{2+}, CO$_3^{2-}$, HCO$_3^-$ and Cl$^-$) using the methodology described by Richards (1954). The sodium adsorption ratio (SAR) [Na$^+$/{(Ca^{2+} + Mg^{2+})/2}$^{1/2}$] and residual sodium carbonate (RSC) [(CO$_3^{2-}$ + HCO$_3^-$)-(Ca^{2+}+Mg^{2+})] values were also calculated for classification of saline and sodic water (Richards, 1954).

2.6. Preparation of the thematic layers for base map and degraded soils

The Survey of India topographical maps at 1:50,000 scales, Universal Transverse Mercator (UTM) projections and ILWIS GIS software were used for geo-referencing and digitizing thematic layers for administrative and political boundaries (state/district), infrastructure (roads/railways), irrigation/drainage (canal/river) and settlements (state/district capitals). These layers were overlaid to prepare a base map for the study area. GPS was used to collect geo-referenced data for soil profile locations, soil and water sampling sites, tube wells and were stored in an attribute table that was linked with the base map. The spatial coverage of interpreted units was delineated using on-screen digitizing and were overlaid on the base map. Distinguishing colors were used for representative map units. These were annotated for scale, north direction, legends, title, boundary coordinates and other cartographic elements (Figures 1, 2 and 6).

2.7. Mapping of degraded soils

An integrated approach of image interpretation, ground truth survey and laboratory analysis data for soil physical and chemical properties was used for mapping degraded soils in the Indo-Gangetic plain of Central Haryana (Figure 5). Legends were developed for mapping degraded soils based on the methodology developed by National Remote Sensing Agency (2007). Categories of salt-affected and waterlogged soils were identified based on the soil physical and chemical properties. The area statistics of map (soil) polygons were used to assess spatial extent of salt-affected and waterlogged soils (Dwivedi, 2006; Mandal & Sharma, 2011, 2012, 2013).

Table 4. Physico-chemical characteristics of soils in Kurukshetra district

Horizon	Depth (cm)	pHs	ECe (dS m⁻¹)	Na⁺ (me L⁻¹)	K⁺ (me L⁻¹)	Ca²⁺ + Mg²⁺ (me L⁻¹)	CO₃HCO₃⁻ (me L⁻¹)	Cl⁻ (me L⁻¹)	SAR	OC (%)	CaCO₃ (%)	ESP (%)	CEC (cmol kg⁻¹)	Sand (%)	Silt (%)	Clay (%)	Texture
P1 29°59'54.3" N 76°25'41.3" E strongly sodic soils (sodic haplustept) and sodic GW, Ghaggar plain, Kikar, Prosopis juliflora plantation																
A1	0–14	9.6	7.4	98.7	0.9	4.0	17.0	31.0	49.3	0.08	2.45	49.6	13.1	71.2	14.9	13.7	sl
AB	14–40	10.0	5.5	80.0	0.2	4.0	21.0	18.0	40.0	0.06	2.45	51.8	8.8	61.9	21.6	16.3	sl
B21t	40–82	10.1	4.6	70.4	0.1	3.0	28.5	15.0	40.6	0.03	0.98	76.9	11.1	58.8	20.0	21.0	scl
B22t	82–109	10.2	5.3	83.9	0.1	3.0	27.5	12.0	48.4	0.02	1.47	54.1	13.9	58.6	18.6	22.6	scl
C	109–139	9.8	4.5	75.8	0.1	3.0	18.5	11.5	43.8	0.02	2.85	43.2	9.5	63.2	20.1	16.7	sl
P2 29°59'02.7" N 76°59'37.6" E strongly sodic (sodic haplustept), good quality (normal) GW, Yamuna plain, forestry plantations																
A1	0–12	9.8	1.2	11.5	0.1	1.0	15.7	3.5	11.5	0.17	1.3	46.8	21.6	59.0	23.0	18.0	sl
B21t	12–28	10.7	5.3	54.0	0.1	1.5	30.2	12.5	44.1	0.11	1.0	56.5	26.3	53.0	22.0	25.0	scl
B22t	28–58	10.6	6.2	60.6	0.1	1.0	27.5	10.5	60.6	0.17	2.0	51.8	22.0	52.0	24.0	24.0	scl
B23t	58–99	10.7	7.6	67.1	0.1	1.0	31.5	12.0	67.1	0.11	3.3	50.0	21.2	52.0	27.0	21.0	scl
Ck	99–142	10.5	8.5	75.4	0.1	1.0	35.5	12.5	75.4	0.11	4.9	55.0	21.0	53.0	28.5	18.5	sl
P3 30°02'33.6" N 76°51'40.7" E moderately sodic soil (sodic haplustept), cropped (rice), good quality (normal) GW, Markanda plain																
Ap	0–15	9.3	2.4	28.6	0.2	3.0	5.0	25.0	16.5	0.18	5.3	50.0	13.2	51.3	25.1	23.4	scl
AB	15–39	9.7	2.6	33.0	0.1	2.0	4.0	20.0	23.3	0.08	1.7	94.6	15.5	43.9	31.1	24.8	l
Bw1	39–76	10.1	4.0	53.9	0.1	2.0	10.0	25.0	38.1	0.08	1.3	94.2	15.9	47.3	26.8	25.8	scl
Bw2	76–105	10.2	5.9	96.7	0.2	2.0	12.5	30.0	68.3	0.08	1.7	83.1	18.4	48.1	24.7	27.0	scl
Bw3	105–149	10.0	6.6	99.5	0.2	2.0	15.5	32.0	70.3	0.05	1.9	89.7	16.8	46.2	25.8	28.0	scl
P4 30°08'56.3" N 76°25'37.2" E strongly sodic soil (typic natrustalf), reclaimed at surface, partially cropped, sodic GW, Ghaggar plain																
Ap	0–18	9.1	1.2	13.3	0.1	1.5	5.0	5.0	10.8	0.6	4.4	53.1	19.2	47.1	28.1	24.8	scl
AB	18–41	9.6	1.4	15.7	0.1	1.0	6.0	4.0	15.7	0.2	8.1	66.7	16.8	48.5	24.8	26.7	scl
B21t	41–68	9.7	1.6	18.5	0.1	1.0	7.0	4.5	18.5	0.1	5.2	69.0	20.0	38.0	23.3	38.8	cl
B22t	68–105	9.6	1.7	19.5	0.1	2.5	6.0	3.0	12.3	0.2	9.5	58.5	21.2	35.9	29.7	34.5	cl
B23k	105–156	9.8	1.8	22.0	0.1	4.5	10.0	8.5	10.3	0.1	14.5	65.8	22.0	36.8	35.5	27.7	l

Note: GW = ground water.

SN	Water sample-location and latitude-longitude, depth (m)	pH_{iw}	EC_{iw} (dS m^{-1})	Na$^+$ (me L^{-1})	K$^+$ (me L^{-1})	Ca^{2+} + Mg^{2+} (me L^{-1})	CO$_3^-$ (me L^{-1})	HCO$_3^-$ (me L^{-1})	Cl$^-$ (me L^{-1})	RSC (me L^{-1})	SAR
T1	Village Macheri (108 m) 29°59'08.4"N76°25'55.0"E	8.8	1.4	13.9	0.1	2.5	2.0	13.2	1.7	12.7	12.4
T2	Village Macheri, (60 m) 29°59'54.3"N76°29'41.3"E	8.7	1.1	9.9	0.2	2.5	1.5	10.0	1.7	9.0	8.9
T3	Village Seonsar (108 m) 29°59'29.9"N76°29'39.4"E	9.5	1.4	13.3	0.1	2.0	3.0	11.0	1.7	12.0	13.3
T4	Village Kheri Daban (100 m) 30°01'19.9"N76°27'36.6"E	9.1	1.3	12.6	0.1	1.5	Tr.	2.5	10.0	6.4	14.5
T5	Village Hansu Majra (100 m) 29°59'36.4"N76°29'55.5"E	9.3	1.4	14.0	0.1	1.0	Tr.	3.0	6.0	2.0	19.7
T6	Village Tatiana (120 m) 30°04'57.5"N76°27'35.8"E	9.1	1.6	16.6	0.1	1.0	0.0	3.0	5.0	2.0	23.4

Table 5. Quality of ground water samples in Kurukshetra district

Notes: RSC = (CO$_3$ + HCO$_3$) − (Ca^{2+} + Mg^{2+}), SAR = [Na$^+$/(Ca^{2+} + Mg^{2+})]$^{1/2}$, T = tube well.

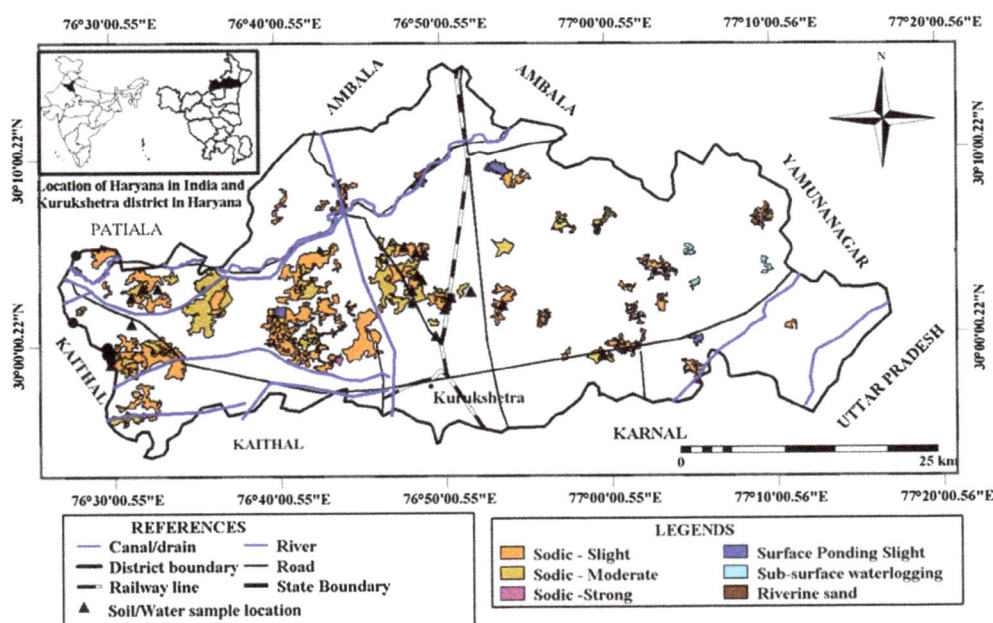

Figure 6. Distribution of degraded soils in Kurukshetra District, Haryana, India.

3. Results and discussion

3.1. Results

3.1.1. Image interpretation and ground truth studies of degraded soils

The spatial characteristics of salt-affected and waterlogged soil; natural vegetation and field crops were presented with soil chemical properties and ground water data (Table 2). The strongly sodic soils (white to yellowish white tone in B321, irregular shape), normal crops (bright red tone, continuous), waterlogged (surface ponding, dark blue to black tone, irregular shape) soils and riverine sands (yellowish white with definite shape along the river bed) were easily detected based on their strong signatures from the visible and infrared bands in IRS data (Figures 1 and 2). The seasonal data showed higher extents of moist salt-affected soils and waterlogged areas (irrigated) during March and October (Figure 1) and dry salts (salt crust) during June. This may possibly be due to similar reflectance of salt and sand and the absence of vegetative cover during the dry season. The normal cropped areas were identified by the distinct (bright) red tones at different growth stages, while stressed vegetation was identified by lighter red tones, patchy occurrence and patchy white tones for salt crust in the saline and moist surface in waterlogged areas (Mandal & Sharma, 2010). Riverine sands were identified by the yellowish white tone and spotted natural vegetation along the river course (Figure 1).

Field studies indicated prominent salt crusts, scanty vegetation, scattered salt-tolerant natural vegetation, scrub and pastures in strongly sodic soils and in places, intercepted with forestry planta-tions for biological reclamation. Moderate and slightly sodic soils appeared as mixed red and gray tones in irrigated areas, the ground truth studies showing patchy salts, scattered crop cover, moist soil surface, low permeability and absence of natural drainage. In dry areas, strongly sodic soils ap-peared as white patches of barren salt crust underlain by sodic ground water. Moderately sodic soils appeared as tiny white patches and red to dark red tones for crops irrigated with good quality ground water (Mandal & Sharma, 2011). Partially reclaimed sodic soils showed moderate crop cover and intermittent salt patches in low lying flats and depressions (Mandal, 2012). Slightly sodic soils showed good to very good crop and vegetative covers (Howari, 2003), though the field study re-ported low productivity due to crop damages in the maturity stage.

In irrigated areas, permanent waterlogged soils (surface ponding) were in the low-lying flats/de-pressions and appeared as gray to dark gray tones in all seasons (Figure 2). Mixed red and reddish gray tones were identified in the irrigated areas supporting vegetation. Field studies indicated high water table depth (sub-surface waterlogging, WT < 1.5 m depth), crop cover and secondary soil sa-linization during the post-monsoon season (Mandal & Sharma, 2001, 2010). However, using moder-ate spatial and spectral resolution of IRS data, the segregation of mixed signatures of water and crop in sub-surface waterlogged areas was difficult (Singh et al., 2010). It was authenticated on the strength of ground truth.

3.1.2. Digital analysis of remote sensing data for spectral properties of degraded soils

Spectral analysis of IRS data identified prominent energy absorption for waterlogged areas (surface ponding, SP_S) during October (B3 > B4 > B2) and March (B3 > B4 > B2) (Mandal & Sharma, 2001). The NDVI values were low (0.1–0.3) as a result of low crop cover (Mandal & Sharma, 2011). In irrigated areas, spectral values ranging from 60 to 148 in B3 and 58 to 66 in B4 indicated high water table depth or sub-surface waterlogging SSW (Mandal & Sharma, 2011). The NDVI values (0.24–0.34) indi-cated the presence of stressed vegetation (Dwivedi & Sreenivas, 2002; Joshi, Toth, & Sari, 2002).

Strongly sodic soils (SS_SGW) with poor quality ground water showed higher reflectance (80–110) of dry salts during June (B2 > B3 > B4) and the reflectance (40–90) of freshly precipitated moist salt surfaces during March (B3 > B4 > B2) and October (B4 > B2 > B3). NDVI values (0.28–0.37) indicated scattered and poor vegetative cover due to high soil sodicity (Figure 3 and 4). Strongly sodic soils with normal ground water showed high reflectance of B3 from salty (carbonate type) surface during October (B3 > B4 > B2), March (B3 > B4 > B2) and June (B3 > B2 > B4) respectively. These areas showed low NDVI values (0.18–0.29) due to poor to very poor crop cover (Figures 3 and 4). Moderately sodic soils irrigated with normal (MS_NGW) and poor quality ground water (MS_SGW) showed similar trends of spectral reflectance (B3 > B4 > B2) during October and March (Mandal & Sharma, 2001). NDVI values (Figure 4) were low (0.18–0.52) for cropped areas irrigated with poor quality ground water and higher (0.48–0.52) in normal ground water zone. Slightly sodic soils also showed higher reflectance of B3 followed by B4 and B2 (Figure 3). Similar data were also reported by Coleman, Agbu, Montgomery, and Prasad (1991).

Matured winter crops showed higher reflectance of B3 during March (B3 > B4 > B2) while crops in moist soil surface showed higher values of B4 during October. NDVI values showed similar trends during March and October, respectively. The spectral reflectance of riverine sand was high (60–100) due to bare surface (Figure 3) and low NDVI values (−0.04 to 0.04) which indicated scanty vegetative cover (Figure 4).

Principal component analysis was performed to homogenize digital data and achieve higher accuracy in classification. The principal component coefficients (PC) showed significant relationship between B1 and PC1 (0.524); B2 and PC4 (0.831); B3 and PC1 (0.707); and PC2 (0.683); B4 and PC3 (0.670). PC1 showed 93.5% variance while PC2, 3 and 4 showed 5.91, 0.30 and 0.09% variance, respectively. An average (AVG 3 × 3) filter was used to enhance sharpness of the images for visual analysis and to reduce noises prior to multi-band image classification. The nearest neighboring nine pixels were calculated to assign the values for central pixel to reduce noises and enhanced interpretation of the images.

Digital classification was performed using a supervised classification based on maximum likelihood classifier. Ground truth, laboratory analysis and land use data (field crop, forestry, urban settlement, road, natural water for pond, river and canal) were included as training sets for digital classification. Legacy data such as digitized maps of salt-affected soils, water table depth and quality and other collateral data including topographical maps of the Survey of India were also used as supporting data (Saxena, 2003; Verma, Singh, Sreenivas, Dwivedi, & Mathur, 2004). The salt-affected soils map at 1:250,000 scale was also consulted as supporting data. Clusters of pixels showing average reflectance for B1–B4 in March data were assigned a class name and the sample statistics (feature space) of the training set was generated to provide a visual overview of the separation of classes for the training pixels using a scatter plot for two bands. The feature spaces for B1 and B2, B1 and B4, and B2 and B4 indicated positive relation while B1 and B3, B2 and B3, and B3 and B4 showed partial or null relationships. An interactive (cross) database was prepared using maps prepared from visual analysis and digital classification. A confusion (error) matrix was prepared to assess the accuracy of digital classification (Table 3). The data showed an overall accuracy of 25.4%, average accuracy of 18.0% and reliability 10.5%, respectively. The highest accuracy was shown for slightly sodic soil (34%, reliability 65%) followed by sub-surface waterlogging (27%, reliability 3%), riverine sand (27%, reliability 21%) and moderately sodic soil (11%, reliability 35%), respectively.

3.1.3. Physical and chemical characteristics of salt-affected soils and waters

The field morphological characteristics of four representative soil profiles ranges from deep to very deep, pale brown to dark yellowish brown, sandy loam to sandy clay loam/clay loam texture, medium to strong, coarse to fine angular/sub-angular blocky structure, sticky, plastic to very sticky, very plastic consistency, presence of few to abundant $CaCO_3$ nodules and moist to wet sub-surface horizons. A few iron and manganese mottles were also found in sub-surface (50 cm) layers of P3 (Markanda plain) and P4 (Ghaggar plain), due to prolonged saturation with water. $CaCO_3$ concretions (2–5 cm, 10–30%) were found at 1 m depth in P2 and P4. The textural changes occurred from sandy loam to sandy clay loam and sandy clay loam to clay loam at P1, P2, and P4 apparently due to clay illuviation. The silt and clay contents were higher than sand content in P3 and P4 possibly due to lower topographic position.

Soil physical and chemical properties of four representative soil profiles are presented in Table 4. The pHs value ranges from 9.1 to 10.7 indicating alkaline reaction. The depth distribution of ECe values of P1 (4.5–7.4 dS m^{-1}) and P3 (2.4–6.6 dS m^{-1}) indicated moderate soil salinity while P2 (1.2–8.5 dS m^{-1}) showed higher salinity at sub-surface depth and soil salinity in P4 (1.2–1.8 dS m^{-1}) is low in general. The carbonate plus bicarbonate content is high in P1 (17.0–28.5 me L^{-1}) and P2 (15.7–35.5 me L^{-1}) and low in P3 (5.0–15.5 me L^{-1}) and P4 (5.0–10.0 me L^{-1}). A significant content of $CaCO_3$ (calcretes) was noted at 99 and 105 cm depths in P2 (1.3–4.9%) and P4 (4.4–14.5%) which caused restricted drainage and caused low permeability. CEC values were low in P1 (8.8–13.1 c mol (p+) kg^{-1}) and P3 (13.2–18.4 c mol (p+) kg^{-1}) due to coarse texture. The higher ESP values in P1 (49.6–76.9), P3 (50.0–94.6), P2 (46.8–56.5), and P4 (53.1–69.0) favored higher alkalinity.

Chemical properties such as pH (8.7–9.5), RSC (9.0–12.7 me L^{-1}), and SAR (12.4–23.4) water samples are presented in Table 5. Among the anions CO_3^{2-} (1.5–3.0 me L^{-1}), and HCO_3^- (2.5–13.2 me L^{-1}) and cations Na$^+$ (9.9–16.6 me L^{-1}) and Ca^{2+} + Mg^{2+} (1.0–2.5 me L^{-1}) and Cl$^-$ (1.7–10.0 me L^{-1}) were dominant. RSC values range from 2.0 to 12.7 me L^{-1} and are critical in T1 (12.7 me L^{-1}), T3 (12.0 me L^{-1}),

Sl no.	Categories of salt-affected soils and associated degradations	Area (ha)	%
	Table 6. Spatial extent of salt-affected soils and associated land degradations		
1	Slightly sodic soil	10,409	61.0
2	Moderately sodic soil	5,697	33.6
3	Strongly sodic soil	34	0.2
4	Surface ponding-slight	363	2.1
5	Sub-surface waterlogging	203	1.2
6	Riverine sand	210	1.2
Total		16,916	

T2 (9.0 me L^{-1}) and T4 (6.0 me L^{-1}) respectively (Richards, 1954). SAR values are higher (>10) in general.

3.1.4. Distribution of salt-affected and waterlogged soils

The spatial distribution of salt-affected and waterlogged soils is shown in Figure 6 and the extents were presented in Table 6. Slightly sodic soils have the largest area (10,409 ha) covering 61% of the total degraded soil in Kurukshetra district. It is followed by moderately sodic soils 5,697 ha covering 33.6% area and strongly sodic soils that occur in 0.2% of the area. Surface ponding occupies 363 ha (2.1%) while sub-surface waterlogging (203 ha) covers 1.2% area. Riverine sand covers 210 ha (2.1%) along the flood plain of the Markanda River.

3.2. Discussion

3.2.1. Remote sensing studies

The digital analysis of remote sensing data revealed mixed surface properties for salts, soil particles during dry (June) season and complex spectral signatures of moist soil surface and moderate crop cover in salt-affected soils (Khan et al., 2005). The similarity of spectral signatures for village settlements (muddy roof top) and barren salt-affected soils caused spectral confusion during digital analysis. Visual analysis revealed definite shape and sizes of rural settlements that differs from irregular pattern in salt-affected soils (Khan et al., 2005). Mixed gray to reddish gray and mottled red tones indicated waterlogging in cropped areas (Mandal & Sharma, 2013), which was authenticated during field studies. The linear shape of canals and typical curvilinear meandering rivers differs from stagnant water bodies (waterlogged surface) though these elements showed similar spectral reflectance. Irrigated areas with poor quality ground water showed mixed spectral signatures for poor crop stand (light to red tone) and moist soil surface (light to gray tones). Ground truth studies showed salt enrichment, unfavorable physical properties and poor drainages in soil profiles (Mandal et al., 2013; Sharma & Mondal, 2006). The low reflectance values of irrigated sodic soils in March data (40–60) appeared to be due to surface moisture. Similar results were reported for carbonate rich salts in visible (0.55–0.77 um) and infrared (0.9–1.3 um) ranges (Csillag, Pasztor, & Biehl, 1993; Khan et al., 2005; Rao et al., 1995). The higher NDVI values of moderately sodic soils (0.29–0.52) may be ascribed to higher vegetative cover and also management interventions at selected locations (Mandal & Sharma, 2011; Raghuwanshi, Tiwari, Jassal, Raghuwanshi, & Umat, 2010). The mixed reddish gray to dark gray tone for sub-surface waterlogged areas indicated scattered crop cover, and higher moisture content at soil surface.

3.2.2. Soil studies

Slight to strong soil alkalinity/sodicity indicated variable and complex chemical properties of sodic soils in the Kurukshetra district. The higher soil pHs (P3) at 40 cm depth indicated unfavorable soil physical properties and development of waterlogging. The high soil pHs of P1 (9.6–10.2) and P2 (9.8–10.7) at surface depth also limited its use for arable cropping. The dominance of $CO_3^{2-} + HCO_3^-$ anions and high Na$^+$ content in P1, P3, P2 and P4 indicated the sodium carbonate and bicarbonate parent materials that favored sodicity development in soils (Bhargava, Sharma, Pal, & Abrol, 1980; Sharma, Mandal, Singh, & Singh, 2011). The low contents of $Ca^{2+} + Mg^{2+}$ are due to precipitation of calcium carbonates in an alkaline medium (Bhargava & Bhattacharjee, 1982). The texture analysis of P4, P3, P1 and P2 indicated higher clay contents in sub-surface layers that caused restricted drainage and favored waterlogging. Higher CEC values in P2 and P4 is attributed due to higher clay content. The high ESP values showed significant saturation with exchangeable Na^+ that favored alkali soil formation. The high $CaCO_3$ contents caused drainage congestion. The soil physical and chemical properties indicated variable alkalinity dominated by alkaline earth metals and poor drainage caused low permeability (Raghuwanshi et al., 2010).

The high pH, RSC and SAR values of water samples indicated their sodic nature dominated by the presence of CO_3^{2-}, HCO_3^- and Na^+ while the presence of $Ca^{2+} + Mg^{2+}$ and Cl^- is also noted. Higher SAR values indicated dominance of Na^+ ion, causing soils unsuitable for agriculture (Richards, 1954). The critical limits of RSC in T1, T3, T2 and T4 indicated the need for treatment with amendments for irrigation in field crops. Treatment with gypsum is required for water samples with high RSC (T1–T4). Samples with moderate alkalinity (T5 and T6) may be used for the growing salt-resistant varieties.

3.3. Reclamation and use potential of salt-affected and waterlogged soils

Sodic soils of the Gangetic plain in Central Haryana are rich in sodium carbonate and bicarbonate salts and showed high ESP and variable soil texture. Strongly sodic soils (P1 and P2) containing high Na_2CO_3 and $NaHCO_3$ salts, coarse soil texture and sodic ground water needs gypsum application @ 8–10 t ha^{-1} to reduce alkalinity in soil and water followed by leaching of excess soluble salts. Moderately sodic soil (P3) containing soluble Na_2CO_3 and $NaHCO_3$ salts and fine soil texture can be reclaimed by addition of 4–6 t ha^{-1} gypsum. Due to high clay content and presence of $CaCO_3$ concretions, P4 (slightly sodic soil) showed drainage restrictions and waterlogging. It may be used for growing salt-tolerant rice and wheat crops. The addition of FYM in soils and cultivation of Dhaincha (*Sesbania* sp.) is suggested to improve physical properties, drainage conditions and reduce waterlogging.

4. Conclusions

Visual and digital analysis of IRS LISS III multi-temporal data was used for identification and delineation of sodic soils and waterlogged areas in the Gangetic plain of Central Haryana. Field validation and laboratory analysis for physical and chemical properties facilitated development of map legends. High values for spectral reflectance were observed from salty surfaces, and higher energy absorption in visible and infrared bands suggested the identification of strongly sodic soils and surface waterlogging. The mixed spectral signatures for salt, scattered crop covers and waterlogging were authenticated by field investigation. Saturation of Na_2CO_3 and $NaHCO_3$ salts in soil and ground water caused alkalization and low soil productivity. Fine soil texture and the presence of concretionary calcium carbonate layer at sub-surface depths tended to produce waterlogging. Sodic soils and sodic water can be reclaimed with suitable amendments such as gypsum or pyrite.

Acknowledgments
The author thanks Director and Head, Soil and Crop Management Division CSSRI Karnal for necessary support and guidance to carry out the work. Sincere thanks are also due to technical personnel involved in soil analysis and cartographic works. Thanks are also due to NRSC Hyderabad for providing satellite images and other analytical supports for image analysis and technical guidance for database generations.

Funding
The authors received no direct funding for this research.

Author details
A.K. Mandal[1]
E-mail: arupkmondal@gmail.com
[1] Department of Soil and Crop Management, Central Soil Salinity Research Institute, Karnal 132001, Haryana, India.

Cover image
Source: Author.

References
Bhargava, G. P., Sharma, R. C., Pal, D. K., & Abrol, I. P. (1980). A case study of the distribution and formation of salt affected soils in Haryana state. In *Proceedings of the International Symposium on Salt Affected Soils* (pp. 83–91). Karnal: CSSRI.
Bhargava, G. P., & Bhattacharjee, J. C. (1982, February, 8–16). Morphology, genesis and classification of salt affected

soils. In *Review of Soil Research in India, 12th International Congress of Soil Science* (pp. 508–528). New Delhi: Indian Society of Soil Science.

Coleman, T. L., Agbu, P. A., Montgomery, O. L., & Prasad, S. (1991). Spectral band selection for quantifying selected properties in highly weathered soils. *Soil Science, 151,* 355–361. http://dx.doi.org/10.1097/00010694-199105000-00005

Csillag, F., Pasztor, L., & Biehl, L. L. (1993). Spectral band selection for the characterization of salinity status of soils. *Remote Sensing of Environment, 43,* 231–242. http://dx.doi.org/10.1016/0034-4257(93)90068-9

Dwivedi, R. S., & Sreenivas, K. (2002). The vegetation and waterlogging dynamics as derived from spaceborne multispectral and multitemporal data. *International Journal of Remote Sensing, 23,* 2729–2740. http://dx.doi.org/10.1080/01431160110076234

Dwivedi, R. S. (2006). Study of salinity and waterlogging in Uttar Pradesh (India) using remote sensing data. *Land Degradation & Development, 5,* 191–199.

FAO/IIASA/ISRIC/ISS-CAS/JRC. (2008). *Harmonized world soil database (version 1.0)*. Rome: FAO.

Gupta, S. K. (2010). *Management of alkali water* (Technical Bulletin CSSRI/Karnal/2010/01, p. 62). Karnal: Central Soil Salinity Research Institute.

Howari, F. M. (2003). The use of remote sensing data to extract information from agricultural land with emphasis on soil salinity. *Australian Journal of Soil Research, 41,* 1243–1253. http://dx.doi.org/10.1071/SR03033

Howari, F. M., Goodell, P. C., & Miyamoto, S. (2002). Spectral properties of salt crust on saline soils. *Journal of Environment Quality, 31,* 1453–1461. http://dx.doi.org/10.2134/jeq2002.1453

Jackson, M. L. (1986). *Advanced soil chemical analysis*. New Delhi: Prentice Hall of India.

Joshi, M. D., & Sahai, B. (1993). Mapping of salt-affected land in Saurashtra coast using Landsat satellite data. *International Journal of Remote Sensing, 14,* 1919–1929. http://dx.doi.org/10.1080/01431169308954012

Joshi, D. C., Toth, T., & Sari, D. (2002). Spectral reflectance characteristics of na-carbonate irrigated arid secondary sodic soils. *Arid Land Research and Management, 16,* 161–176. http://dx.doi.org/10.1080/153249802317304459

Kapoor, B. S., Singh, H. B., Goswami, S. C., Abrol, I. P., Bhargava, G. P., & Pal, D. K. (1981). Weathering of micaceous minerals in some salt affected soils. *Journal of the Indian Society of Soil Science, 29,* 486–492.

Khan, N. M., Rastoskuev, V. V., Sato, Y., & Shiozawa, S. (2005). Assessment of hydrosaline land degradation by using a simple approach of remote sensing indicators. *Agricultural Water Management, 77,* 96–109. http://dx.doi.org/10.1016/j.agwat.2004.09.038

Manchanda, H. R. (1976). *Quality of ground waters in Haryana* (p. 160). Hisar: Haryana Agriculture University.

Manchanda, M. L., & Iyer, H. S. (1983). Use of Landsat imagery and aerial photographs for delineation and categorization of salt-affected soils of part of North-West India. *Journal of the Indian Society of Soil Science, 31,* 263–271.

Mandal, A. K. (2012). Delineation and characterization of salt affected and waterlogged areas in the Indo-Gangetic plain of central Haryana (District Kurukshetra) for reclamation and management. *Journal of Soil Salinity and Water Quality, 4,* 21–25.

Mandal, A. K., & Sharma, R. C. (2001). Mapping of waterlogged areas and salt affected soils in the IGNP command area. *Journal of the Indian Society of Remote Sensing, 29,* 229–235. http://dx.doi.org/10.1007/BF02995728

Mandal, A. K., & Sharma, R. C. (2005). Computerized database on salt affected soils in Haryana State. *Journal of the Indian Society of Remote Sensing, 33,* 447–455.

Mandal, A. K., & Sharma, R. C. (2006). Computerized database on salt affected soils for agro-climatic regions in the Indo-Gangetic plain of India using GIS. *Geocarto International, 21,* 47–57. http://dx.doi.org/10.1080/10106040608542383

Mandal, A. K., & Sharma, R. C. (2010). Delineation and characterization of waterlogged and salt Affected areas in IGNP command, Rajasthan for reclamation and management. *Journal of the Indian Society of Soil Science, 58,* 449–454.

Mandal, A. K., & Sharma, R. C. (2011). Delineation and characterization of waterlogged salt affected soils in IGNP using remote sensing and GIS. *Journal of the Indian Society of Remote Sensing, 39,* 39–50. http://dx.doi.org/10.1007/s12524-010-0051-5

Mandal, A. K., & Sharma, R. C. (2012). Description and characterization of typical soil monoliths from salt affected areas in Rajasthan. *Journal of the Indian Society of Soil Science, 60,* 299–303.

Mandal, A. K., & Sharma, R. C. (2013). Mapping and characterization of waterlogged and salt Affected soils in Loonkaransar area of Indira Gandhi Nahar Pariyojona for reclamation and management. *Journal of the Indian Society of Soil Science, 61,* 29–33.

Mandal, A. K., Sethi, M., Yaduvanshi, N. P. S., Yadav, R. K., Bundela, D. S., Chaudhari, S. K., ... Sharma, D. K. (2013). *Salt affected soils of Nain experimental farm: Site characteristics, reclaimability and potential use* (Technical Bulletin: CSSRI/Karnal/2013/03, p. 34). Karnal: Central Soil Salinity Research Institute (CSSRI).

Mandal, A. K., Obi Reddy, G. P., & Ravisankar, T. (2011). Digital database of salt affected soils in India using geographic information system. *Journal of Soil Salinity and Water Quality, 3,* 16–29.

Metternicht, G., & Zinck, J. A. (1997). Spatial discrimination of salt- and sodium-affected soil surfaces. *International Journal of Remote Sensing, 18,* 2571–2586. http://dx.doi.org/10.1080/014311697217486

Mougenot, B., Pouget, M., & Epema, G. F. (1993). Remote sensing of salt affected soils. *Remote Sensing Reviews, 7,* 241–259. http://dx.doi.org/10.1080/02757259309532180

National Remote Sensing Agency. (1997). *Salt affected soils*. Hyderabad: Author.

National Remote Sensing Agency. (2007). *Manual: Nationwide mapping of land degradation using multi-temporal satellite data*. Hyderabad: Author.

National Remote Sensing Agency. (2008). *Mapping salt-affected soils in India* (p. 54). Hyderabad: Author.

Phogat, V., Satyavan, K. S., & Sharma, S. K. (2011). Effects of cyclic and blending uses of saline andgood quality waters on soil salinization and crop yields under pearl millet-wheat rotation. *Journal of the Indian society of Soil Science, 59,* 94–96.

Raj kumar, Ghabru, S. K., Singh, N. T., Ahuja, R. L., & Sharma, B. D. (2010). Origin of salinity in the Indus plains of sub-continent. *Journal of Soil Salinity and Water Quality, 2,* 24–33.

Raghuwanshi, S. R. S., Tiwari, S. C., Jassal, H. S., Raghuwanshi, O. P. S., & Umat, R. (2010, October 8–10). *Clay mineralogy study of salt affected soils of Bhind district, Madhya Pradesh* (p. 29). National Seminar on "Issues of Land Resource Management: Land Degradation, Climate Change, and Land Use Diversification", Nagpur.

Rao, B. R. M., Sharma, R. C., Ravi Sankar, T., Das, S. N., Dwivedi, R. S., Thammappa, S. S., & Venkataratnam, L. (1995). Spectral behaviour of salt-affected soils. *International Journal of Remote Sensing, 16,* 2125–2136. http://dx.doi.org/10.1080/01431169508954546

Rao, B. R. M., Dwivedi, R. S., Sreenivas, K., Khan, Q. I., Ramana, K. V., Thammappa, S. S., & Fyzee, M. A. (1998). An inventory of salt-affected soils and waterlogged areas in the Nagarjunsagar Canal Command Area of southern India, using space-borne multispectral data. *Land Degradation & Development, 9*, 357–367. http://dx.doi.org/10.1002/(ISSN)1099-145X

Richards, L. A. (Ed.). (1954). *Diagnosis and improvement of saline and alkali soils.* Washington, DC: United States Department of Agriculture.

Sachdev, C. B., Lal, T., Rana, K. P. C., & Sehgal, J. (1995). *Soils of Haryana for optimizing land use* (Vol. 44, p. 59). Nagpur: National Bureau of Soil Survey and Land Use Planning.

Samra, J. S., Singh, G., & Ramakrishna, Y. S. (2006). *Drought management strategies in India* (p. 277). New Delhi: ICAR.

Saxena, R. K. (2003). Application of remote sensing in soils and agriculture. *Journal of the Indian Society of Soil Science, 51*, 431–447.

Saxena, R. K., Sharma, R. C., Verma, K. S., Pal, D. K., & Mandal, A. K. (2004). *Salt affected soils, Etah district (Uttar Pradesh)* (Vol. 108, p. 85). Nagpur: NBSS-CSSRI Publ., NBSS&LUP.

Sharma, R. C., Mandal, A. K., Singh, R., & Singh, Y. P. (2011). Characteristics and use potential of sodic and associated soils in CSSRI experimental farm, Lucknow, Uttar Pradesh. *Journal of the Indian Society of Soil Science, 59*, 381–387.

Sharma, R. C., & Mandal, A. K. (2006). Mapping of soil salinity and sodicity using digital image analysis and GIS in irrigated lands of the Indo-Gangetic plain. *Agropedology, 16*, 71–76.

Sharma, R. C., Saxena, R. K., & Verma, K. S. (2000). Reconnaissance mapping & management of salt affected soils using satellite images. *International Journal of Remote Sensing, 21*, 3209–3218. http://dx.doi.org/10.1080/014311600750019831

Sharma, R. C., Saxena, R. K., Verma, K. S., & Mandal, A. K. (2008). Characteristics and reclaimability of sodic soils in the alluvial plain of Etah district, Uttar Pradesh. *Agropedology, 18*, 76–82.

Shrestha, R. P. (2006). Relating soil electrical conductivity to remote sensing and other soil properties for assessing soil salinity in northeast Thailand. *Land Degradation & Development, 17*, 677–689. http://dx.doi.org/10.1002/(ISSN)1099-145X

Singh, G. B. (2009). Salinity-related desertification and management strategies: Indian experience. *Land Degradation & Development, 20*, 367–385. http://dx.doi.org/10.1002/ldr.v20:4

Singh, G. B., Bundela, D. S., Sethi, M., Lal, K., & Kamra, S. K. (2010). Remote sensing and geographical information system for appraisal of salt-affected soils in India. *Journal of Environment Quality, 39*, 5–15. http://dx.doi.org/10.2134/jeq2009.0032

Verma, D., Singh, A. N., Sreenivas, K., Dwivedi, R. S., & Mathur, A. (2004, February 9–14). Evaluation of image fusion techniques on IRS multi-sensor data for delineation of sodic lands. In *International Conference on "Sustainable Management of Sodic Lands- Extended Summaries* (p. 553). Lucknow: Uttar Pradesh Council of Agricultural Research (UPCAR).

Verma, K. S., Saxena, R. K., Barthwal, A. K., & Deshmukh, S. N. (1994). Remote sensing technique for mapping salt affected soils. *International Journal of Remote Sensing, 15*, 1901–1914. http://dx.doi.org/10.1080/01431169408954215

Verma, K. S., Saxena, R. K., & Bhargava, G. P. (2007). Anomalies in classification of salt affected soils under USDA Soil Taxonomy. *Journal of the Indian Society of Soil Science, 55*, 1–9.

Verma, T. P., Singh, S. P., Gopal, R., Dhankar, R. P., Rao, R. V. S., & Lal, T. (2012). Characterization and evaluation of soils of Trans-Yamuna area in Etawah district, Uttar Pradesh for sustainable land use. *Agropedology, 22*, 26–34.

PERMISSIONS

All chapters in this book were first published in CG, by Cogent OA; hereby published with permission under the Creative Commons Attribution License or equivalent. Every chapter published in this book has been scrutinized by our experts. Their significance has been extensively debated. The topics covered herein carry significant findings which will fuel the growth of the discipline. They may even be implemented as practical applications or may be referred to as a beginning point for another development.

The contributors of this book come from diverse backgrounds, making this book a truly international effort. This book will bring forth new frontiers with its revolutionizing research information and detailed analysis of the nascent developments around the world.

We would like to thank all the contributing authors for lending their expertise to make the book truly unique. They have played a crucial role in the development of this book. Without their invaluable contributions this book wouldn't have been possible. They have made vital efforts to compile up to date information on the varied aspects of this subject to make this book a valuable addition to the collection of many professionals and students.

This book was conceptualized with the vision of imparting up-to-date information and advanced data in this field. To ensure the same, a matchless editorial board was set up. Every individual on the board went through rigorous rounds of assessment to prove their worth. After which they invested a large part of their time researching and compiling the most relevant data for our readers.

The editorial board has been involved in producing this book since its inception. They have spent rigorous hours researching and exploring the diverse topics which have resulted in the successful publishing of this book. They have passed on their knowledge of decades through this book. To expedite this challenging task, the publisher supported the team at every step. A small team of assistant editors was also appointed to further simplify the editing procedure and attain best results for the readers.

Apart from the editorial board, the designing team has also invested a significant amount of their time in understanding the subject and creating the most relevant covers. They scrutinized every image to scout for the most suitable representation of the subject and create an appropriate cover for the book.

The publishing team has been an ardent support to the editorial, designing and production team. Their endless efforts to recruit the best for this project, has resulted in the accomplishment of this book. They are a veteran in the field of academics and their pool of knowledge is as vast as their experience in printing. Their expertise and guidance has proved useful at every step. Their uncompromising quality standards have made this book an exceptional effort. Their encouragement from time to time has been an inspiration for everyone.

The publisher and the editorial board hope that this book will prove to be a valuable piece of knowledge for researchers, students, practitioners and scholars across the globe.

LIST OF CONTRIBUTORS

Kousik Das and Prabir Kumar Paul
Department of Mining Engineering, Indian Institute of Engineering Science and Technology, Shibpur, Howrah, India

Akram Afifi and Ahmed El-Rabbany
Department of Civil Engineering, Ryerson University, Toronto, Ontario, Canada

Bambo Dubula, Solomon Gebremariam Tesfamichael and Isaac Tebogo Rampedi
Dubula, Department of Geography, Environmental Management and Energy Studies, University of Johannesburg, Johannesburg, South Africa

Brahima Koné, Sehi Zokagon Sylvain and Kouassi Kouassi Jacques
Earth Science Unit, Soil Science Department, Felix Houphouet-Boigny University, 22 BP 582, Abidjan, Côte d'Ivoire

Traoré Lassane
Department of Economic Sciences, Biology Sciences, Peleforo Gon Coulibaly University, BP 1328 Korhogo, Côte d'Ivoire

Leilei Xiao
Jiangsu Key Laboratory for Microbes and Functional Genomics, Jiangsu Engineering and Technology Research Center for Microbiology, College of Life Sciences, Nanjing Normal University, Nanjing 210023, China
Key Laboratory of Coastal Biology and Utilization, Yantai Institute of Coastal Zone Research, Chinese Academy of Sciences, Yantai 264003, China

Bin Lian, Qibiao Sun, Huatao Yuan, Xiaoxiao Li, Changmei Lu and Yue Chu
Jiangsu Key Laboratory for Microbes and Functional Genomics, Jiangsu Engineering and Technology Research Center for Microbiology, College of Life Sciences, Nanjing Normal University, Nanjing 210023, China

Yulong Ruan
Key Laboratory of Karst Environment and Geological Hazard Prevention, Ministry of Education, Guizhou University, Guiyang 550003, China

Enoch Bessah and Appollonia A. Okhimamhe
Department of WASCAL, Federal University of Technology, Bosso Campus, Minna, PMB 65, Niger State, Nigeria

Abdullahi Bala
Department of Soil Science, Federal University of Technology, Bosso Campus, Minna, PMB 65, Niger State, Nigeria

Sampson K. Agodzo
Department of Agricultural Engineering, Kwame Nkrumah University of Science and Technology, PMB, Kumasi, Ghana

Zdena Dobesova
Dobesova, Faculty of Science, Department of Geoinformatics, Palacký University, 17 listopadu 50, 771 46 Olomouc, Czech Republic

Mary Antwi
Department of Crop and Soil Sciences, Kwame Nkrumah University of Science and Technology, Kumasi, Ghana

Alfred Allan Duker
Department of Geomatic Engineering, Kwame Nkrumah University of Science and Technology, Kumasi, Ghana

Mathias Fosu
Council for Scientific and Industrial Research, Savanna Agriculture Research Institute, Tamale, Ghana

Robert Clement Abaidoo
College of Agriculture and Natural Resources, Kwame Nkrumah University of Science and Technology, Kumasi, Ghana
International Institute of Tropical Agriculture, Ibadan, Nigeria

Friday Uchenna Ochege
Laboratory for Cartography and GIS, Department of Geography & Environmental Management, University of Port Harcourt, Choba, Rivers State, Nigeria
Faculty of Environmental Studies, Department of Surveying and Geoinformatics, University of Nigeria, Enugu, Nigeria

Chukwunonyelum Okpala-Okaka
Faculty of Environmental Studies, Department of Surveying and Geoinformatics, University of Nigeria, Enugu, Nigeria

Hussien ElKobtan
Nile Research Institute, National Water Research Center, Cairo, Egypt

Mohamed Salem and Sayed Ahmed
Geology Department, Benha University, Benha, Egypt

Karima Attia
Water Resources Research Institute, National Water Research Center, Cairo, Egypt

Islam Abou El-Magd
Environmental Studies Department, National Authority for Remote Sensing and Space Sciences, 23 Josef Tito St., El- Nozha El-Gedida, P.O. Box: 1564 Alf-Maskan, Cairo, Egypt

Geeta Singh
Division of Microbiology, Indian Agricultural Research Institute, New Delhi 110012, India

D. Kumar
Division of Agronomy, Indian Agricultural Research Institute, New Delhi 110012, India

Pankaj Sharma
NRC-DNA (Fingerprinting), NBPGR, Pusa campus, New Delhi, India

R. Hewson
Faculty of Geo-Information Science & Earth Observation (ITC), University of Twente, P.O. Box 217, 7500 AE, Enschede, The Netherlands

D. Robson, A. Carlton and P. Gilmore
Geological Survey of NSW, NSW Trade & Investment, Division of Resources & Energy, P.O. Box 344, Hunter Region Mail Centre, 2310, Sydney, NSW, Australia

Snehal Rajeev Pathak
Department of Civil Engineering, College of Engineering Pune, Pune, Maharashtra, India

Asita Nilesh Dalvi
Department of Civil Engineering, SITS, Pune, Maharashtra, India

S. Minu and Amba Shetty
Department of Applied Mechanics and Hydraulics, National Institute of Technology Karnataka, Surathkal, Mangalore 575025, Karnataka, India

Binny Gopal
Department of Agronomy, University of Agricultural and Horticultural Sciences, Navile, Shimoga, Karnataka 577225, India

A.K. Mandal
Department of Soil and Crop Management, Central Soil Salinity Research Institute, Karnal 132001, Haryana, India

Index